DESIGNING WITH

SOLAR POWER

A SOURCE BOOK
FOR BUILDING INTEGRATED
PHOTOVOLTAICS (BiPV)

First published by Earthscan in the UK and USA in 2005

This edition published 2013 by Earthscan

For a full list of publications please contact:

Earthscan
2 Park Square, Milton Park, Abingdon, Oxon OX14 4RN
Simultaneously published in the USA and Canada by Earthscan
52 Vanderbilt Avenue, New York, NY 10017

First issued in paperback 2020

Earthscan is an imprint of the Taylor & Francis Group, an informa business

National Library of Australia
Cataloguing-in-Publication entry:

Designing with solar power: a source book for building integrated photovoltaics.

Bibliography.
Includes index.

1. Building-integrated photovoltaic systems. 2. Buildings – Energy conservation.
621.31244

Coordinating Editor: Robyn Beaver

Designed by The Graphic Image Studio Pty Ltd, Mulgrave, Australia

Film by Mission Productions Limited, Hong Kong

IMAGES has included on its website a page for special notices in relation to this and our other publications.

ISBN 13: 978-0-367-57808-4 (pbk)
ISBN 13: 978-1-84407-147-0 (hbk)

DESIGNING WITH
SOLAR POWER

A SOURCE BOOK FOR BUILDING INTEGRATED PHOTOVOLTAICS (BiPV)

EDITORS

DEO PRASAD & MARK SNOW

SOLARCH Group, Centre for a Sustainable Built Environment
University of New South Wales, Australia

earthscan
from Routledge

CONTENTS

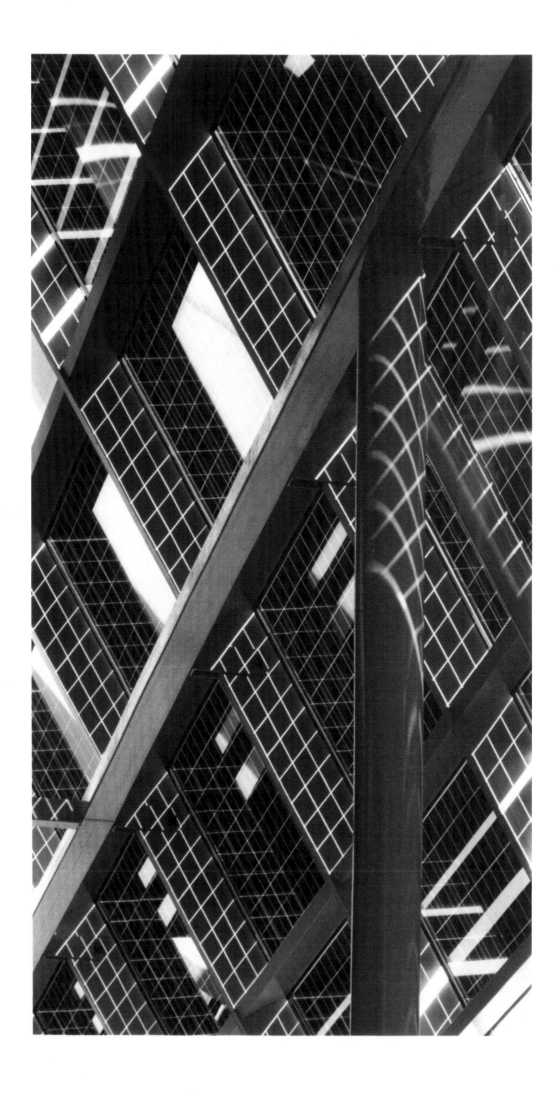

PREFACE

This book on designing with solar electric power in buildings is the result of international collaborative research and development work carried out within the framework of the International Energy Agency's Photovoltaic Power Systems Programme (IEA PVPS), and performed within its Task 7 on 'Photovoltaic power systems in the built environment'. Within this international framework we have brought together world-wide and interdisciplinary expert experience on building-integrated photovoltaics in order to provide state-of-the-art information on technology and design issues. The objectives reflect the general mission of the IEA's collaborative work aimed at performing profound analysis and providing recommended best practice in new energy technologies.

In 1994, a number of government officials from the Netherlands and Switzerland came together to speak about building-integrated photovoltaics and the opportunities that international collaboration could offer to help develop this energy technology of the future. Switzerland had been working on a growing number of projects, demonstrating the potential of PV as an energy-producing building material. The Netherlands was commencing a national PV programme focusing on building integration. The meeting resulted in an international collaborative programme, Task 7, together with a consortium of other countries, a mix of experienced countries including Germany, Austria, the United States and Japan and newcomers such as Australia, Italy and Canada.

From the beginning, it was clear to the international consortium that architects play an important role in developing the technology and language of solar buildings. Particularly at the early stages of development, solar buildings should be positive 'advocates' of the technology. The installation of solar buildings continues to provide increased understanding of bottlenecks and barriers that still exist within the marketplace.

In the years following the first meetings in 1994, solar experts worked on building integration together with architects, researchers and utility representatives. These participants closely followed a number of solar projects. Experiences from these projects, with a focus on the architectural and building-related aspects, have been collected in this book, to provide examples and inspiration for the designers of tomorrow's buildings, to reflect the enthusiasm today's PV experts put into their design work, and to document the diverse potential of solar for future buildings.

Continuing from the 1996 *Photovoltaics in Buildings* book, developed under the IEA Solar Heating and Cooling Programme, Task 16, this new publication captures the dramatic advancement of the technology. Design guides are available and many well-conceived projects are producing solar power every day. We hope this new book will help to develop responses to the questions that still need to be answered. Our sincere gratitude is extended to the co-editors, Deo Prasad and Mark Snow from the University of New South Wales, Sydney and all the Task 7 experts that have contributed.

Over the past years, grid-connected, distributed photovoltaic power systems have become the fastest growing market segment of this technology, the majority as systems in the built environment. This trend is expected to continue in the future and solar electric systems may well become a common feature of future building practice. To capture the main issues of this development, the IEA PVPS Programme has started a new project: Task 10 on 'Urban Scale Photovoltaic Applications'. The aim of this project is developing the means to enhance the opportunities for wide-scale, solution-oriented application of photovoltaic power electricity production in the urban environment.

This book will have achieved its mission if it can attract the interest of its target audience, namely architects, designers and engineers and inspire them to further ideas in this exciting area. We hope that many new building professionals will join these efforts and that this book will be another milestone in the widespread deployment of solar designed buildings.

Opposite: PV façade on Solar Office, Doxford International Business Park, UK
Source: Dennis Gilbert Photographer

Stefan Nowak
Chairman
IEA PVPS Programme

Tony Schoen
Operating Agent Task 7
Photovoltaics in the Built Environment

INTRODUCTION

Photovoltaic (PV) power's potential for wide distribution makes it a unique and novel energy source that can be embedded within the fabric of individual buildings, while shifting power generation away from being large-scale and regionally located. As a consequence, a free, clean and silent electrical supply can be introduced into cities, towns and built-up areas.

Building-integrated photovoltaics (BiPV) involves combining solar photovoltaic electricity technologies with those of building construction. This subject is of great interest to those in the fields of energy conservation and building design. Its significance, however, cannot be underestimated in the context of the more familiar notion of sustainable development. The concept of sustainability is more relevant than ever; it is a dynamic process that enables all people to realise their potential, and improve their quality of life in ways which simultaneously protect and enhance the Earth's life-support systems. BiPV addresses these essential aspects.

The current level of fossil fuel power generation is by far the greatest barrier to a state of sustainable equilibrium. Photovoltaic energy is already making a significant contribution towards the transition to 'renewable' sources – the key to achieving a sustainable society.

The relevance of BiPV in a sustainable world becomes clear after close examination of the three words in the acronym:

Building(s) protect against the extremes of climate. They evoke mood; they can excite, delight, and create a sense of wellbeing and repose. In the developed world, buildings also account for approximately one half of all energy consumed. The current energy generation process and reliance on fossil fuel sources is one of the major threats to achieving sustainable goals for the building sector.

Integrated means interdependence and interaction. In a sustainable world, it is recognised that each action has consequences beyond the specific end at which it is aimed. The construction of a building may successfully accommodate and enhance the activities contained within it, but it will also have an effect on its immediate physical setting, the climate, the neighbours and local community, the region and, ultimately, the globe itself. Sustainable development endeavours to anticipate these effects and, through integration with one another, ensures that adverse effects are minimised and, if unavoidable, are in some way balanced by those that are benign or of equal value.

Integration in sustainable development therefore seeks to reduce and harmonise detrimental environmental impacts. It also seeks economy of means and of materials in developing new industry opportunities. This, in the right hands, promotes elegant design. The goals of integration in BiPV are just those of sustainable development.

Photovoltaics is a technology whereby sunshine is converted into electricity. Just as plants use chlorophyll to photosynthesise the sun's irradiation in order to provide energy for their growth, a building can use particular composite solar components to meet the energy needs of its occupants. Only 14.4 per cent of sunshine survives filtering from the Earth's atmosphere and falls on land where it can be harvested. This is, however, 2,800 times more than our energy needs!

This book encapsulates five years of work by more than 30 international experts from the fields of research, development and design. In addition to greatly assisting the spread of the knowledge of BiPV, it is hoped that the book will encourage governments and agencies to support its deployment and raise confidence in its use by architects, engineers, environmentalists, planners and their clients.

Opposite: PV incorporated into roof of ECN building 42, the Netherlands
Source: Het Hout Blad
Photographer: John Lewis Marshall

CHAPTER ONE

BUILDING DESIGN AND ENVIRONMENTAL CONCEPTS

A renewable future

Imagine a world where petroleum is not the dominant source of energy. Over the next two decades, the burning of oil and gas is forecast to peak and the use of renewable energy to accelerate, so much so that by 2050, renewable energy could supply half of the world's energy requirements. The solar community has been predicting possibilities of this nature for some time[1] but now they are also coming from the oil industry.[2] There is no doubt that the world will undergo a major transition in energy sources and infrastructure systems over the next few decades. Oil industry analysts expect increased price pressure on petroleum as demand outstrips supply. The global distribution of fossil fuels creates strategic ramifications for the security of energy supply. Without a secure and affordable energy supply, cities become vulnerable to economic, environmental and societal decay. An 'affordable energy supply' is now recognised as one that neither compromises the ability to pay for energy nor jeopardises the ecosystems on which humans depend. 'Security of energy supply' is crucial to maintaining independence; nations that depend heavily on importing energy sources from overseas risk security of supply.

As the era of cheap oil draws to a close, other energy options must be developed to power buildings as well as transportation, agriculture and industry. These options must be more sustainable and have minimum ecological impact. It may come as a surprise to many architects and their clients, but every building they design that relies on fossil fuel may become obsolete within its lifetime.

Buildings and their processes account for roughly one half of all energy consumption. When the energy required to mine, produce, deliver and assemble materials for the construction of buildings is included, the total impact far surpasses that amount. For developed countries to continue to enjoy the comforts that are taken for granted, and for the developing world to ever hope to attain them, sustainability must become the cornerstone of our design philosophy. Rather than merely using less of the non-renewable fuels and creating less pollution, we need to design sustainable buildings that rely on renewable resources to provide most, and eventually all, of their own energy needs and eliminate pollution.

One of the most promising renewable energy technologies is photovoltaic (PV) power. PV power is a truly elegant means of producing on-site electricity from the sun, without concern for energy supply or environmental harm. It is estimated that just one hour of solar energy received by the earth is equal to the total amount of energy consumed by humans in one year. Photovoltaics are solid-state devices that simply make electricity out of sunlight, silently and with little to no maintenance, no pollution and no significant depletion of material resources.

Growth in PV power, originally in the form of stand-alone or off-grid markets (not connected to the electricity grid system) but now in grid-connected applications, has grown appreciably. Between 1992 and 1999, the growth rate of total installed capacity was between 20 per cent and 31 per cent per annum. This rate of increase rose to 34 per cent between 2001 and 2002. It can be seen from figure 2 that the majority of this rise was due to the continued dramatic increases in Japan (184 MW) and Germany (82.6 MW). In fact, of the 1,328 MW installed in 2002, almost 80 per cent was installed in Japan and Germany alone. In 2002, Japan achieved the highest installed power per capita (5 W/$_{capita}$), above Germany (3.37 W/$_{capita}$) and Switzerland (2.67 W/$_{capita}$).

PV is practical on a number of fronts. For example, the same technology that can power water pumps, thresh grain and provide communication and village electrification in the developing world can also produce electricity for the buildings and distribution grids of industrialised countries. It is clear that designing with solar power will become the norm for the future.

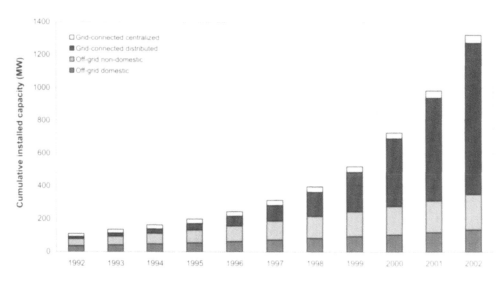

Fig 1 Cumulative installed PV power by application area between 1992–2002 for International Energy Agency (IEA)-reporting countries
Source: IEA Photovoltaic Power Systems Program (PVPS); www.iea-pvps.org

Country	1992	1993	1994	1995	1996	1997	1998	1999	2000	2001	2002
Australia	7.3	8.9	10.7	12.7	15.7	18.7	22.5	25.3	29.2	33.6	39.1
Austria	0.6	0.8	1.1	1.4	1.7	2.2	2.9	3.7	4.9	6.6	9.0
Canada	1.0	1.2	1.5	1.9	2.6	3.4	4.5	5.8	7.2	8.8	10.0
Switzerland	4.7	5.8	6.7	7.5	8.4	9.7	11.5	13.4	15.3	17.6	19.5
Denmark		0.1	0.1	0.1	0.2	0.4	0.5	1.1	1.5	1.5	1.6
Germany	5.6	8.9	12.4	17.8	27.9	41.9	53.9	69.5	113.8	194.7	277.3[1]
Spain	4.0	4.6	5.7	6.5	6.9	7.1	8.0	9.1	9.1[2]	16.0[3]	16.0[4]
Finland	0.9	1.0	1.2	1.3	1.5	2.0	2.2	2.3	2.6	2.7	3.1
France	1.8	2.1	2.4	2.9	4.4	6.1	7.6	9.1	11.3	13.9	17.2
United Kingdom	0.2	0.3	0.3	0.4	0.4	0.6	0.7	1.1	1.9	2.7	4.1
Israel	0.1	0.1	0.2	0.2	0.2	0.3	0.3	0.4	0.4	0.5	0.5
Italy	8.5	12.1	14.1	15.8	16.0	16.7	17.7	18.5	19.0	20.0	22.0
Japan	19.0	24.3	31.2	43.4	59.6	91.3	133.4	208.6	330.2	452.8	636.8
South Korea	1.5	1.6	1.7	1.8	2.1	2.5	3.0	3.5	4.0	4.8	5.4
Mexico	5.4	7.1	8.8	9.2	10.0	11.0	12.0	12.9	13.9	15.0	16.2
Netherlands	1.3	1.6	2.0	2.4	3.3	4.0	6.5	9.2	12.8	20.5	26.3
Norway	3.8	4.1	4.4	4.7	4.9	5.2	5.4	5.7	6.0	6.2	6.4
Portugal	0.2	0.2	0.3	0.3	0.4	0.5	0.6	0.9	1.1	1.2	1.7
Sweden	0.8	1.0	1.3	1.6	1.8	2.1	2.4	2.6	2.8	3.0	3.3
USA	43.5	50.3	57.8	66.8	76.5	88.2	100.1	117.3	138.8	167.8	212.2
Total [5]	109.9	136.2	163.9	198.6	244.7	314.0	395.7	520.0	725.8	990.0	1,327.7

Fig 2 Cumulative installed PV at end 2002 for participating IEA countries
Source: IEA PVPS; www.iea-pvps.org

1 Includes off-grid data estimate
2 Data not provided. Installed power for end 1999
3 Data from IEA statistics
4 Data from IEA statistics (previous year, 2002 information not available)
5 Totals reflect conservative 'best estimates' based on the latest information made available to the IEA–PVPS programme from all countries for previous years, and are updated as required

The opportunity for building-integrated photovoltaics

Energy planners have long envisioned large utility-scale solar power plants covering large expanses of desert. While this vision has many favourable attributes, the economics require careful investigation. Ground-mounted PV systems require the allocation of land, which must be acquired and prepared to accept the PV system. The cost of land and the cost of site work can be considerable. In Europe, Japan and many other countries, the lack of available large open tracts of land has effectively precluded the large-scale stand-alone PV option as afforded to Sacramento in California (figure 3).

As interest in solar electricity increases, there is a growing consensus that distributed PV systems that provide electricity at or near the point of use will be the first to reach widespread commercialisation. Distributed PV systems have considerable logic. They:

- Provide grid support, particularly in areas of summer peak loads.
- Eliminate costs and losses in transmission and distribution.
- Create a diverse and resilient energy system.
- Typically require no special approvals or permits.
- Can be deployed very rapidly.

Fig 3 Sacramento Municipal Utility District's (SMUD) photovoltaic array with nuclear cooling towers in the background
Source: Warren Gretz

The most attractive distributed applications are PV power systems for individual buildings, which have the following compelling benefits:

- The buildings and the processes they house consume the majority of electricity.
- The real estate comes 'free' with the building.
- There is no real estate tax on land to support the PV system or on the PV itself.
- There are no site development costs – they are part of the building construction.
- The utility interconnection already exists to serve the building (in most cases).
- PV can displace electrical power on the customer's side of the meter at the retail rate.
- PV can provide demand charge reductions and very favourable economics under time-of-use rates.
- PV can provide significant sectoral greenhouse gas (GHG) offsets in line with GHG emission reduction targets.

Interest in the building of integrated photovoltaics (known as BiPV), where the PV elements are integral to the building, often serving as the exterior weathering skin, is growing worldwide. PV specialists from some 14 countries have worked within the International Energy Agency's Photovoltaic Power Systems Implementing Agreement over the past several years to optimise these systems, and architects in Europe, Japan, the United States and Australia are now beginning to explore innovative ways of incorporating solar electricity into their building designs.

BiPV has many additional benefits:

- The building itself becomes the PV support structure.
- System electrical interface is easy – just connect to a distribution panel.
- BiPV components displace conventional building materials and labour, reducing the net installed cost of the PV system.
- On-site generation of electricity offsets imported and often more carbon-intensive energy.
- Architecturally elegant, well-integrated systems will increase market acceptance.
- BiPV systems provide the building owners with a highly visible public expression of their environmental commitment.

With BiPV systems, building owners are already paying for façade and/or roofing materials and the labour to install them. The land is already paid for, the support structure is already in place, the building is already wired, the utility is already connected, and developers can finance the PV as part of the overall project. Another benefit comes from distributing the BiPV installations over a broad geographic area and a large number of buildings, mitigating the effects of local weather conditions on the aggregate and producing a resilient source of supply.

Low-energy design, the prerequisite to BiPV

It is clear that the integration of PV into the built environment offers considerable scope for energy-demand offsets and reduction of greenhouse gas (GHG) emissions. It is also clear that the manner in which new buildings are designed and existing buildings are refurbished, altered and extended can significantly influence energy consumption and environmental impacts.

Many government-supported agencies in the developed world have carried out studies to determine benchmarks for energy use in various types of buildings. These benchmarks mostly point out the opportunities for significant energy demand reductions and consequent GHG emission reductions. Buildings have, in the past century, been designed with little consideration for their energy use and environmental impact. It is no surprise, therefore, that studies show that significant efficiencies are attainable in existing buildings at very cost-effective rates. In Australia, a study sponsored by the Australian Greenhouse Office and conducted by the SOLARCH Group and EMET Consultants showed that in order to meet the sectoral reductions in GHG emissions in the non-

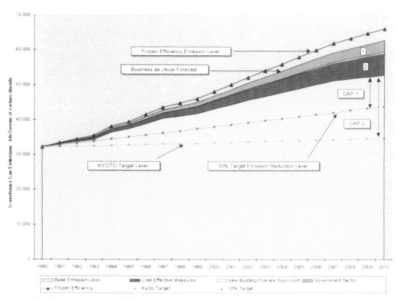

Fig 4 Forecast targets in GHG emission reductions for the Australian commercial sector 1990–2010
Source: EMET Consultants Pty Ltd and SOLARCH Group, New South Wales, Australia

residential building sector alone, a greater than 30 per cent per building reduction in energy use would be needed. More is necessary in order to achieve the 8 per cent increase in GHG emissions set for Australia under the Kyoto Protocol over the 1990 to 2010 period. Figure 4 highlights this finding.

Low-energy design is a clear prerequisite for all sustainable buildings, especially those with BiPV. It makes little sense to waste energy, particularly when it is generated using premium innovative technologies such as PV.

The passive solar office and the energy-efficient workplace

Passive solar design involves utilising natural forces such as the sun and wind for the heating, cooling and lighting of living spaces. Well-designed buildings take advantage of the natural energy characteristics in materials and air created by exposure to the sun that reduce the need to purchase utility energy sources to control, for example, the temperature and lighting of a building.

Recently, the Energy Technical Support Unit (ETSU) in the United Kingdom commissioned studies of three new out-of-town office developments. Consultants were asked to design three low-energy office buildings in the context of a specific site and client. An analysis of the design and its energy performance was then compared with that of an existing conventionally designed, air-conditioned 'reference' building. The average reduction in energy consumption between the low-energy design and the reference office building was 52 per cent, giving an energy cost reduction of 49 per cent. The capital costs for construction were similar to the conventional building, and the carbon dioxide emissions were reduced by 15 per cent.

	Units	Passive Solar Office	Reference Building	Energy Self-sufficient Workplace
		Estimated	Actual	Estimated
Gross floor area	m²	5,511	4,900	5,500
Net to gross floor area	%	82	77	82
Window to wall area	%	30	43	
Roof light to roof area	%	30	0	
Heating demand	kWh/m²	22.5	16.6	11.54
Lighting demand	kWh/m²	10.4	28.0	8.84
Cooling energy	kWh/m²	0	25.5	0
Incidental loads	kWh/m²	–	–	28.0
Totals	kWh/m²	32.9	70.1	48.38
Costs				
Capital cost (1992)	US$/m²	1,240	1,571	–
Heating cost	US$/m²	0.68	0.5	–
Lighting cost	US$/m²	0.74	1.97	–
Cooling cost	US$/m²	0	0.99	–
Electricity demand charges	US$/m²	0.56	0.98	55.04
Total running cost	US$/m²	1.98	4.43	–
Area available for PV				
Pitched roof	m²	0		1,740
Façade, including sunbreakers	m²	0		390
Total	m²	0		2,050

Table 1 Summary of information relating to the Passive Solar Office

Summary information relating to the Passive Solar Office (figure 5 shows a model representation) at Milton Park for Landsdown Estates is shown in table 1. This project was selected as it was subsequently redesigned as the Energy Self-sufficient Workplace (figure 6), to show the additional energy benefits and design implications of incorporating an extensive PV installation. This suggested that, in a location in the southeast of England, with an additional 29 per cent to the cost of construction, the building could be energy self-sufficient. The details of the PV installation are shown in table 2.

The assessment covered the design of a speculative office building of 5,510 square metres for Landsdown Estates on the Milton Park business estate, in Abingdon, United Kingdom. It was designed by Studio E Architects, and services engineers SVM. Comparison was made with an existing office building of the same rental value designed on a conventional sealed, air-conditioned basis.

PV sizing:
Annual average solar radiation: 2.5 sun hours/day
PV power rating: 120 Wp/m²
Energy requirement per day: 5,500 (floor area) x 34.4 (estimated energy demand)/365 (days) = 518.4 kWh/day
Required PV generation: 518.4 x 2.5 (Average solar radiation) = 207.36 kWp
PV area required: 207,360 (Wp) x 120 (PV power rating) = 1,726 m²
Available area on building: 2,050 m²

Table 2 PV installation details, Energy Self-sufficient Workplace

Fig 5 View of the south façade of the Passive Solar Office
Source: Studio E Architects, UK

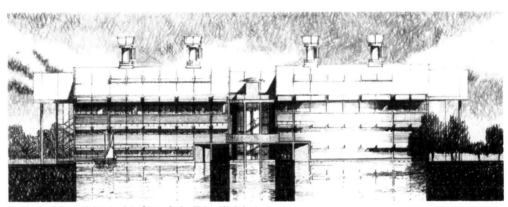

Fig 6 View of the south façade of the Energy Self-sufficient Workplace
Source: Studio E Architects, UK

13 •

Other significant points of the Passive Solar Office include:

- If capital depreciation and maintenance costs were included, the Passive Solar Office would benefit from a further 25 per cent reduction in running costs.
- Overheating is contained at an acceptable level if a natural ventilation rate of 8 ach (air changes per hour) is achieved and night time venting is implemented.
- The Passive Solar Office would require more knowledgeable and more sensitive comfort control procedures than the reference building.
- Some elements of cost are relatively high due to passive solar or other energy efficiency measures. These include two wind scoops, low-emissivity glazing and external sunbreakers/light shelves.
- The Passive Solar Office has fewer and less costly windows than the reference building.

The Energy Self-Sufficient Workplace took the Passive Solar Office, refined its energy saving and environmental measures and then applied BiPV with a view to providing a net-zero-energy commercial office building and an elegant architectural solution.

Other significant points in respect to the Energy Self-Sufficient Workplace include:

- Capital cost of the building would be an additional 29 per cent of the cost of the Passive Solar Office.
- Running costs would be 15 per cent less than the Passive Solar Office.

If a contribution towards the energy consumption of the building is to be supplied from an integrated PV installation, the first consideration must be to limit, as far as possible, the demand for power in the building. In so doing, the consumption of grid-supplied electricity is minimised and the use of PV-generated supply maximised.

If, for example, a building located in a temperate or continental climate is poorly insulated and the external envelope is poorly sealed, a considerable proportion of the energy used to heat the building in winter is wasted. If the first call for power is from a PV installation, that 'clean' power is also wasted. Electrical energy used for heating the building would in that case draw on fossil fuel, grid-connected power and would add to an increased GHG emission.

The approach required for energy reductions

It is outside the scope of this book to describe in detail the design and operational measures that have been developed to reduce building energy use – there are many publications that do so. Where BiPV installations are being considered, however, it is important that the following aspects are addressed, whether the design is for a new building or for modifications to an existing one:

- Pursue 'passive' systems, that is, those that exploit natural forces such as wind and sun and the effective use of environmentally sustainable materials, before calling on 'active' systems – such as artificial cooling and lighting – to augment or replace the passive systems. This requires wide-ranging considerations such as solar access and shading, orientation, use of thermal mass, effective daylighting, natural ventilation and thermal insulation.
- Consider the construction of the building as a whole, including: energy used to extract, manufacture and transport materials and components; the extent of construction waste; the transport and environmental implications of extensive earthworks; safety during construction; and the efficiency of construction and commissioning.
- Carefully assess the climate of the site. This should be undertaken in respect to both the climatic region and the immediate locality. Clearly, different parts of the world require very different measures. Building form, layout and massing should reflect the local climate.
- Minimise the need for artificial lighting and promote daylighting.
- Minimise the need for wasteful heating and cooling. This may mean investigating the possibilities of 'mixed-mode' systems where passive and active measures are combined to provide comfort effectively and efficiently.
- Ensure an approach that can deal with complexity effectively. Determine, for example, the split and interface in the building between centralised comfort control by the building manager and local control by the occupants. The more devolved control is to the latter, the happier they will usually be. Two basic strategies that usually work are low technology/low management input, and advanced technology/high management input. Avoid the other two possibilities: high technology/low management input (which perpetuates the myth that technology does not need to be managed), and low technology/high management input (impractical for most organisations).

- Ensure that control systems default to low-energy use; for example, put systems in place to ensure lights do not remain on unnecessarily and blinds are not closed when it is overcast.
- Ensure that the building occupants and those who run it 'buy into' the design approach and the practicalities of its operation. This can be assisted by presenting designs for comment, organising tours of the partially completed building, explaining the building's characteristics prior to occupation and providing feedback on the building's performance after occupation.

The list is not exhaustive – every building will have unique, as well as generic, requirements. A complex series of decisions is necessary to find the right balance between the functional demands of occupants and processes, energy restraint measures and the capacity to stimulate and delight residents and visitors. Finding this balance is the key to architectural proficiency.

This state should be the goal of all building design. Unfortunately, in BiPV projects, due to preoccupation with aspects particularly connected to PV – its novelty, technology and the need to cover the additional cost – it is, so far, rarely found. However as PV becomes better understood and more commonplace, BiPV buildings of elegance and confidence will increase.

Principles of holistic building design

Designing and constructing a low-energy building that is both efficient and stimulating requires a 'holistic' approach; that is, one that is comprehensive in its consideration of design issues, weighs them up carefully and finds a particular harmony between them.

A holistic approach to building design involves:

- *A fully integrated multidisciplinary design team.* Although the architect may lead the team, creative decisions are needed from the start, in particular, from structural and building services engineers. Engineers should contribute key information and ideas about energy and comfort systems and structural design. They need to impress upon the client that the end product will be economically, environmentally and socially responsible while providing a delightful setting for work or home.
- *An emotionally and intellectually compatible design team.* Members must have the right skills and understanding of the issues (or at least be ready and quick to learn) and be able to work together harmoniously.

- *The provision to call on specialists* when required and, at least in the early stages of the deployment of new technologies or concepts, to undertake research and development where appropriate and where it can be accommodated within the project budget and programme.
- *The early development of energy consumption targets* for the range of consumption categories in the building. In fact, the whole sustainable building strategy may be developed at this stage.
- *The capability to carry out physical and computer modelling* of aspects of the building ranging from massing and form to thermal, ventilation, daylight and acoustic computations to ensure targets are being met.
- *Effective cost-planning, cost-checking and cost-benefit evaluating* procedures including direct and indirect costs and value-added benefits over the short term and the building lifecycle.
- *Structured audits and reviews of the design* as it develops – a value engineering system that works.
- *The full involvement of the client in the design process*; engagement of the occupants at key stages and, where possible, the involvement of the builder in the detailed design stages. Consideration should also be given to the post-handover 'tuning' period when the design intentions are matched with actual operational characteristics.
- *A determination to retain and protect the principles behind the design*, with the foresight and agility to adapt and reconstruct them should circumstances change (which they often do) during the course of a project.

Many already adopt these strategies, but more do not, assuming that the buildings they are designing are readily understood and of a type that they have tackled before. However, to make an impact on environmental damage, environmentally sound design must apply to all construction and this requires a shift in perception of the designers, their clients and any intermediary agencies. While this shift is being made, new values have to be embraced, new technologies understood and new techniques assimilated. A holistic approach offers the means by which this can be realised. The extra investment in design time should recoup benefits in the manner in which the building responds to the demands of the occupants. It has also been noted that, in those buildings which are the product of this approach and which are significantly slanted towards passive rather than active systems, the very measures that have been used to address environmental depredation are usually those that make the building a stimulating and life-enhancing place to be.

The opportunity provided by integrated renewable energy

Thus far we have concentrated on the need for low-energy design, and the means and the processes associated with its realisation. The introduction of integrated renewable energy makes the holistic approach even more important. It also, as has been pointed out earlier in this chapter, allows buildings to be not only consumers, but also suppliers of energy. It therefore raises the vision of many small, low-technology contributors in urban or suburban locations supplying to the national power need and, in respect to the building within which this renewable source is installed, the possibility of energy self-sufficiency and carbon neutrality. For pioneers of solar energy in buildings, zero emissions – meaning that, over the life of the building, carbon dioxide produced by its construction and occupation is balanced by the savings in carbon dioxide made possible by the supply of renewable energy – has been the Holy Grail.

In theory, zero emission is possible and it is useful to have an ultimate goal, but in reality a 50 per cent cut would be satisfactory and is still difficult to achieve in most buildings. BiPV does, however, offer the energy supply side of the equation that is essential to significant carbon dioxide reductions.

The building integration attributes of photovoltaics

Of the various renewable energy sources, photovoltaic power is the only one whose hardware lends itself to composite manufacture with conventional building materials such as glass, metal and plastics. The composition, physics and performance characteristics of the photovoltaic process are described in Chapter 2. From the point of view of the architect or engineer considering the incorporation of PV, the product is available in the same way as any other building component, but also in a variety of forms and with very particular installation requirements.

Essentially, the product in its manufactured state is in the form of micron-thin saucer-sized wafers, each faced with a grid of filaments wired to each other and sandwiched between two sheet materials, the outside facing one being transparent (mono- and poly-crystalline modules), or as a fine deposition on a sheet material with the active areas again wired together (amorphous or thin film modules). The latter are about half as efficient as the former, but about half the price per unit area. Since both are manufactured or produced in sheet form and once sandwiched or deposited are relatively robust, they lend themselves to incorporation into the external envelope of the building, either to replace a conventional component or as an additional one. It is this ability to offer the same attributes as a conventional building component and its versatility in application that makes PV uniquely suitable for building integration.

The choice of proprietary photovoltaic-composite systems is wide. It includes:

- Curtain wall systems for vertical and inclined façades.
- Rainscreen cladding systems.
- Fixed and motorised solar shading louvre systems.
- Integrated roof cladding, sheeting and tiling systems.
- Pitched and flat roof mounted systems.
- Roof light systems (semi-transparent).

A number of manufacturers are also developing combined solar-electric and solar-thermal roof integrated systems – these are discussed in later chapters.

The designer, therefore, has considerable technical scope for incorporating PV into buildings, and the choice will increase over the next decade. However, like any new technology that is both inherently complex and has wide-ranging non-technical ramifications, introduction of PV installations has to be carefully conceived and effectively developed. A detailed overview of PV technologies and integration techniques is provided in Chapter 2.

Principles of photovoltaic integration

As we have noted, what has come to be called the holistic approach to building design is essential for successful BiPV. Unlike any other building installations, BiPV can affect every aspect of the design process. In their fully integrated form they affect the:

- Layout and orientation.
- Form and massing.
- Layout and height of surrounding buildings and planting.
- Energy strategy.
- Building structure and modularity.
- Selection and assembly of other building materials, components and systems.
- Capital and running costs.
- Construction integrity and detail.
- Appearance and architectural expression.
- Manner in which the building owner and its occupants are perceived.

Even where a PV array is retrofitted or applied to an existing building (as in the instance of an array mounted on a flat roof) it can affect many of these aspects.

Design and installation of a BiPV system, unlike most other building systems, do not reside in the realm of one particular skill within the consultant team. The architect sees the installation as an element of the building envelope (one that not nly generates power, but keeps the weather and burglars out nd allows daylight in); the structural engineer sees it as a articular load requiring a bespoke supporting structure; and he building services engineer sees it as an electronic system equiring expert design, specification, commissioning and, in nany cases, monitoring. This extended parentage clearly equires multidisciplinary collaboration from the outset. The ollaboration is facilitated by carefully structured design levelopment, moving from the strategic to the specific, with lear benchmarks identified along the route. Shared modelling ools and CAD systems can also help to bring together team nembers and their output.

Developing the energy strategy for a building

A priority at the outset of a BiPV project is to determine its energy strategy. This involves:

- Reviewing the activities planned for the building and determining future activities.
- Determining diurnal and seasonal use patterns.
- Determining and agreeing on the extent to which the design should exploit and rely on passive systems.
- Establishing local renewable energy sources.
- Establishing the cost of mains supply energy.
- Estimating the energy consumption of the building (given the assumptions on the extent of use of passive systems) and identifying the different types of energy use (such as heating, cooling, lighting, incidental loads).
- Setting a total energy consumption target from estimates relating to the different energy use types.
- Determining the extent to which the estimated energy consumption can be met by a BiPV installation. Unless already determined by a particular national programme, this is usually decided on the basis of cost and/or the area that can be made available within the building development. This latter condition may be largely determined by site conditions such as orientation, the extent of overshading by adjoining properties, and townscape requirements. There are tools available to assist this critical determination. These and the criteria for evaluation are covered in Chapter 5.
- Estimating likely capital costs, the extent to which PV components can offset the cost of conventional ones, payback periods and running costs. Estimating the likely savings in energy and operational costs arising from the low-energy design and the 'free' PV electricity compared to a conventional energy use equivalent. In addition, determining the 'value-added' benefits of BiPV such as shading impact on cooling loads and thermal comfort.
- The formulation of energy strategy scenarios for development of building design options by the design team.

Fig 7 Interior of conservatory with transparent modules at ECN in Petten, the Netherlands
Source: Houtblad, the Netherlands

17 •

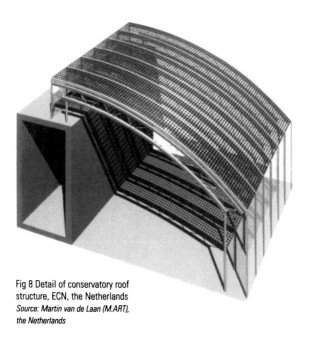

Fig 8 Detail of conservatory roof structure, ECN, the Netherlands
Source: Martin van de Laan (M.ART), the Netherlands

In passive solar buildings, design attention is very much focused on the external envelope: the roof, the walls and the ground slab. Issues of insulation, thermal capacity, impermeability, daylight, solar and glare control, solar gain, perimeter heating, views out, natural ventilation, security, noise control, atmospheric pollution and insect ingress all come together in the design of the external envelope and most particularly in the design of the glazing and fenestration. PV must also be integrated into the external envelope, and in many cases, large expanses of it. This brings issues of accommodation of cable runs and electrical kit, and possibly cooling the array and/or reclaiming free heat (a product of PV conversion). Design of the envelope can, accordingly, become complex.

Fig 9 Interior of Solar Office Doxford International, Sunderland, UK
Source: Denis Gilbert Photographer

Design measures that would benefit the optimisation of PV power can run counter to those that would assist the optimisation of passive solar measures. For instance, in the design of a BiPV façade, a design team may be looking to maximise electrical output while at the same time ensuring sufficient daylight reaches the spaces behind the façade, to avoid reliance on artificial lighting during the day. Many PV cells and depositions are opaque. Accordingly, power output from opaque areas of the façade would need to be carefully weighed against the energy-saving and psychological benefits of good daylight, mitigation of solar gains and glare, and views out.

In resolving this issue, the design team will have recourse to semitransparent PV modules; those of glass–glass mono- and poly-crystalline modules, where the cells can be spaced out to allow daylight through, and those of the laser etched amorphous modules where patterns of transparency are applied to the surface. Semitransparent PV technologies, such as those based on titanium dioxide may also be available soon. Figures 7–9 illustrate the use of interior effects in atrium designs at the ECN Building in the Netherlands, and at the Solar Office at Doxford International, United Kingdom.

Another example is the situation where the design team may be looking to maximise the thermal capacity of the fabric of the building in order to even out internal temperatures in summer while at the same time incorporating a BiPV pitched roof and glazing system. The modules would normally be mounted in a metal framework and supporting structure that has low thermal capacity. This is not to say that BiPV installation is incompatible with low-energy design. The point is that careful thought needs to be put to integration so that the measures are reinforced and, where not possible, compensatory design measures are incorporated.

The critical aspects of a BiPV installation, in so far as influencing layout and form, are the need to maximise irradiation on the array through orientation and slope, and the need to avoid overshadowing of the array. Equally important is the need to meet these two requirements while satisfying, first, any planning, heritage or aesthetic demands imposed by statutory bodies; second, the client brief; and third, creating illuminating and stimulating architecture through the manipulation of form, light and space.

Orientation of the array is a question of ensuring that it receives the maximum amount of direct sunshine possible given the fixed circumstances of the site and the density at which it is being developed. Actual orientation and inclination of the array will, of course, depend on the global location of the site. There is considerable latitude in deviating from the optimum orientation before output falls off significantly. Figure 10 shows daily insolation (kWh/m²) relative to orientation for Sydney, Australia.

Overshadowing (or overshading as it is also referred to in the industry) is a more tricky aspect of the design. For instance, if part of the array is overshadowed by another building for any part of the day, the loss of output can be significantly greater than the output of the actual modules overshadowed. This is because the modules are typically wired together in groups (known as strings) and therefore, overshadowing part of the group knocks the whole string out (see Chapter 6). Ideally, no part of the array should be overshadowed during any part of the day on any day of the year. This is a very onerous design aim, particularly in respect to BiPV buildings in a town or city. The best arrangement, given the particular circumstances, is usually determined by a succession of studies. These can be carried out by computer modelling (see Chapter 5). Overshadowing by a nearby object may at times be unavoidable, but overshadowing by elements of the host building, although common, should be avoidable. If the array is on a façade, it is very easy to inadvertently introduce projecting elements that overcast the array. This may be by projecting accommodation, but it can also be by small projections adjacent to the modules such as the framing in which they are set (self shading). The problem of overshadowing, along with PV output, are the main reasons why so many installations are placed on, or incorporated within, the roof, where the possibility of shading is minimised. It is also the rationale for consideration of 'AC modules', that is, modules with individual inverters, such that shading of one module does not impact on the output from the rest of the array. This is discussed further in Chapter 6.

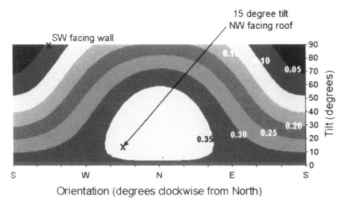

Fig 10 Daily insolation (kWh/m²) relative to orientation for Sydney, Australia
Source: PV Centre, University of New South Wales, Australia

Fig 11 There is no evidence of the 180 kWp PV installation on the Suglio Building, Zurich when seen from the surrounding spaces. The installation does not appear to have any significant influence on layout and form.
Source: Studio E Architects, UK

A BiPV aesthetic?

Another reason why PV installations end up sitting on a flat roof or embedded in a pitched one is the inability or unpreparedness of the designers to come to terms with PV as an expressive architectural element. One of the most positive attributes of PV is its appearance; it conveys the sophisticated coolness of 'high-tech' with the social responsibility of mitigating environmental depredation. In other words, it can support and enhance a highly contemporary architectural idiom while countering the sense of profligacy with which this type of architecture is often associated. The potential architectural expression that PV contains has yet to be fully realised. We need many more architect champions to embrace this and demonstrate its architectural potential and versatility.

When PV is an element in a truly low-energy, holistic building design, not only is its own expressive potential available to the designer, but its low-energy elements – sunshields, wind towers, solar scoops – to name the most overt, are also available. The question of whether eco- or bio-climatic architecture has evolved its own aesthetic is the subject of much debate. It is clear that many recent bio-climatic buildings are recognisably so. It has also been demonstrated that the low-energy measures in buildings can be completely suppressed without impairing their effectiveness (figure 11). It may be that with the necessary close relationship between building and the environment, the resulting architecture will need to clearly demonstrate the principles upon which it is based. One day, BiPV and low-energy measures will no doubt be as commonplace as other revolutionary innovations, such as lift installations and central heating, thus fulfilling a tenet of the Modern movement of the 1930s. This was based on a manifesto of social awareness and responsibility. Unfortunately its purist aesthetic was often hijacked by international commerce, and the low-cost housing and workplaces for which the aesthetic evolved are now largely expressed in quasi-vernacular garb. The green movement, if nothing else, has arisen in order to protect and enhance humanity. Architecture that takes its credo from principles of sustainability can attempt to reintroduce the social dimension that was denied to Modernism.

Figures 12 to 14 demonstrate the visible nature of PV and its considerable influence on layout and form.

Fig 12 Fire Station Houten, the Netherlands, designed by architect Philippe Samyn
Source: Novem, Hans Pattist

Fig 13 University of Melbourne PV, Australia
Source: Sustainable Technologies International, Australia

Fig 14 The aesthetic effect of photovoltaics at twilight at the Solar Office, Doxford International
Source: Denis Gilbert Photographer

The BiPV marketplace

With decreasing capital and installation costs, improved aesthetics and all the benefits of distributed generation, BiPV systems are a prime candidate for early widespread market adoption. Innovative architects are now beginning to integrate PV into their designs, and PV manufacturers are responding with modules specifically for BiPV applications, including integral roof modules, roofing tiles and shingles, and modules for vertical curtain wall façades, sloped glazing systems and skylights. Other architectural module designs, discussed in detail in Chapter 2, employ glass-superstrate, crystalline modules with space between the cells and opaque backings, to provide diffuse daylighting along with their electricity production.

In the past, incorporating PV into a building design required tradeoffs and concessions in the architectural design process. Today, as PV manufacturers are matching products to building-industry standards and architects' requirements, companies in the United States, Japan, Europe and Australia are actively pursuing new module designs that displace traditional building materials. Such BiPV components are facilitating a new generation of solar architecture. With the right design, sunlight falling on a building and/or its site can provide much or all of the power it requires. In urban areas, the power that could be generated by incorporating PV in the process of renewal, change and expansion can only be imagined. If BiPV is considered for incorporation as a matter of course for all rebuilding, new development and property upgrading, the vision of cities becoming overall energy suppliers rather than profligate energy consumers becomes an achievable goal.

Notes

1 Browne, J 1997, Speech presented by the chief executive of BP Solar to Stanford University.

2 World Energy Outlook: www.worldenergyoutlook.org.

CHAPTER TWO

TECHNOLOGIES AND
INTEGRATION CONCEPTS

Fig 1 Colt PV system optimised for shading
Source: BEAR Architecten, Netherlands

Fig 2 Shade design optimised for PV, general and detail views
Netherlands Energy Research Foundation (ECN) – Building 31
Source: BEAR Architecten, Netherlands

Introduction

Photovoltaic (PV) product technologies and techniques for integrating systems as an intrinsic part of buildings are rapidly evolving as market deployment increases. This section introduces PV technology basics, and includes a detailed summary of integration concepts in use for all building typologies, with comments on architectural aspects and appropriateness.

Holistic PV design

The application of PV systems must be part of a whole (holistic) approach to building design and construction. A high-quality PV system can provide a substantial part of the building's energy needs if, in the first instance, the building has been designed to accommodate PV appropriately and to be energy efficient. The more energy consumption is reduced, the more significant PV becomes as a supply option.

In a holistic approach, the integration of a PV system is not simply the replacement of building materials and components or the resolution of formal aesthetic concerns in the design. The integration also embraces other functions of the building envelope. Mounted on a sloped roof, a PV system can be part of the weatherproof skin using glass, or it can be mounted above a watertight foil protecting it against direct and ultraviolet (UV) sunlight, thereby extending the life of the foil. These kinds of systems are available for flat roofs as well. PV systems mounted on extruded polystyrene insulation material as a thermal roof system are available and are well-suited to the renovation of large flat roofs. PV systems can also be integrated on building elements such as canopies and exterior shading systems. The designer must carefully examine the details of shading and PV technology operation to fully understand how to integrate PV systems effectively and elegantly. One of the first things that a designer will discover is that a good PV system is not automatically a good shading system, nor is the reverse the case, as depicted in figures 1 and 2.

Orientation is one of the major design considerations for buildings, and is particularly important in low-energy buildings. A building's heating load, the need for solar shading and the design of façades are all influenced by the orientation. Orientation is important for the performance of PV systems. Façade systems might be more suitable in certain countries, especially at a northern (above 50° N) or southern (below 50° S) latitude. For countries between these latitudes, sloped surfaces facing the sun or even horizontal surfaces might be more appropriate. These are discussed in detail in Chapter 5. The final choice made by the designer will be influenced by restrictions on orientation, the amount of irradiation in the region, climatic operating constraints (such as temperature, relative humidity and snowfall), any shading by surrounding buildings and obstructions, and the required aesthetics of the design.

leating and cooling loads and daylight control systems can be combined with the integration of PV systems. Moreover, detailed performance studies by a designer will discover that PV can also be an effective and integral part of the thermal envelope or thermal system. The most important issue for the architect is to become fully conversant with the capabilities of the PV cell typologies and comfortable in finding creative integration possibilities at the early stages of design. A PV system may not be easily or cheaply added to a building that was not initially designed with that intention. A PV system is a design element of a building and thus, should be considered in the very early stages of building concept design planning.

PV basics

Solar power is the most reliable source of electricity in the world today. Photovoltaic modules generate electricity when they are exposed to sunlight. The actual creation of usable electrical current in a solar cell takes place at the atomic level. The most commonly available solar cells are made from high-grade silicon that is treated with negatively and positively charged semi-conductors, phosphorous and boron. This process is called 'doping'. When light energy (photons) strikes the face of the cell, it excites the electrons within the cell. This flow of electrons (current) from the negative semi-conductor (phosphorous) to the positive semi-conductor (boron) is what we call the photovoltaic (PV) effect.

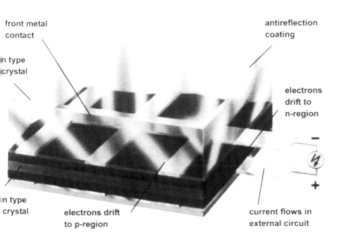

Fig 3 Solar cell operating principles
Source: Demosite, Switzerland

Solar cells can be separated into the following categories, according to their crystalline structure:

Mono-crystalline silicon cells

These cells are made from pure mono-crystalline silicon. The silicon has a single and continuous crystal lattice structure with almost no defects or impurities. The principal advantage of mono-crystalline cells is their high efficiency, typically around 15 per cent, although the manufacturing process required to produce mono-crystalline silicon is complicated, resulting in slightly higher costs than other technologies. Different manufacturing methods are used, depending largely upon the Czochralski method of growing, or pulling a perfect crystal that

has a solid, cylindrical shape. EFG (Edge-defined Film-fed Growth) has become popular, where the cells are cut from an octagon, to allow higher packing densities in modules. Another approach deposits grown films of crystalline silicon onto a low-cost substrate. The cost of silicon ingot sawing is eliminated and the quantity of silicon per solar module can be reduced significantly. A third approach is a string ribbon technique, where two high temperature strings are pulled vertically through a shallow silicon melt and the molten silicon expands and freezes between the strings.

Fig 4 Mono-crystalline silicon cell
Source: Enecolo, Switzerland

Poly- or multi-crystalline silicon cells

Poly-crystalline (also called multi-crystalline) cells are produced using ingots of multi-crystalline silicon. In the manufacturing process, molten silicon is cast into ingots, which are square or rectangular in shape, and allowed to cool so as to form large crystals. These ingots are then cut into very thin wafers and assembled into complete cells. New manufacturing methods also use the approach of grown films of poly-crystalline silicon on a low-cost substrate. Such substrates have included a metallurgical-grade silicon sheet, stainless steel, ceramics and quartz glass, using a variety of growth techniques to deposit silicon films onto these substrates. Poly-crystalline cells are cheaper to produce than mono-crystalline cells, due to the simpler manufacturing process. They tend to be slightly less efficient however, with average efficiencies of around 12 per cent.

Fig 5 Poly-crystalline silicon cell
Source: Enecolo, Switzerland

Amorphous silicon cells

Amorphous silicon cells are composed of silicon atoms in a thin homogenous layer, rather than a crystal structure. Amorphous silicon is produced by deposition onto a substrate, rather than wafer sawing, so the cells can be thinner. For this reason, amorphous silicon is also known as a 'thin film' PV technology. Amorphous silicon can be deposited on a wide range of substrates, both rigid and flexible, which makes it ideal for curved surfaces and 'fold-away' modules. Amorphous cells are, however, less efficient than crystalline-based cells, with typical efficiencies of around 6 per cent, but they require less material and are therefore cheaper to produce. Their low cost makes them ideally suited for many applications where high efficiency is not required and low cost is important. For instance, their early market has been in appliances such as calculators and watches.

Fig 6 Amorphous silicon cell
Source: Enecolo, Switzerland

Other thin films

A number of other promising materials such as copper indium diselenide (CIS) (figure 7), and cadmium telluride (CdTe), are now being used for PV modules. The attraction of these technologies is that they can be manufactured using relatively inexpensive industrial processes, certainly in comparison to crystalline silicon technologies, and typically offer higher module efficiencies than amorphous silicon. Some of the raw materials required, however, are less abundant than silicon and there are lingering concerns over the environmental toxicity of some of the elements used, although it appears possible to overcome these, with careful manufacturing, recycling and disposal processes.

Dye-sensitised solar cell (DSC)

Dye-sensitised Solar Cell (DSC) technology is best considered as artificial photosynthesis. It performs well under indirect radiation, during cloudy conditions, and when temporarily or permanently partially shaded. DSC technology has been dominated by the Grätzel titanium dioxide (TiO$_2$) cell. Particles of titanium dioxide are coated with a photosensitive dye and suspended between two electrodes in a solution containing iodine ions. When this dye is exposed to light energy, some of its electrons jump on to the titanium dioxide particles, which are then attracted to one of the electrodes. At the same time, the iodine ions transport electrons back from the other electrode to replenish the dye particles. This creates a flow of electrons around the circuit. Efficiencies over time are still to be established but technically could achieve around 10 per cent or more, and are very effective over a wide range of sunlight conditions.

Fig 7 Copper indium diselenide (CIS)
Source: Enecolo, Switzerland

Fig 8 Grätzel titanium dioxide (TiO$_2$) cell
Source: Sustainable Technologies Australia

Table 1 compares the typical efficiencies of PV technologies on the market and looks into the future, by seeing what can be achieved in the laboratory.

	Typical efficiencies %	Maximum recorded outdoors	Maximum recorded laboratory efficiency %
Mono-crystalline (m-Si)	12–15	22.7% ± 0.6	24.7
Poly-crystalline (p-Si)	11–14	15.3% ± 0.4	19.8
Amorphous silicon (a-Si)	6–8	10.4% ± 0.5	12.7
Cadmium telluride (CdTe)	7–10	10.7% ± 0.5	16.0
Copper indium diselenide (CIS)	8–12	13.0% ± 0.6	18.8
Gallium Arsenide (GaAs)		25.1% ± 0.8	
Titanium Dioxide Grätzel cell (TiO$_2$)	10	n/a	

Table 1 Comparison of solar cell efficiencies
Source: IEA Task 1 (2003); www.iea-pvps.org

From solar cells to PV modules and arrays

The solar cell is the basic unit in a PV system. An individual solar cell can vary in size from about 1 cm (0.4 inch) to about 15 cm (6 inches) across and typically produces between 1 and 2 watts, hardly enough power for the great majority of applications. The power is increased by connecting cells together to form larger units called modules.

The cells are welded in series to a string of several solar cells. Standard modules use around 36 solar cells and have a peak rating (Wp) of around 60 watts. For large modules (150 Wp), two cell strings are employed and can be connected at the back to electrical junction boxes. Thin-film materials such as amorphous silicon, CIS and cadmium telluride can be made directly into modules. The cell material is sputtered onto a substrate, either glass, polyamide or stainless steel, and interconnected to a module by laser.

Fig 10 PV module layered structure
Source: Martin Van Der Laan (M. ART)

Fig 9 Amorphous silicon flexible module structure
Source: Unisolar

A PV module is composed of interconnected cells that are encapsulated between a transparent cover and weatherproof backing. The modules are typically framed in aluminium for mounting, although frames may not be required for building applications. The PV module is the basic building block of any PV power system. The term 'solar panel' is often used to refer to a PV module. However, the same expression is also used in reference to solar water heating systems, so to avoid confusion, 'photovoltaic module' is preferred.

Solar cells are laminated to protect them from the external environment. On the front, a tempered, low iron-content glass is usually used. This type of glass is relatively cheap, strong and stable. Furthermore, it has high transparency, good self-cleaning properties and prevents the penetration of water, water vapour and gases. On the rear side, a thin polymer sheet is usually used. The sheet should also prevent the penetration of undesirable vapours and gases. For bi-facial modules, which can generate electricity from front and rear, or when extra strength or semi-transparency is required, glass is used at the rear. To provide adhesion between the different components of the module, the cells are sandwiched between thin sheets of ethyl vinyl acetate (EVA). The encapsulant should be stable at elevated temperatures and under UV exposure. The stability of the encapsulant is one of the major contributors to the expected lifetime of the module. To improve the strength and rigidity of the module, it can be framed using aluminium. Some of the crystalline silicon PV module manufacturers now guarantee a lifetime of 20 years for their modules.

Typical module sizes are 0.5 x 1.0 metre and 0.33 x 1.33 metres. However, modules of any desired size can be produced. Because of the previously discussed properties of amorphous silicon, modules of this material can have various forms and sizes, though most of the commercially available modules are rectangular and are similar in size to the crystalline modules. Furthermore, amorphous silicon can be deposited directly onto building components such as window glass, metal sheets, plastics and roof tiles.

Standard rectangular modules can be delivered with or without frames. Frameless modules, or laminates, can essentially be processed as normal glass panes. The thickness of glass–Tedlar laminates is generally 8 mm. Glass–glass laminates are typically at least 10 mm thick. Tedlar is used to provide back reflection and a high transmissivity toughened glass acts as a superstrate.

A major cost component of a conventional photovoltaic module is the processing of the silicon wafers, electrical interconnection and encapsulation. There are many methods under development, aiming to reduce these costs. A new Sliver Cell™ developed at the Australian National University, is produced using special micro-machining techniques, requiring the equivalent of two (rather than 60) silicon wafers to convert sunlight to 140 watts of power. The thin slivers are bi-facial and can be spaced out as required within a glass encapsulant to alter its translucent properties.

Fig 11 Sliver Cell™ (left) and components (right)
Source: Centre for Sustainable Energy Systems, Australian National University

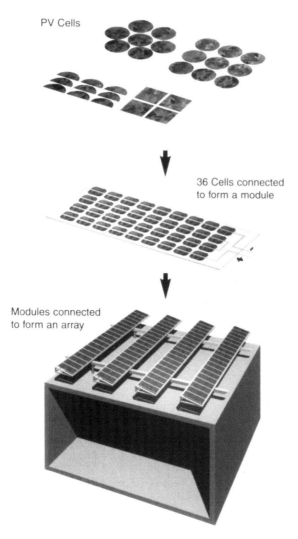

PV Cells

36 Cells connected
to form a module

Modules connected
to form an array

Fig 12 Diagram illustrating the cell to module to array relationship
Source: Martin Van Der Laan (M. ART)

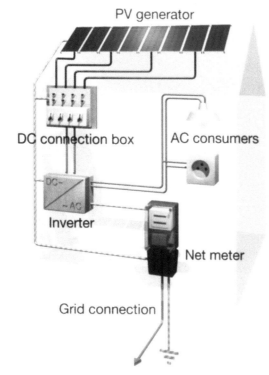

PV generator

DC connection box

AC consumers

Inverter

Net meter

Grid connection

Fig 13 Grid-connected PV installation
Source: Enecolo, Switzerland

Modules can be connected in series to strings and then in parallel to form even larger units known as arrays. Figure 12 illustrates the cell to module to array relationship.

Modules or arrays, by themselves, do not constitute a PV system. Also needed are structures oriented towards the sun on which to fix them, and components that transport and convert the DC electricity produced into alternating current (AC), for use in building applications. These structures and components are referred to as the balance of system (BOS) and are illustrated in figure 13.

BOS components include support structures, electrical junction boxes and inverters (also referred to as power conditioners) for DC to AC conversion. These elements account for approximately 40 per cent of the total investment cost for a PV installation.

The electricity meter, which is already a component of all grid-connected buildings, is not considered part of the BOS. Interconnection of solar energy to the electricity grid network varies by country and by power distribution utility. It is now commonplace that the solar power produced may be fed into the grid, either using a separate production meter or a bi-directional net meter.

When using a separate PV production meter, the power utility is able to buy the solar power at a different rate than conventional tariff electricity. This concept is currently not well-developed and varies depending on the power utility's policy. Bi-directional net metering is becoming more accepted, where the PV owner receives the same energy rate for solar power produced as is charged for conventional utility-sourced energy. Differential rates and the use of separate production meters may become more common in coming years, as more power utilities develop renewable energy labelling and pay a higher energy rate for renewable sources. Building designers should check this issue at the early stages of the BiPV project, as it can substantially improve grid connection planning and heighten the incentive to invest in PV. More technical details pertaining to metering issues can be found in Chapter 6.

Key design issues for architects and associated practitioners
Building and PV design interaction

A successful BiPV solution requires interaction between building design and PV system design. The approach can be to fully integrate the PV system in the building, displacing a conventional external building material, such as tiles on a roof or cladding against a façade. An alternative, but equally valid approach is to see the PV system not as an intrinsic building design issue, and to place it onto a building element, such as a roof or other fixture. The integration of PV systems in architecture can be divided into five approaches. It can be:

- applied seamlessly;
- added to the design;
- added to the architectural image;
- used to determine the architectural image;
- used to explore new architectural concepts.

These categories have been classified according to the increasing extent of architectural integration. However, a project does not necessarily have to be of a lesser quality just because PV modules have been applied seamlessly. A highly visible PV system is not always appropriate, especially in renovation projects with historic architectural styles. The challenge for architects is to integrate PV modules into buildings properly. PV modules are new building materials that offer new design options. Applying PV modules in architecture should therefore lead to new designs. A selection of projects described below provides further explanation.

1 The PV system is **applied seamlessly** and is therefore not architecturally 'disturbing'. The PV system on a dwelling in Tokyo (figure 15) harmonises with the total project by maintaining roof tile dimensions and a complimentary colour sequence. Another example is in Maryland, USA (figure 14) where the PV is laminated on a Spandek roof system and is barely visible. This solution was chosen because the entire project is of historical significance. A modern high-tech material would clearly not be appropriate for this architectural style.

2 The PV system was **added to the design**. The building may be missing a design function that PV can fill in the form of, for example a practical PV shading device, as shown in a building in Madrid in figure 16. This can occur if the intended purpose of internal spaces within the building changes or the comfort levels required need to be improved. PV provides both an active and passive solar shading solution and invariably, a well-designed PV eave (figure 17), awning or louvre retrofit can avoid the need for a mechanical cooling system upgrade. Often, the PV additions do not necessarily mean that architectural integration is lacking as the 'added' PV system is not always highly visible.

Fig 14 PV standing seam metal roof on dwellings in Maryland, USA
Source: United Solar System Corp

Fig 16 IES building in Madrid, Spain
Source: BEAR Architecten, Netherlands

Fig 15 Dwelling in Tokyo, Japan
Source: New Energy and Industrial Technology Development Organisation (NEDO), Japan

Fig 17 Solar canopy house design in Nieuwland Amersfoort, Netherlands
Source: BEAR Architecten, Netherlands

3 The PV system **adds to the architectural image** by being integrated into the total design of the building without dominating the project. In other words, the contextual integration is excellent. PV provides a visual statement that can either offer subtle or substantial changes to the architectural dynamics of the building. The EMPA building's PV façade and roof canopy in figure 18 gives visual acuity to the design without overpowering the original building form.
A stylish PV eave system can effectively modernise an older office building (see case study, Netherlands: Le Donjon) with previously predictable flat roofline features. This can effectively create a visual edge to the building that could lead to a positive influence on its market value.

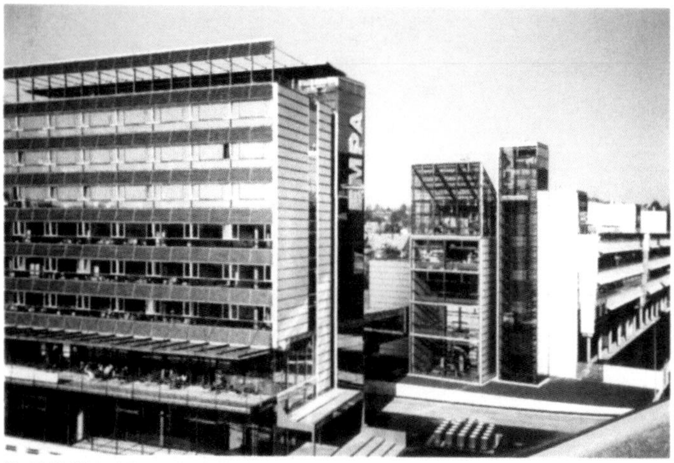

4 The PV system **determines the architectural image**. The PV system is used as an integral part of the building envelope and thus espouses a core building characteristic. Figure 19 shows dwellings in the Dutch development at Langedijk using PV to dominate the roofscape and aesthetic feel of the area. The blue-coloured PV roof, while unconventional, blends effectively with the water and sky vista. In other examples, such as the office building shown in figure 20, PV revitalises what might have been a stereotypical glazed corridor into a prominent feature, playing an important role in the total image of the building.

5 The PV system **leads to new architectural concepts**. The application of PV modules, particularly in combination with passive solar design concepts, leads to new designs and new architecture. Chapter 1 figure 7 shows the translucent properties of PV, creating curved or dynamic surfaces as a fundamental construct of the building. Architecturally, this presents new design options, working with a variety of support structure materials and complementary building textures, such as wood and steel. Importantly, the architect can control and experiment with the natural lighting dynamic within the building and transform the colour and feel of the internal spaces as the sun's position alters during daylight hours. Similarly, the use of artificial light on different textures and PV laminates, including fluorescent backings as used for the Sydney Olympic Boulevard lighting pylons (chapter 4 figures 39 and 40), can re-energise the solar panels when natural light at night has disappeared. This creates a new appeal and innovative architectural building form.

![Fig 18 EMPA building in St. Gallen, Switzerland]

Fig 18 EMPA building in St. Gallen, Switzerland
Source: Enecolo, Switzerland

Fig 19 Dwellings on the 5 MW site Langedijk, Netherlands
Source: BEAR Architecten, Netherlands

Fig 20 IMEC building in Leuven, Belgium
Source: IMEC, Belgium

ow can a good BiPV design be determined?

f photovoltaic applications are integrated in an elegant and aesthetically pleasing way, high-quality examples can be used to convince clients, architects and the public of the positive role that PV systems can play, both in terms of performance and as an ntegral building element. To identify high-quality projects, some criteria are needed. Although most built PV projects aim to achieve good technical performance, PV systems have tended not to be very well-integrated and the architectural quality can often be disappointing.

To qualify a project as 'well-integrated', the architectural quality has to be of a high standard, and of course, both the building quality itself and the technical performance of the PV system have to be met. A poorly integrated PV system on a well-designed building might be disturbing, but a poorly integrated PV system on a poorly designed building is clearly worse. Similarly, an elegant PV system will not necessarily improve the overall design.

Architectural criteria that have been formulated by the architects in IEA PVPS Task 7 are given below and attempt to define and characterise what a good BiPV design might entail. These criteria are not mutually exclusive but provide a basis from which good questions and comparisons concerning the architectural quality of a BiPV project can be made. These architectural criteria need discussion, particularly with non-architects and manufacturers of photovoltaic systems for integration in roofs and façades, who often believe that their systems always fit perfectly. In many instances, architects are not aware of the attractiveness or expressive opportunities of integrating PV systems into their buildings.

Naturally integrated

Natural integration refers to the way that the PV system forms a logical, natural part of the building and how, without a PV system, something will appear to be missing. The system completes the building. The PV system does not have to be that obvious. In renovation situations, the result should look as though the PV system was there before the renovation.

Architecturally pleasing

Based on a good design, does the PV system add eye-catching features to the design?

The building should look attractive and the PV system should noticeably improve the design. This is a very subjective issue, but there is no doubt that people find some buildings more pleasing than others.

Well-composed

The colour and texture of the PV array should be in harmony with the other exterior finish materials. In many cases the PV system will intentionally be produced in a certain way, for example, frameless instead of framed modules. Specific PV technologies can be chosen to achieve a suitable colour, transparency, shape or texture.

Grid, harmony and composition

The sizing of the PV system matches the sizing and grid of the building. This will determine the dimensions of the modules and the building grid lines used. 'Grid' refers to the modular system of lines and dimensions used to structure the building.

Well-contextualised

The total image of the building should be in harmony with the PV system and match the context of the building. The entire appearance of the building should be consistent with the PV system used and vice versa. In a historic building, a tile-type system will look better than large modules. A high-tech PV system, however, would fit better in a high-tech building.

Fig 21 The PV roof at De Kleine Aarde, Boxtel, is architecturally pleasing
Source: BEAR Architecten, Netherlands

Fig 22 80.5 kWp amorphous silicon tiles. The Atlantis Sunslates are part of a good composition in the National Horse Stables in Bern
Source: Atlantis Solar Systems AG, Bern, Switzerland

Fig 23 Amorphous silicon solar fascia at London Waterloo's
new community youth project, The Centre Lambeth, UK
Source: Solar Century

Fig 25 This structural glazing detail is well-engineered and innovative
Source: Officine di Architettura, Cinzia Abbate

Well-engineered

This does not only concern the waterproofing or reliability of the construction. It includes the elegance of the details. Have details been well conceived? Did the designers pay attention to detail? Has the amount of material been minimised? Are details convincing? These considerations will determine the influence of the working details.

Innovative new design

PV is an innovative technology, asking for innovative creative thinking by architects. New ideas can enhance the PV market and add value to buildings.

PV systems can be used in many ways and we still do not know all of them. The IEA Task 7 international competition to design PV products for use in the built environment has helped stimulate new design concepts and innovation. Winning designs included a solar sail feature in front of an office complex, and a PV ventilated rainscreen stressed-skin timber construction, as well as careful consideration of ways in which the different passive, active and mechanical energy systems interact. Competitions such as the University Solar Decathlon in America, solar car racing events, the virtual World Solar Challenge and many others are inspiring new PV design concepts and technological advancements.

BiPV project scale

The way photovoltaics is dealt with in architecture can differ from country to country. It can also depend on the scale, culture and type of financing for PV building projects. In countries such as Denmark and the Netherlands, where public housing is common, large-scale housing developments are apparent. Professionals, such as project developers and architects, who implement the housing construction process, are presented with numerous opportunities for PV roof integration in single-family houses and for façade and roof integration in apartment buildings. This allows for the integration of a large number of PV systems during construction, such as those seen in Nieuwlands, Amersfoort and the Sydney Olympic Village. Integration of PV systems in residential homes can be carried out on an ad hoc basis but often the motivation comes from the private owner. There is a significant market for private homeowners who buy 1–4 kWp small-scale PV systems and mount them somewhere on their house. In the private sector, more building-integrated PV systems are found in commercial and industrial buildings, where building professionals are closely involved. With these types of buildings, PV systems are more readily integrated both into façades and roofs.

Fig 24 Sainsbury's Supermarket petrol station
Source: Solar Century, UK

The first issue to clarify when designing with photovoltaics is the primary reason for integrating PV in the building. For example, is it for general energy supply or is it to make the building more independent of conventional utility power? For general energy supply, large PV systems will be necessary. Key design considerations will be the architectural treatment of large areas of PV and selecting from different types of modules, different shapes, different colours or different textures of modules to be used to face off the building. For buildings aspiring to be energy independent, the actual size of the PV system will depend on the efficiency of the system and the generated yearly output. The designer will have to accommodate a specific amount of modules and will have to design the building around the integrated system.

Core design issues for BiPV

The next critical step in the design will be to establish the number of modules, the dimensions of the modules and the total dimensions of the system to be integrated into a roof or façade. Shadowing of the modules is an important consideration, particularly when they are connected together in a string. A module that is partly shaded will lose more efficiency than is expected since when one row of cells in a module is covered or heavily shaded the output of the whole string of modules can be affected. Small objects that can cause shading are less important. The shade will move during the day and there is typically a significant amount of indirect light available. Some modules have integrated diodes to make a short cut when a row of cells is covered or shaded. AC modules also help to isolate the impact of shading as each module's DC power output is converted to AC and drawn individually. Other photovoltaic technologies, such as amorphous silicon and titanium dioxide cells, are less impacted by irregular shading effects, due to different electrical connection characteristics and better performance in low light. Inverter characteristics are also critical, since most have a cut-off point. In general, shading should be avoided as much as possible.

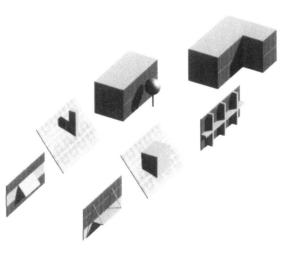

Fig 26 Shadow impacts
Source: Martin Van Der Laan (M.ART)

Space is also needed for the inverters to be housed safely and effectively. The modules have junction boxes on the back that will be connected to the inverters. Improved efficiency is gained by placing the inverters near the modules. From the inverters, an AC cable has to be fed into the grid via the meter. The efficiency of crystalline silicon cells diminishes when the temperature rises and hence the backs of the modules must be ventilated. Since space for a junction box is also needed on the back of the module, a gap of between 20–50 mm (depending on the size of the junction box) can be used for both functions. Safety switches will be needed near the inverters to work safely on the PV system. BiPV electrical issues are discussed in detail in Chapter 6.

Maintenance and cleaning

Modules, particularly those laid flat, will collect dirt, especially in urban environments or dusty rural environments, and this can decrease the efficiency by approximately 4 per cent or more. Modules with an angle over 20° will self-clean through the action of rain. Special treatment with PV Guard can help to keep the modules clean.

Fig 27 Soiling of panels
Source: ECN, Netherlands

Depending on the application, the PV module must meet appropriate building standards. As with hail, fire, wind, structural load and other building regulations, the developer of PV installations should consider the applicable national and local council codes. In general, flat or sloped roof applications are less affected by building codes than overhead or façade installations. Specific compliance requirements are provided in Chapter 6.

Fig 28 Hosing down panels
Source: National Renewable Energy Laboratory (NREL)

Urban aspect

PV integrated systems in buildings offer added functionality beyond power generation. PV structures can effectively displace conventional building materials and assist in reducing overall system costs. To generate maximum power from building-integrated systems, several urban and architectural aspects are important. The starting point is the maximum power that can be produced from a system. Hindrances can occur from the part shadowing of a system by other buildings or objects and, very occasionally, by reflective glare.

Orientation and angle

The amount of irradiation depends on the latitude of the building and the local climate. The maximum irradiation depends on the orientation and the angle of the collection surfaces. For latitudes 52° north good results (over 90 per cent) can be achieved between southeast and southwest with system tilt angles between 30° and 50°. Orientations between east and southeast and between southwest and west are acceptable for tilt angles between 10° and 30°, since the irradiation will only be reduced by approximately 15 per cent from optimum. Figure 29 depicts annual irradiation in relation to orientation and tilt for Freiburg, Germany (northern hemisphere) and for Sydney, Australia (southern hemisphere).

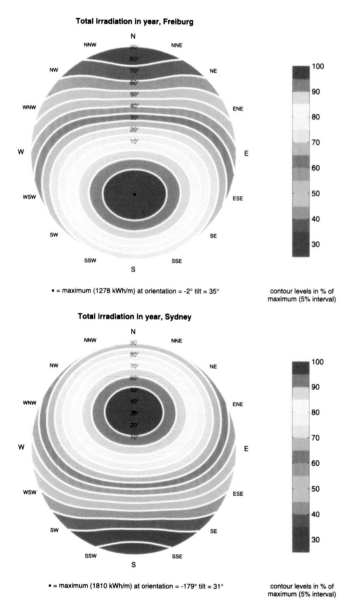

Total irradiation in year, Freiburg

• = maximum (1278 kWh/m) at orientation = -2° tilt = 35°

contour levels in % of maximum (5% interval)

Total irradiation in year, Sydney

• = maximum (1810 kWh/m) at orientation = -179° tilt = 31°

contour levels in % of maximum (5% interval)

Fig 29 Total irradiation depending on tilt and orientation for Freiburg, Germany and Sydney, Australia
Source: ECN, Netherlands

Flat roof systems with very low angles (between 5° and 10°) can be a good solution should orientation prove difficult. The loss of irradiation for the Freiburg example is estimated to be between 5 per cent orientated to the south and 20 per cent for a northerly orientation. Detailed discussion on orientation is provided in Chapter 5.

Distance between buildings

Shadowing between buildings should be avoided. For low-rise settlements the distance between the buildings can be easily calculated and shadowing is a relatively easy problem to solve. For mixed settlements it will be more difficult. A high-rise apartment building in a low-rise neighbourhood, for example, can generate a considerable amount of unwanted shading. The building density of an area is critical. In high-density areas such as cities and urban centres, the distances between buildings will be limited and shadowing will occur for significant periods of the year. Typically, façade systems are the most susceptible to shading and require larger distances between buildings than roof systems.

Trees

Greening the area around buildings offers a number of positive qualities, including improved appearance and a microclimate that will be comfortable for the inhabitants. However, the shadowing effect of trees is very important, especially as their crowns will be at a maximum during the summer period. In addition, in the winter period, when trees lose their leaves, the branches can also generate significant shade.

The rate and extent of tree growth is also typically underestimated. Since every tree will grow a little more each year, planning is very important to avoid problems a few years after the building is constructed or the PV system is installed. Possible solutions include:

- Only planting trees on the north side of buildings (northern hemisphere), or the south side for southern hemisphere buildings.
- Only planting small trees that have a growth limit up to roof height.
- Yearly pruning of trees to avoid shading of the PV collection surfaces.

Zoning

PV systems in urban areas may require special solar zoning. Three-dimensional maps and computer models can be used to identify rights to sunlight and to establish the borders of building areas to prevent future shading problems.

Glare

Although not typically a major problem, unwanted glare can occur under certain circumstances. There are typically few problems in low-rise settlements, but in a mixed development of low- and high-rise buildings with increased façade surface area, there might be hindrance from solar reflection off the PV to surrounding buildings. The extent of this depends on the textural properties of the PV and mounting system, the sun's position and direct irradiation intensity during specific time periods of the year. Complying with required distances between buildings to avoid shadowing would also eliminate most of the potential glare problems.

orm and colour

olar cells are typically blue, dark blue or almost black in colour.)ifferent colours – grey, green, red, orange and yellow – are ossible, but are not typically standard manufacture and may be gnificantly more expensive than standard modules. Blue oloured cells have the highest efficiency.

he type of module also has an impact on colour. Frameless nodules give a very harmonious impression as the roof is not isturbed by frame patterns of different colour or material to that f the cells. The colour of the cells will be the colour of the total urface and the roof will look like one large surface.

ramed modules give another visual accent. The frames can look eavy and profoundly influence the overall visual impact of the V array and its integration with the building. Smaller frames in he same colour as the cells are almost invisible at the surface. ometimes the frames can be used to make a specific mpression. Framed modules will emphasise the dimensions of ach module and will influence the mounting profile of the roof.

he colour of the frames and of the encapsulant used as the rear urface of the module can also be different, thus providing more pportunity for design interest.

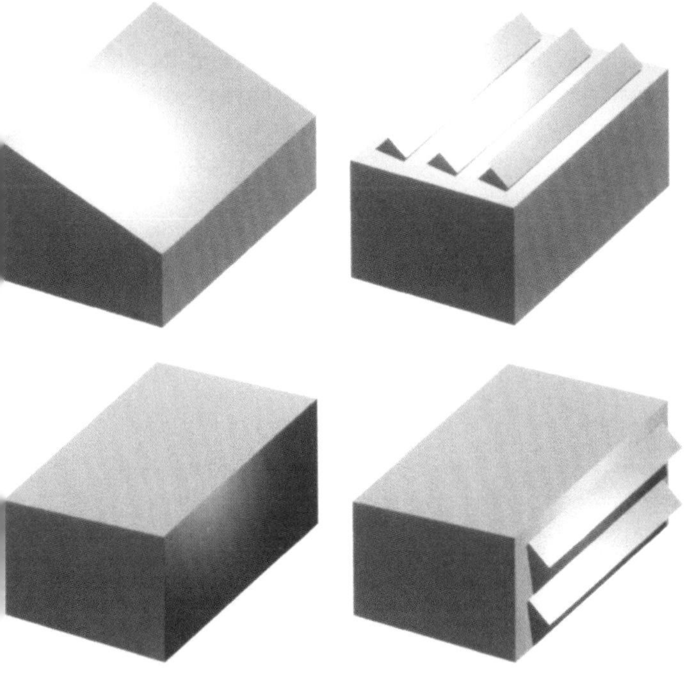

BiPV integration techniques

The building envelope provides a number of possibilities for the integration of PV. The four main options are:

- Sloped roof
- Flat roof
- Façade special applications
- Shading system

Sloped roof applications have the advantage of using the inclined roof as a platform, whereas flat roof applications need a special mounting structure to provide the required orientation angle for the modules. Façade installations have a high profile, due to the high visibility of the installation. The technical requirements are mostly higher than for a flat or sloped roof installation, because of the wiring and the junction boxes, which have to be hidden, and the increased difficulty in fixing the array to the building. The fourth category relates to special applications, for example, shading elements, atriums, louvres, balconies and skylights. This section focuses on the four categories and presents different typologies of integration. A comprehensive list of product typologies can be found on an online International Energy Agency (IEA) internet database at www.pvdatabase.com.

Sloped roof

Sloped roof constructions are very common for residential buildings and are most suited for a PV installation if orientated approximately towards the equator. Different typologies of mounting systems are readily available on the market for sloped roof applications. One of the cheapest typologies is the mounting of a profile system off the roof, above the tiles (figure 31).

Fig 31 Concept of sloped roof mounting structure
Source: Ha Wi Energietechnik, Germany

Fig 30 Basic building envelope PV integration typologies
Source: Martin Van Der Laan (M. ART)

A common installation approach is to fix special roof clamps onto battens. Vertical profiles are installed to form a base, followed by horizontal aluminium profiles. The modules are placed on those profiles and fixed either by screws or clamps.

A second approach (figure 32) replaces some standard tiles with special ground plates, onto which the horizontal profiles are fixed directly.

Fig 32 Klöberprofile system
Source: Klöber, Germany

The main advantage of such mounting techniques is that an air gap exists between the rear of the module and the roof, providing a cooling effect that can be achieved on existing buildings with no large effort or additional costs. This may attract differing visual appeal with some preferring a more integrated solution. Some examples of stand-off mounting techniques, which differ mainly in the method of fixing the modules to the profile system, are given in figures 33–35.

Fig 33 Sun Stick
Source: RegEN, Germany

The Sun Stick system prefabricates the modules with a hook at the back of the modules to snap them onto the profiles during construction. Other systems use standard modules or laminates that are directly mounted onto the profiles and then fixed by a clamp or by a so-called diamond clip.

Another system, the Plug & Power™ product, is also manufactured with rear metal clips in advance. The modules are clicked onto the profiles, plugged in and provide power in a relatively straightforward installation process.

Fig 34 Plug & Power™, Pacific Solar, Australia
Source: Pacific Solar, Australia

Fig 35 Plug & Power™,
Pacific Solar, Australia
Source: Pacific Solar, Australia

Fig 36 Alu-Tec
Source: Solarmarkt, Switzerland

Several systems are available for 'real' integration of standard PV laminates or modules into the roof profile itself. The key issue, other than the need for water-tightness, is that these systems need to be suitable for standard modules as well as for custom-made ones, if needed. Thus, the overall product design is slightly more expensive than on-roof mounted designs, but in some applications, they are cost competitive. As for the roof-mounted systems, most integrated systems apply profiles, which are mounted onto the roof batten structure. The laminates are fixed to the profiles either by rubber solutions or mechanically. EPDM (ethylene propylene diene terpolymer) rubber is often used as a malleable product with high resistance to punctures, UV radiation, heat, weathering and microbial attack.

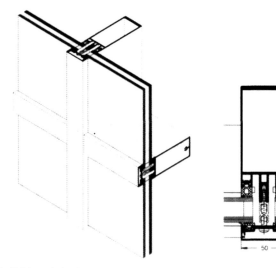

Fig 37 Schematic drawing
Source: Schüco, Germany

Fig 38 Installation with Schüco system
Source: Schüco, Germany

Fig 39 Detail of BOAL system
Source: BEAR Architecten, Netherlands

The BOAL profile system is a vertical profile fixed to the existing roof construction using regular screws. The number of screws required is determined by the maximum local wind load. The PV laminates slide into the vertical profiles and are glued in place. The top cover is clipped into the vertical profile.

Integrated systems typically show a better aesthetic than on-roof mounted installations, but have the disadvantage of less cooling at the back of the modules. Experience shows that the temperature of the module may reach up to 60 °C for mid-European conditions, thus resulting in a power reduction of up to 15 per cent during ambient temperature peaks. Calculating the annual energy reduction because of a poorer cooling effect, a value of approximately 2–5 per cent can be noticed for installations in mid-European locations. This will be even higher under very high ambient temperature conditions. It is therefore important to consider a ventilation inlet at the bottom of the PV installation and an outlet at the top, which would facilitate a natural ventilation effect and bring the high temperature during summer down to approximately 50 °C.

Sloped roof: shingle roof tile

A very dynamic field is the shingle and PV roof tile. Several new developments have been introduced into the market. Important issues include aesthetics, ease of handling and installation time. The category covers a wide spectrum from a simple tile, equipped with a very small PV module, to custom-made laminates glued onto a building substrate, and to the direct lamination of PV cells with, for example, reinforced fibre plates. The following section presents some products and provides more information about the integration typologies.

All systems have one aspect in common, that of replacing conventional tiles and being mounted directly on the roof structure (battens or water barrier foil). Most products use custom-made laminates, with the PV capacity varying from just 7 Wp up to 100 Wp, either placed on a building substrate and then integrated, or framed suitably for direct integration. The products are usually easy to connect to the existing or new tiles and are based either on a shingle system or have an overlapping area that will be covered by profiles. Some systems provide a nearly watertight installation, whereas others have the same properties as standard tiles and provide protection against rain and normal storm conditions.

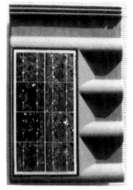

Fig 40 Sunny Tile
Source: Star Unity, Switzerland

Fig 41 Sunslate
Source: Enecolo, Switzerland

Fig 42 SED Dachziegel
Source: SED, Austria

Sloped-roof profiles

Figure 40 shows the Sunny Tile, made in Switzerland by Star Unity. It is a PV laminate (7 Wp) integrated in a plastic tile, with the same properties as a conventional tile. The result is a very aesthetic appearance, thanks to the small units. These are relatively costly, in the range of approximately EUR10–13/Wp for the tile only, depending on the capacity. Another example, which is based on an existing building element with the same dimensions, is shown in figure 41. Sunslate is a Swiss product and consists of an approximately 14 Wp laminate, which is glued to a reinforced fibre plate. Figure 42 shows a product from Austria called SED Dachziegel.

One advantage of these products is that normal roofers (used to tiles and shingles) can mount the tiles and interconnect them without the need of specially trained personnel. Consequently, many systems are similar in approach and design to existing building products, to ensure simple shingle or tile mounting.

Some module products are screwed to the wooden batten (Terra Piatta®, figure 45) and others are mounted on prefixed clamps on top of prefabricated plastic moulds (Braas, figure 44). One of the oldest known solar tile products, seen in figures 46 and 47, is produced by Phönix, and was invented and distributed by Newtec, Switzerland.

Fig 43 Solar tiles, Peter Erling Pty Ltd
Source: Peter Erling

Fig 44 Braas tile, SRT 35
Source: Braas, Germany

Fig 45 Terra Piatta®
Source: Pfleiderer, Germany

Fig 46 Solardachziegel Newtec
Source: Enecolo, Switzerland

Fig 47 Mounting of the Solardachziegel
Source: Newtec, Switzerland

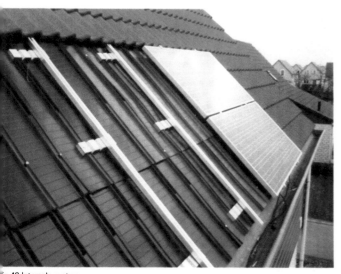

Fig 48 Intersole system
Source: Ecofys, Netherlands

Fig 49 Installation with Intersole
Source: Ecofys, Netherlands

Fig 50 SOLRIF module
with a BP Solar laminate
Source: Enecolo, Switzerland

Fig 51 Installation with large roof tile (SOLRIF)
Source: Isofoton, Spain

Fig 52 UniSolar shingle
Source: USSC, USA

All the systems above use custom-made PV laminates and are therefore more expensive than profile-integrated installations. In attempts to reduce mounting structure costs, new products were developed, based on standard laminates or modules where the mounting structure is adjusted to fit the dimensions of the PV element. Figures 48 and 49 show a product from the Netherlands called Intersole.

SOLRIF (Solar Roof Integration Frame) (figures 50 and 51) is a mounting system for inclined roofs and consists of special profiles for the direct integration of PV modules into the roof. SOLRIF creates a watertight roof comparable to a clay tile. The profiles are cast from aluminium by extrusion and are independent of any PV laminate. An innovative feature is the aspect of the lower edge where the frame is fixed on the back, thus allowing free water drainage. The problem of soiling along the lower edge of the module is also avoided.

Thin film roof systems

New mounting structures can be explored if using thin film technologies. The PV cells are deposited directly onto a metal substrate, which can be produced in a roll-to-roll technique of metal sheets up to 5.8 metres in length if vertically mounted, or up to 2.5 metres when horizontally mounted.

Another product based on thin film technology is a flexible and lightweight shingle system shown in figure 52, produced by UniSolar, USA. The shingles (2.2 x 0.31 metres) are horizontally mounted on the roof. The modules can withstand the bodyweight of roof maintenance workers and the unobstructed lower edge shingle design helps reduce soiling of the modules.

Another special shingle system from Germany is produced by a company called Rathscheck. One module has a capacity of 50 Wp and can be walked on without concern for damage. The disadvantage is that, in case of failure of the solar module, the neighbouring shingles have to be removed first.

This product idea (figures 54 and 55) illustrates the major advantage of BiPV products. PV laminates do not just need to be glued onto a building substrate. The cells can be laminated (integrated) into the building material, preferably as a part of the manufacturing process. So far only demonstration projects have been realised with such products but it is only a question of time before the building industry accepts the PV cells and incorporates them in their materials, leading to 'real' BiPV products.

Some ideas are: integrated in a plate of concrete (figure 57), into reinforced fibre (figure 58) and another PV cell laminated on a plastic substrate and covered with a plastic sheet (figure 59). These are prototypes and are not commercially available products as yet.

It is predicted that some healthy cost reductions are possible if the PV cells are integrated into well-known building materials. Furthermore, one can assume that the aesthetics of such an element will be optimised with minimal barriers to its application. This however, requires further research into appropriate lamination materials and a better understanding of the building industry.

Even so, all the systems presented bring the disadvantage of less than optimal cooling at the back of the modules. It is important to consider a ventilation inlet at the lower part of the PV installation and an outlet at the top, to encourage a natural ventilation cooling effect.

Fig 53 Vertical mounting of Thyssen Solartec®
Source: Thyssen, Germany

Fig 54 Metal sheets are linked using clip-on profiles
Source: Thyssen, Germany

Fig 55 Rathscheck shingles being installed
Source: Rathscheck, Germany

Fig 56 Completed roof with Rathscheck shingles
Source: Rathscheck, Germany

Fig 57 PV cells integrated in a plate of concrete
Source: Enecolo, Switzerland

Fig 58 PV cells integrated into reinforced fibre
Source: Enecolo, Switzerland

Fig 59 PV cells laminated on a plastic substrate and covered with a plastic sheet
Source: Enecolo, Switzerland

Flat roofs

Flat roofs present a large potential of suitable areas for installations of PV plants. In a study carried out by IEA Task 74 of potential in industry, it was argued that large factories, corporate offices and apartment complexes could offer, on average, roofing equivalent to a quarter of the ground floor area for suitable PV siting. For example, if the ground floor area is 100 square metres, the area suitable for PV is approximately 25 square metres of flat roof. Sloped roof potential is approximately 15 square metres and façades 15 square metres.

Flat roof PV installations have the advantage of being able to be optimally positioned with support structures, and the inclination angle can be adjusted to meet specific demand and the location. The installations can track to the optimal sun angle either as one- or two-axis systems. Different systems for mounting can be separated into three categories: mechanically fixed to the roof structure; based on weight foundation; and an integrated solution, which incorporates insulation properties and forms the watertight layer.

In general, a flat roof PV generator helps to reduce the thermal load of a building due to the shading of the modules, but it is important to ensure there is space for maintenance visits and thus ensure the functionality of a flat roof over its operation lifetime.

Installations that are mechanically fixed to the roof structure by bolts and screws were the first application of PV in buildings. Nowadays this is less so, due to the possibility that any connection to the roof may leak over time and also because of the metal-intensive nature of such installations. Figures 60 and 61 are examples of systems that are available on the market. Many more appear as custom-made applications, depending on the roof structure.

Designers have also experimented with lower mounting support structures, fixed only by weight. Although this does not always have an aesthetically pleasing outcome, bringing added weight to the roof can pose demands on the load-bearing ability of the building and this needs to be assessed with care. At present, several systems are available and more are under development. The research for new solutions, less cost and less material is comparable to the approach for BiPV products in the inclined roof sector. The systems based on weight foundation have the advantage of being flexible, as they can be placed almost at any place and can be easily extended and are very suitable for a modular concept. The SOFREL design shown in figure 62 is an example. Some of these systems present low-cost solutions and have been applied in installations of hundreds of kilowatts.

Fig 62 SOFREL® Type 98
Source: Enecolo, Switzerland

Concrete elements are placed in a metal pan as weight foundation. Horizontal profiles are fixed to the base element according to the inclination angle needed.

Other concepts use the existing gravel on the roof as ballast. The structure is made of reinforced fibre or PE-plastic, which can be recycled. It is important that these systems are flexible and can take different sizes of modules in the range of 70–130 Wp. As the mounting structure is made of more expensive material, costs for the structure are higher, but can be compensated by less cost for the installation overall.

Fig 60 SolarFamulus
Source: RegEN, Germany

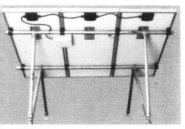

Fig 61 T Solar array structure
Source: BP Solar, UK

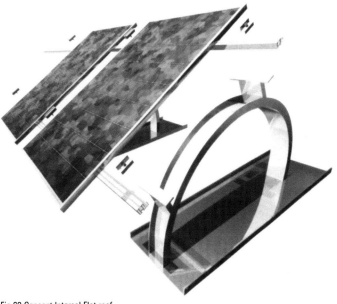

Fig 63 Concept Intersol Flat roof
Source: Donauer Solartechnik, Germany

A mounting structure for PV modules and laminates called SOLGREEN has been especially developed for green roofs. The aim is to develop a lightweight system, which adds almost no load to the green roof. The structure is fixed by the gravel foundation of the roof and the substrate for the roof vegetation. Thanks to this approach, solar arrays can be combined with vegetated roofs in an economical and aesthetically pleasing way. An important design issue is to give as much space between the ground and the module as possible, since the vegetation can grow up to 40 cm and the green roof requires maintenance, such as grass mowing.

Fig 64 Ansicht Schlet-Top Flach, Germany
Source: Enecolo, Switzerland

Fig 65 SOLBAC
Source: Enecolo, Switzerland

Fig 66 ConSole
Source: Ecofys, Netherlands

Fig 67 SOLMAX
Source: Solstis, Switzerland

Fig 68 Installation SOLMAX in Bielefeld, Germany
Source: Solstis, Switzerland

Fig 69 SOLGREEN small
Source: Solstis, Switzerland

Fig 70 SOLGREEN large
Source: Solstis, Switzerland

Fig 71 Green roof application
Source: Solstis, Switzerland

n a third category are systems that incorporate the properties of a roofing element, such as a watertight barrier and an insulation ayer. These products add very little extra weight to the roof. Although more expensive than standard flat roof solutions, there are gains in synergies concerning roof properties, and in overall aesthetics.

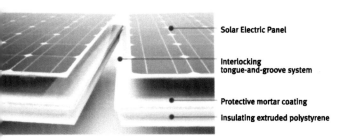

Solar Electric Panel

Interlocking
tongue-and-groove system

Protective mortar coating

Insulating extruded polystyrene

ig 72 Detail PowerGuard®
Source: PowerLight, USA

Fig 73 15 kWp installation in Kailua-Kona, Hawaii
Source: Enecolo, Switzerland

PowerGuard® is a practical solution that delivers solar electricity while insulating and protecting the roof. Each element interconnects easily, with no roofing penetration, and works as a retrofit over existing roofs as well as for new buildings. The system incorporates PV backed with extruded polystyrene foam. The elements are applied un-adhered over built-up, mechanically attached or fully adhered membranes.

Another interesting product is a roofing foil that includes Uni-Solar's triple-junction PV cell. The product, Evalon®-Solar, combines two products: an amorphous PV module, developed by UniSolar in the US, and a traditional roofing foil, made by Alwitra in Germany. It forms a watertight barrier and can be installed in a similar way to existing roofing foils. The product offers a new (and lightweight) aesthetic, due to the horizontal mounting. The amorphous PV module is produced with a 'roll to roll' process on a metal substrate and is so flexible that it can be rolled. The dark blue modules are up to 1.55 metres in width and 6 or 12 metres in length and are encapsulated in protecting plastic. Each module has an installed capacity of 128 Wp. In northern Europe, each module is expected to produce about 100 kWh of solar electricity each year. Alwitra foil consists of EVA (Ethyl-Vinyl-Acetate) and PVC (Polyvinylchloride).

Fig 74 Evalon® system
Source: Alwitra, Germany

Fig 75 PV installation with Evalon®
Source: Alwitra, Germany

All flat roof mounting systems, except the roof-integrated products, show a good performance, thanks to an improved cooling effect at the back of the modules. As the module is mounted from 10 cm up to 100 cm above roof level, airflow at the back of the module is guaranteed. Thus, the flat roof PV generators usually provide a higher energy yield than sloped roof or façade installations of equivalent size.

Other flat roof PV constructions, such as the Williams Building in Boston (figure 76) have developed effective mounting systems, which are fastened to the roof using epoxy-embedded anchors into the concrete roof deck. In this installation, the amorphous silicon modules from PowerLight provide 37 kWp of DC power to assist running a chiller system, and 28 kWp of AC power that feeds directly into the grid.

Sustainable houses designed at Etten-Leur, in the Netherlands, (figure 77) supply all their own energy. To accomplish this, they use passive solar energy, photovoltaic solar energy, heat pumps, and heat recovery. During the summer, heat is stored in the ground, to be used during the winter. The houses are situated in the district known as 'de Keen', a newly developed area with zero-energy houses being built to help reduce CO_2 emissions. These zero-energy houses are covered with an 'energy-roof' containing 6.2 kWp BP Solar PV modules (around 45 square metres) for each house. This roof construction overhangs a row of houses, which means that the energy systems can be easily reached, maintained and if necessary, expanded. Good ventilation improves the performance of the PV system. The orientation of the houses is more or less independent of the solar systems. The system can be further optimised in the future. Because the use of solar energy is a major part of the design process, new designs arise which are architecturally very interesting. The project shows how new concepts enrich the architectural value of the (flat) roof.

Other examples illustrate the ability of PV to blend into the existing cultural building context and remain aesthetically unobtrusive. This is evident with the 150 kWp metal-lined roof tile system on a high school gymnasium in Japan (figure 78).

Fig 76 Flat roof PV in the CBD of Boston
Source: PowerLight, USA

Fig 78 High school gymnasium, Kagoshima, Japan
Source: Jiro Ohno

Fig 77 Etten-Leur, The Netherlands
Source: BEAR Architecten, Martin van der Laan (M. ART)

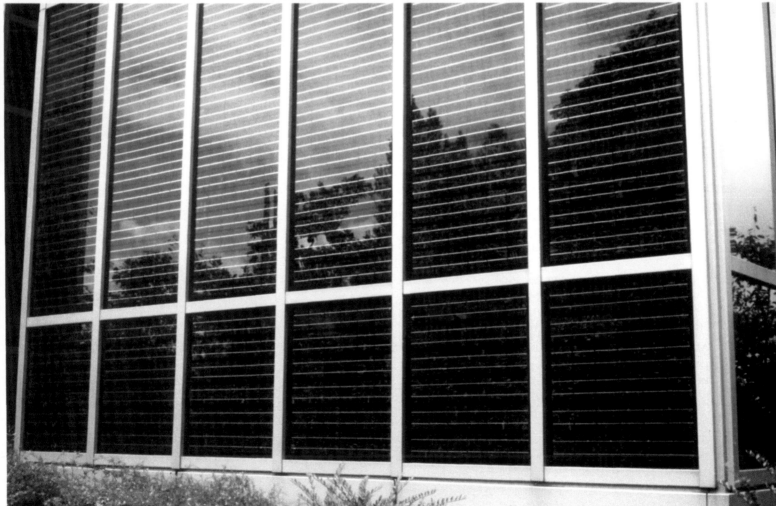

Fig 79 PV façade of Schüco International HQ, Bielefeld, Germany
Source: BEAR Architecten, Netherlands

Façades

The use of PV modules in façades seems to be obvious. Many façades have glass or tiles as a skin – PV modules can replace these materials. Often, façades present large surface areas for PV use, but under a typical vertical profile are usually sub-optimal in orientation. The extent of this very much depends on latitude, though there are a number of benefits to be gained by using a PV façade approach, particularly for east- or west-facing vertical building surfaces that require protection from sometimes very harsh morning or afternoon sun. Façades are, however, more prone to external shading effects so careful site evaluations and shade modelling are recommended to determine solar access.

Façade constructions can be separated into two main groups: ventilated and not ventilated. Because of the heat build-up behind the PV modules, it is important to know whether or not a construction is ventilated. Ventilated façades are suitable for the integration of crystalline silicon PV modules whose output efficiency is sensitive to high temperature. Non-ventilated façades require technologies that can tolerate much warmer ambient temperatures and unventilated conditions, such as amorphous silicon PV modules.

Façades are basically constructed using in-situ bricklaying or concrete constructions, prefab elements or structural metal façades that are mounted in place. Concrete constructions form the structural layer and are covered with insulation and a protective skin. This skin will be a cladding of different materials.

There are three basic system types:

- Cladding with open joints, hidden mounting, mechanical or glued. Materials: concrete or ceramic tiles, natural stone, metal or plastic panels and laminates, glass sheets and wood;
- Curtain walls, window frames and sheeting mounted in profiles, closed air gap. Materials: metal or plastic panels and laminates, glass sheets;
- Structural glazing, glass mounted on high-tech steel construction, gaskets between the glass.

For luxury office buildings, which often have expensive cladding, cladding with PV modules is not more expensive than other commonly used materials, such as natural stone, marble and expensive specialty glass. This cladding costs around $US1,000 per square metre, comparable to the cost of today's PV module.

Façades – integrated cladding

Curtain walls are well-known and used on a large scale. When these constructions cover the whole building there is often a second layer on the inside with insulation and an inner skin. To prevent condensation, this layer can be almost airtight, and hence any airgap is closed and the array is not ventilated.

Structural glazing or structural façades are constructed using highly developed mounting systems, which can be filled with all types of sheeting, such as glass or frameless PV modules. Gaskets or profiles are used to close the gap between the sheets of glass.

Examples of integrated cladding façade designs include a bank in Suglio, Italy (figure 80), and an office in Austria (figure 81). PV façade projects are becoming an increasingly popular alternative to conventional cladding materials, forming a distinctive feature and practical point-of-use power generation source.

Poly-crystalline solar cells can also be integrated into reflective glass panels. One such façade system was designed and patented by FLABEG (formerly Flabeg/Pilkington Solar) for Schuler and Jatzlau Architects who first used it in 1993 on the colourful 'Oekotec 3' Berlin office façade (figure 80). Modules are made as resin-filled glass–glass panels. The rear glass sheets of the PV panels contain a reflective coating around the cell area to achieve a unified design with the other glass panels in the façade.

Fig 80 'Oekotec 3', Berlin office façade
Source: BEAR Architecten, Netherlands

Fig 81 UBS bank, Suglio, Italy
Source: Enecolo, Switzerland

Fig 82 Kyocera commercial office building, 20 kWp poly-crystalline façade, Austria
Source: Stromaufwärts, Austria

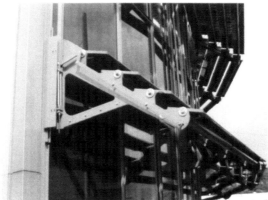

Fig 83 Colt International shadovoltaic PV louvre, SBL Office, Austria
Source: H. Wilk

Fig 84 Transmitting station in Tabarz Inselberg, Germany
Source: BEAR Architecten, Netherlands

V modules are easily configured to create curved façade profiles. renovated radio and TV transmitting station in Tabarz Inselberg, ermany (figure 84) used a special façade framing system so that 9.8 kWp of poly-crystalline silicon modules, manufactured by rSolarstrom GmbH, could be mounted from the inside.

Sloped façade profiles

Sloped façade profiles are particularly practical solutions for optimising the available PV surface area. They can often also achieve architectural novelty, both from an external building form perspective and through the control and experimentation of passive lighting and active PV structures, visible from within the building. The Solar Office at Doxford International (see case study, Chapter 3) is a well-known example of this for commercial office spaces, but the concept is also practical for residential applications (figures 85 and 86) and as a sloped framing louvre construction (figure 87). Façades used in this way can optimise orientation of large building surface areas to maximise the solar power generation potential.

g 85 3.2 kWp poly-crystalline residential V façade design, Kitzmantel, Austria ource: H. Wilk

g 86 Sloped façade design
ource: Ecofys, Netherlands

Fig 87 Stadtwerke Konstanz with louvre system, Germany
Source: BEAR Architecten, Netherlands

Façades – vertical fixtures

PVs integrated as balustrades, handrails and ornamental façade fixtures may be applied in both new and retrofit designs. This type of integration can revitalise the external building composition. Using PV as an eco-sensitive design solution, a Japanese PV balustrade (figure 88) creates a sophisticated finish to what could have been a bland and uninspiring balcony design.

Fig 88 PV balcony balustrade, Yokahama Media Tower, Japan
Source: NEDO

Eaves

PV modules can be effectively incorporated into eaves as a continuation of a roofline or as an overhead screening device. When used in conjunction with glazed façades, this can produce vibrant PV mirror reflections from the streetscape, visually appealing effects from within the office space, or eye-catching external features.

Sunscreens and louvres

Façades are also suitable for the addition of all types of screens, sunshades, louvres and canopies. There is a logical combination of shading a building in summer and producing electricity at the same time. Architects recognize this and many examples of PV shading systems are being constructed in an increasing number of countries.

The Schüco International KG, Top Sky I product (figure 90) from Germany uses an aluminium sunscreen design as a PV canopy. For safety reasons, Schüco recommends the use of compound safety glass or wired glass with these canopy roofs. The system works just as well with wood, stone or concrete façades as it does with aluminium or steel.

Schüco's Top Sky II product increases the maximum profile length to 1500 mm, and the angle can be fixed between 30° and 45°. A range of standard aluminium powder-coated colours can be used on the product.

At de Zonnegolven, Boxtel (figure 91) 36 Dutch houses, with a total of 43 kWp integrated PV systems, have a variety of roof and façade BiPV elements. Of particular interest is a cleverly distributed row of PV shading devices that provides an impressive and changing lighting effect against a residential red brick façade.

Fig 90 Schüco Top Sky I with a 30° tilt angle
Source: Schüco

Fig 89 4 kW PV shading system, TIZ Office Building, St. Florian, Austria
Source: H. Wilk

Fig 91 De Zonnegolven, Boxtel, The Netherlands
Source: NOVEM Riesjard Schropp

Figure 92 shows the PV façade integration on 22 houses in Dordrecht. These houses are located near a noisy highway and consequently have no openings on that side (northeast), except the entrance door. The south and west façades have large windows and glass doors. To prevent overheating in summer, shading devices are needed. The living room façade has a fixed shading device and the PV panels are part of these elements. From the point of view of the PV system, this solution is not optimum, but in the context of the design of the house and the façades, the architectural integration looks effective.

Piping-shaft blind louvres incorporating 36 kWp PV have been used extensively at a research centre building in Tokyo (figure 93), with ventilation gaps to minimise temperature increase of the solar cells. The gaps are shaped so as not to create noise during windy conditions.

The entrance to the Baglan eco-factory in Wales demonstrates the use of both cantilevered crystalline silicon awnings and a vertical amorphous silicon wall system (figure 94).

PV sunscreen devices are usually exposed structures and thus benefit from being well-ventilated. Examples are shown in figures 96–99.

Fig 92 22.8 kWp amorphous silicon façade sunscreens on houses in Dordrecht
Source: BEAR Architecten, Netherlands

Fig 93 36 kWp NTT Musasashino R & D Centre, Tokyo, Japan
Source: Jiro Ohno

Fig 94 Baglan Eco-Park, designed by the Welsh School of Architecture, Cardiff University, Wales
Source: Newport City Council

Fig 95 Close-up of BP Solar cantilevered awning and amorphous silicon wall
Source: Newport City Council

Fig 96 PV glass canopy detailing by Gregory Kiss
Source: Kiss and Carthart Architects, USA

Fig 97 University of Erlangen, Research Centre for Molecular Biology, Erlangen, Germany
Source: Solon AG, Berlin

Fig 98 10 kWp Colt sunshades on the AMAG Centre, Lausanne, Switzerland
Source: BEAR Architecten, Netherlands

Fig 99 Shop with 10 kWp poly-crystalline PV lamella shading, Lausanne, Switzerland
Source: BEAR Architecten, Netherlands

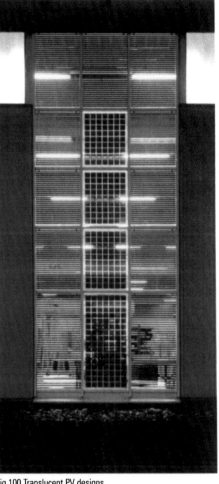

Fig 101 8.1 kWp
mono-crystalline
silicon Church spire
Potsdam, Germany
*Source: Solon AG,
Berlin, Germany*

Fig 100 Translucent PV designs
Source: Solon AG, Berlin, Germany

Translucent PV

Translucent PV modules used as roofing materials serve as water and sun protection as well as transmitting daylight. In glass-covered areas, such as sunrooms and atriums, sun protection in the roof is necessary in order to avoid overheating in summer. The distance between the cells is between 5–20 mm, depending on the amount of transmitted light required. The PV cells absorb 70–80 per cent of the sun's radiation. The space between the cells transmits enough diffuse daylight to achieve a pleasant lighting level in the area. PV cells were used in this way at the Centre for Sustainability De Kleine Aarde, in Boxtel, The Netherlands and at the Brundtland Centre in Denmark (see case study). The use of semi-transparent PV modules in the Solar Office in Doxford has resulted in a similar contrast in the façade. In order to control the amount of daylight to the workplaces, semi-transparent PV modules have been used instead of glass. The modules serve as sun protection, while also allowing daylight to pass through.

Stunning internal lighting effects can be created by modifying the translucent properties of the PVs (figure 100) and selecting cells that resonate colour, which is often associated with poly-crystalline wafer structures. For example, the south, east and west oriented sides of the church roof spire in Potsdam, Germany have been installed with trapezoid custom-sized PV panels to create a stunning effect (figure 101).

Designs using highly translucent Grätzel cell technology, with its red-brown colouring and flexibility, have been put forward as innovative façade structure concepts (figure 102), where PV glazing adaptability can be put to stunning architectural effect.

Fig 102 Grätzel cell façade design
Source: Sustainable Technologies Australia

Fig 103 12 kWp translucent PV atrium, Rohm Co., Kyoto, Japan
Source: Jiro Ohno

Fig 104 Floriade PV canopy
Source: BEAR Architecten, Netherlands

Fig 105 Digital Equipment Corporation (DEC),
Colt shadovoltaic louvre system, Switzerland
Source: Colt International www.coltgroup.com

Atria and skylight PV designs

Possibly one of the most elaborate and architecturally invigorating applications of PV has been in atria or skylighting systems. Examples, such as ECN and Doxford (chapter 1 and case studies) have become well-known BiPV building icons. Combined PV glazing structures (such as those in figures 103–105) are being used to gain publicity in solar power design circles as benchmark architectural BiPV applications.

At the Dutch Floriade 2002 exhibition in Haarlemmermeer, the NUON electricity company built the largest roof-integrated PV system in the world. 28,000 square metres of semi-transparent mono-crystalline silicon Siemens PV panels yielded 2.3 MWp of installed PV power (figure 104).

At the Netherlands Energy Research Foundation (ECN) in Petten, Building 42 has a conservatory with 43 kWp BP solar roof-integrated transparent laminates, which reduce light and sun transmission by around 70 per cent compared to glass. The conservatory therefore acts as a big parasol over the offices, protecting them from the sun while still providing enough daylight (see case study). In this way, it was possible for a passive cooling system to be used for the building, instead of a mechanical cooling system.

The cost of translucent PV modules is typically 20–30 per cent higher than for standard modules. However, the integration possibilities, the multifunctional use (daylight, shading, passive cooling) and the possibility of reducing or avoiding costs for mechanical cooling systems, make them well worth investigating, as the total building cost may be lower. In Building 42 at ECN, the application of translucent PV modules was a neutral investment (refer to case study, chapter 3).

PV thermal co-generation applications

The negative temperature coefficient associated with crystalline silicon PV cell technologies raises interesting obstacles and opportunities for BiPV applications. As described earlier and in more detail in Chapter 6, there is an inverse relationship between temperature and PV cell efficiency primarily due to band-gap properties in crystalline PV cells. It affects the open circuit voltage most dramatically. There is an associated slight increase in short circuit current, but this is of such small magnitude that it is essentially negligible.

Efficiency for crystalline silicon is typically reduced by 0.4–0.6 per cent for each degree increase in temperature above 25 °C; these are generally accepted figures for this phenomenon. The effect of increased temperature on voltage open circuit (Voc) is shown in figure 106. Amorphous silicon, cadmium telluride (CdTe), copper indium diselenide (CIS) and other PV technologies show differing relationships with increased temperature.

uilding-integrated PV modules generally operate at higher temperatures than those arrays mounted outside the building envelope. This is due to increased natural convection heat loss that is possible due to wind and buoyancy effects when both sides of the modules are exposed. Mounting techniques, framing, and encapsulation of cells also impact on the operating temperatures of the cells. Since the vast majority of BiPV products available use crystalline technologies, they will be the primary focus of the discussion. It should be noted, however, that amorphous PV cells have been shown to have almost no degradation due to increased temperatures, and, in some cases, a positive temperature coefficient.

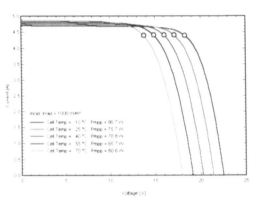

Fig 106 IV curves versus cell temperature
Source: University of New South Wales, Sydney, Australia

PV cells convert approximately 6–18 per cent of incoming irradiance into electrical energy; the rest is reflected, re-radiated, or lost as low-temperature heat. This, along with the temperature coefficient described above, serves as an impetus to remove unwanted heat from behind PV arrays and utilise the fluid flow to create thermal energy. There has been considerable interest in these concepts, which are known as combined PV/thermal (PV/T) systems or PV co-generation systems. In any BiPV co-generation system, the usefulness and timing of the thermal energy produced is crucial. The thermal energy can be:

- transferred via a heat exchanger to hot water systems;
- used in conjunction with air source heat pumps;
- used to heat mass;
- stored in underground pebble beds or phase change materials;
- used to preheat incoming air in cold seasons for buildings with high ventilation requirements.

The heat produced from a BiPV cogeneration system is low-temperature. Thus it can be used directly for the processes named above, but is not useful in generating electricity or in any high-temperature process applications in industry. There is a large variation between system types and climatic conditions in which the systems run. Within these large variations, ratios of thermal output to electrical power production are in the order of 1–3:1.

Built examples

PV/T systems in buildings can span a wide range of technologies:

PV/T modules

In this configuration PV cells are generally pasted to a typical flat plate solar thermal collector and act as the absorber. These modules can utilise either air, or heat transfer liquids as heat transfer mechanisms.

Concentrator technologies

Here, small arrays of PV cells are exposed to high sun concentrations. These cells require substantial cooling and can create higher temperatures than flat plate applications.

PV/daylighting systems

In this category pin-holes are made in the PV cells, using lasers (in the case of thin film applications) or cells are simply spaced apart in order to achieve a desired light transmission. Since larger amounts of incoming radiation penetrate these systems, they can also act as combined systems.

BiPV/thermal applications

These have generally been placed in vertical façade installations where the PV array acts as a second-skin building façade. Forced or natural convective airflows can be used in the cavity that is formed.

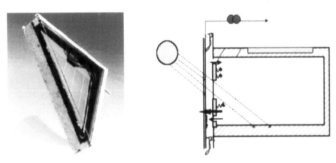

Fig 107 PV/T façade cross-section schematic
Source: University of New South Wales, Sydney, Australia

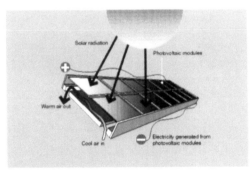

Fig 108 BiPV co-generation schematic
Source: University of New South Wales, Sydney, Australia

Some BiPV-façade buildings utilise the airflow in the cavity behind the PV modules. This can help to increase the stack effect in a commercial building or alternatively, the thermal energy can be captured and used as ventilation air preheating or for direct space heating. Two excellent examples of this type of system are in place on the ELSA (European Laboratory for Structural Assessment) building façade in Ispra, Italy, and in the Doxford solar building in the UK. These buildings have both used holistic design practices to incorporate the benefits of PV façades for co-generation. The test cell schematic (figure 108) shows the basic heat transfer mechanisms involved in the process.

The incoming thermal energy can be vented to the outside or captured as a fluid or in the mass of the building structure. These systems have also been used experimentally in smaller commercial and residential roofs. The decreased installation angles serve to decrease the velocity of the natural convection found behind the modules. In most latitudes, the increased insolation on tilted surfaces can serve to offset this deficiency in energy production.

The CHAPS (combined heat and power system) collectors (figure 109), developed at the Australian National University (ANU) Centre for Sustainable Energy Systems, combine hot water and electricity generation into a single unit, maximising the energy available from the roof space. A system planned for a new campus residence comprises eight rows of 24-metre-long parabolic mirrors that track the sun on a single axis and reflect light onto strips of high efficiency mono-crystalline silicon solar cells at about 35 times the normal solar intensity. The solar cells covering 7.5 square metres convert around 15 per cent of the sunlight into electricity that is delivered to the building and to the electricity network through a 40 kW grid-connected inverter. The balance of the solar energy creates heat which, rather than being wasted, is collected as hot water in 13-cubic-metre thermal storage tanks. It will be used to provide about 80 per cent of the hot water needs of the residential campus building's bathrooms, kitchens and laundry, and about 30 per cent of the winter heating requirement via a hydronic floor heating system.

Summary

This chapter has described the basic principles of building-integrated PV systems and shown examples of the exciting opportunities available to architects in using these new technologies. The following chapters provide more detail of the technical aspects of system installation as well as more detailed descriptions of some of the innovative systems that have been built around the world.

Fig 109 CHAPS (combined heat and power system)
Source: Centre for Sustainable Energy Systems, Australian National University, Canberra, Australia

CHAPTER THREE

INTERNATIONAL BiPV CASE STUDIES

AUSTRALIA: **SYDNEY OLYMPIC VILLAGE**

BIODATA

PROJECT:	623 x 1 kWp Olympic Village
LOCATION/CITY:	Newington, Homebush Bay, Sydney
COUNTRY:	Australia
TYPE OF PV BUILDING:	Roof-integrated
BUILDING TYPE:	Residential
NEW/RETROFIT:	New

CLIMATIC CHARACTERISTICS

LATITUDE:	33° 87′S
LONGITUDE:	151° 12′E
ALTITUDE:	40 metres above sea level
CLIMATIC TYPE:	Sub-tropical (Temperature: winter average = 17 °C; summer average = 26 °C)
SUNSHINE HOURS:	Yearly average = 5.5 hours per day

Fig 1 Olympic village housing with olympic stadium in background
Source: Mirvac LendLease Village consortium

General project background

Newington is a low-rise, inner-city suburb of around 90 hectares. It is located approximately 15 kilometres west of the Sydney CBD, on a site of approximately 262 hectares, which encompasses the Olympic Village. The site originally consisted of salt marshes, wetlands and open grasslands and had been extensively used for industrial purposes. The brownfield site previously housed saltworks, flour and tweed mills, a government asylum and hospital and most recently, a navy ammunition depot.

Project brief

The New South Wales government Olympic Co-ordination Authority (OCA) requested a sustainable approach, including the promise of delivering a 'green' Olympics. The Olympic Solar Village is part of this, with the ambitious goal of changing the world's view of solar energy and energy efficiency, demonstrating to Olympic viewers and overseas visitors the commercial capacity of renewable energy technologies in providing electrical energy to an entire urban residential development.

The project is the world's largest solar village. It was a showcase for the Olympics and is part of a sustainable inner suburb, exemplifying innovative approaches to ESD principles. Its requirements included:

- housing for 15,300 athletes and officials and future Newington occupants (2,000 homes for approximately 5,000 people);
- strict implementation of energy-efficient and demand-side management best practices;
- an environmentally benign community, built on a brownfield site;
- renewable energy integration as a demonstration for future replication;
- BiPV, which satisfies the architectural requirements for a visually acceptable solution, without compromising the technical performance of the roof and the solar power system;
- cost-effectiveness in delivering a clean, green suburb.

Sustainable housing design context

New energy-efficient, passive and active design guidelines were written to commit the design and development of the Village to conform to best practices. The houses follow exemplar design strategies and energy-efficient best practice through passive and active solar application and the maximising of natural ventilation. The energy-smart designs are complemented by gas-boosted solar hot water, gas heating and cooktops and energy-efficient lighting and appliances. The National Housing Energy Rating Scheme (NatHERS) software was used to calculate the energy performance of home designs. The Solar Village designs have achieved an average 4-star NatHERS rating. The overall result is a halving of greenhouse gas emissions when compared to a typical new dwelling in Sydney.

The use of environmentally benign construction methods and materials was also followed. These include minimisation of PVC use by choosing alternative cabling materials, low-allergenic paints, wool instead of fibreglass roof insulation, timber and ceramic tile flooring, fibre cement stormwater piping, and 90 per cent recycling of hard waste during the construction period. Water minimisation strategies, such as reclaimed water used for toilet cisterns and external taps reduce the use of potable water by 50 per cent over conventional homes. The energy initiatives are estimated to reduce non-renewable energy consumption by around 50 per cent compared with standard project housing, which will be equivalent to a saving of 7,000 tonnes CO_2 per annum once the total development is complete in 2005.

BiPV design process
Design issues

Architectural design issues that were considered include:

- Balancing the incorporation of photovoltaics with the desire to create a 'low-tech' streetscape. The visibility of the PV is varied over the site depending on the house design concept, orientation and urban design goals;
- Matching the BiPV system to the different architectural styles of each architect;
- Site planning and roof design so that the majority of roofs lie within the range of 20° west of north and 30° east of north;
- Provision of about 80 per cent of roofs with a 25° pitch to optimise outputs;
- Positioning of the solar hot water units in relation to the PV laminates;
- Controlling the visual appearance of non-integrated systems where roof orientation was not optimal (minority of houses).

Various designs from the eight commissioned architects, co-ordinated by Henry Pollack Architects, ensured a unique variety of PV building-integrated systems with maximum active and passive solar gain given the constraints imposed by the geometric characteristics of the urban plan.

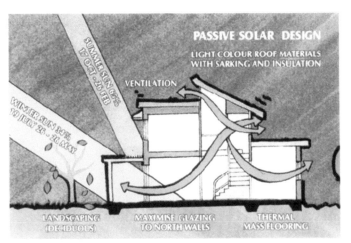

Fig 2 Passive solar energy model
Source: Mirvac LendLease Village consortium

PV system design

Following extensive prototyping and full-scale mock ups of various roof-integrated (PV) system design options, the final product consisted of a metal sub-tray waterproof roof sheet on which frameless BP Solar high-efficiency Saturn PV module laminates were fastened (figure 3) using six custom-designed diamond-shaped mounting clips (figure 4).

Fig 3 Module installation onto prefabricated tray
Source: BP Solar

Fig 4 Diamond block clip design
Source: BP Solar

Fig 5 Module installation onto prefabricated tray alongside a roof integrated solar hot water system
Source: BP Solar

An inverted U-joint between laminates encourages natural convection cooling. A mesh was used to surround the PVs to prevent leaves and animals entering behind the panels. Off-cuts from the sub-tray sheeting acted as weatherproof joints between trays. The design was expected to satisfy a number of requirements. These included each system's delivery requirement of 1,600 kWh per annum, compliance with Australian building load, health and safety specifications and electrical standards, developing and verifying best practices where standards did not exist and providing a simple product for rapid deployment of the PV systems to avoid compromising the overall construction activities of each dwelling.

Fig 6 Customised BP Solar 1200 W inverter
Source: BP Solar

PV performance was marginally compromised through aesthetic requirements to use a dark-coloured backing layer rather than an optimal white layer. This change in the laminate effected a small reduction in efficiency compared with standard test conditions (STC). The darker backing material increases thermal gains and is a slightly less effective internal reflectant of incident light (Collins et al, 2000). System reliability is not considered to be unduly compromised in meeting the required overall PV performance output of the system.

Each PV system is wired to junction boxes mounted in the cavity of the roof, which is eventually brought to a DC isolation switch in a meter box. This is wired to a compact 1200–watt BP Solar inverter (figure 6) that converts the generated PV power into usable AC electricity and feeds directly into the grid. Systems are monitored using telecommunication technology to track performance output and pinpoint any outages that might arise.

More highly-engineered BiPV approaches from overseas were rejected in favour of a design that could meet market needs and be readily accepted by the housing industry for simplicity and ease of installation. Typically, the tray installation would take half an hour and PV wiring a little over two hours. A record number of nine roofs were installed in one day by two skilled labourers. Pacific Power provided third-party indemnity and the systems were covered by a 10-year warranty on deterioration in output due to faulty workmanship or materials. The frameless laminate design and diamond tray mounting clips lower lifecycle energy costs and help to achieve a pleasing balance of cost versus thermal performance and energy yield.

Fig 7 Solar Village roofs
Source: BP Solar

COMPONENT CHARACTERISTICS

PV system power	623 x 1 kWp
Type of building integration	Roof integration
Type of cell technology	Mono-crystalline silicon
Inverter	623 x customised BP Solar 1200 W

Performance characteristics

Some calibration of the weather monitoring equipment caused delay in the release of performance results. Early indicators confirm average PV performance of around 1400–1500 kWh/year, with a number of arrays achieving generation greater than 1600 kWh/year. Post-Olympic reconstruction and development has impacted on the systems. A drop in efficiency was noted following consecutive dry days when construction site dust was suspected of covering the PV systems. This was confirmed as post-rain generation and self-cleaning of the PV laminates caused increased power output relative to irradiance and ambient temperature.

Lessons learnt from the project

While Sydney insolation levels are favourable, temperature extremes prevail in summer periods and place uncomfortable thermal conditions on PV systems. Standard Test Conditions rate PV panels at 25 °C ambient temperature, which is notably below typical Australian summer climatic conditions. Further, the roof area designs of the houses were comparatively small (for Australia), being of terrace or semi-detached style in urban configuration. The design strategy and flexible arrangements of the 12 laminates provided an effective, yet simple solution to overcome these issues.

A small number of inverter locations, which were sited on the side of the house, due to architectural design and site orientation, suffered from direct sunlight heat gain and consequently shut down frequently. This experience supports the importance of good inverter positioning: siting inverters in shade, either by using additional shading components or lee-side façades.

Pacific Power allocated a full-time site manager during the construction phase. This was the building transfer stage once the PV roof was ready to accommodate the installation of the BP laminates. Electricians were sub-contracted to complete the rough wiring and fixing of desirable terminal points. Once the roof and sub-tray system was installed and water tightness approved, the same scaffolding was used, in collaboration with the building construction team (Mirvac LendLease Village consortium), to complete the PV installation. BP Solar completed this work in around three man-hours per installation (usually two workers) and supplied the PV laminates, inverters and BOS using reusable wooden pallets to transport the parts from factory to site. Post-system commissioning, contract spares (24 laminates and 3–4 inverters) are stored by Pacific Power but are yet to be used.

Fig 8 Solar Village
Source: Mirvac LendLease Village consortium

Further installations

Since the completion of the Sydney Olympic Games, Mirvac commissioned 79 additional 1 kWp systems on larger roofscapes under Stage Two of the Newington development. Each home incorporates 12 tiles using BP Solar's 85 Wp Saturn cell PV laminates in the solar tile frame designed and produced by PV Solar Energy Pty Ltd.

Fig 9 Erling solar tiles
Source: Peter Erling

Post-installation feedback
Conclusions

While the BiPV project designs may err on the side of simplicity, they remain sensitive to market reality and are clever in that regard. The houses, along with the PV systems, are there to be sold at an affordable price. In pioneering around 840 kWp of PV projects around the Olympics site, including the nineteen 6.8 kWp PV Boulevard pylons and a 70.5 kWp Superdome amorphous PV installation, an invaluable knowledge base of BiPV applications under Australian conditions is evolving. Experience has been gained in collaborating with architects, developers, the construction industry and real estate operators throughout the PV project process. PV technologies have been trialled, tested, installed and monitored, allowing comprehensive tracking and recording of data to verify system components and commissioning results. Work has been carried out with the inverter manufacturer to improve designs as necessary to meet stringent power quality and safety requirements. Important grid connection issues have been resolved in collaboration between the University of New South Wales, energyAustralia and Pacific Power in respect of system safety and power quality. Detailed on-site training and supervision has been organised to enable a large number of builders and tradespersons to achieve a quality result while working with an unfamiliar product and system. Also, appropriate knowledge has filtered to the sales staff and home-owners at the commercial end of the stream.

While technical issues of BiPV projects have to be satisfied, similar importance must be placed on achieving commercial viability through absorbing the higher PV roof component cost per square metre into the overall marketing of the building as a single package. Added values from PV as a distributed generator and in offsetting greenhouse gas emissions will help to facilitate the wider market acceptance of PV products. Achieving successful integration of systems, cost-effective interconnection with external infrastructure and acceptance by all stakeholders is also a challenging task. However, the Solar Village has required pioneering procedures to ensure that security and safety issues are compliant to Australian requirements and also as rigorous

as current overseas standards. Australian PV grid integration best practices, that avoid poor distribution network harmonics and islanding effects, have resulted. Knowledge gained from the Solar Village thus far has already facilitated a rapid increase in the number of developments proposing to integrate solar technology. As Lord John Browne, CEO of BP Solar, stated in 2000, 'Change begins with the first step and the development of successful systems here [Solar Village] will set a standard which will spread.' It is comforting to note this rhetoric is supported by action, as BP Solar relocated and expanded its Australian operations onto the Homebush site post-Olympics, a move that is expected to accelerate its business growth 'down under'.

Project team

The Mirvac LendLease Village consortium was responsible for co-ordinating the development of a three-stage housing project designed, under strict environmental criteria, by eight local architects. The joint development team consisted of Mirvac and LendLease. Sydney-based electricity generation company Pacific Power owns, manages and maintains the solar electricity component for each house and commissioned BP Solar to supply the grid-connected PV systems.

The NSW government Sustainable Energy Development Authority (SEDA) has supported the sustainable dimensions of the project and checked the quality of work of the consortium delivered to the OCA as client. EnergyAustralia, the local electricity retail utility, has been directly involved in grid connection issues and the services of the University of NSW were sought to assist with ensuring compliance to high levels of safety and power quality standards.

The Mirvac LendLease Village consortium co-ordinated the integration of the PV systems into the building envelope, producing a schedule of procedures for facilitating the use of PV technology by house builders, including retail of the properties. SEDA offered a one-off AUD$500 rebate to entice buyers, while the Olympic Co-ordination Authority ensured compliance and meeting of contractual obligations to ensure the Village was successfully completed.

AUSTRIA: ENERGIEPARK WEST

I O D A T A

PROJECT:	Production hall for thermal collectors and photovoltaic modules, and offices
LOCATION/CITY:	Satteins/Feldkirch, Vorarlberg
COUNTRY:	Austria
TYPE OF PV BUILDING:	Façade-integrated, roof-mounted
BUILDING TYPE:	Commercial
NEW/RETROFIT:	New

CLIMATIC CHARACTERISTICS

LATITUDE:	47° 14'N
LONGITUDE:	9° 40'E
ALTITUDE:	495 metres above sea level
CLIMATIC TYPE:	Alpine, peaking at 3,400 metres above sea level
SUNSHINE HOURS:	Yearly average= 3.9 hours per day

General project background

Energiepark West is located in Satteins near Feldkirch, Vorarlberg, in the most western part of Austria. Satteins is a village with a population of about 2,550, situated on the sunny side of the Walgau valley. The site offered freely available and well-priced real estate. Further, favourable solar energy conditions prevail. The number of sunny days, especially in autumn and in spring, is much greater there than in the industrial Rheintal region.

The owner of the building, DOMA Solartechnik, is one of the leading producers of thermal collectors in Austria. In 1999, the company decided to construct a new production plant and office building. The building was to be large enough to accommodate other firms concerned with renewable technologies, and be easily extendable. Thus, Energiepark West was developed as a new centre for renewables.

With an ecologically sustainable premise in mind, the builder aimed to minimise overall building energy demand requirements and maximise the supply of energy for heating and electricity from renewable sources. Consequently, the heating demand of the building is covered by a solar thermal façade (floor and wall heating) and by two combined heat and power (CHP) systems, working with biodiesel as a backup energy source. Electricity is provided by a PV façade on the south face. More PV modules on the roof are connected to the grid and serve as a power plant (SAG Sorarstrom AG).

The western part of Austria is well known for its innovative and environmentally friendly architecture. It has become common to build low-energy houses, most of which use passive and active solar strategies to reduce greenhouse gas emissions. The shape of the chosen plot did not allow the building to optimise passive solar gains, yet offered the possibility to mount photovoltaic modules on a perfectly south-oriented construction, serving as a wind shelter for the entrance to the offices and to the production hall at the same time.

Fig 1 West façade of Energiepark West: general view and detail
Source: Stromaufwärts

Environmentally benign construction methods and materials were used including minimisation of PVC use by selecting alternative cabling materials. Besides the massive ceilings and load-carrying steel structure necessary for the stability of the building, mostly well-insulated light timber constructions with rock wool were used. There is a long tradition of building with timber in Vorarlberg. Energiepark West also has electric car-charging facilities linked to the photovoltaic power plant on top of the roof. There is an electric charge station directly opposite the entrance and office workers use company electric cars for local business trips.

BiPV design process
Planning process, alternative designs

The aims of the planning process were clearly defined from the start. Sufficient building surface suitable for BiPV and for warm water collectors was provided by the planners, MHM and Gruppo Sportive Architects. They had no need to convince others of the benefits of the photovoltaic components, but the building owner had definite aims, which included:

- the energy demand of the building should be covered by renewables;
- photovoltaics and warm water collectors should be completely integrated into the building envelope;
- pre-heated water from the collectors was to supply heating to the massive concrete floor in the production hall and the interior surface of the walls in the offices;
- the chosen backup system must be environmentally friendly (CHP working with biodiesel). All technical components must work together easily and contribute to the overall energy design strategies;
- the building must be easily expanded if necessary.

Planning strategies were defined within the planning team in co-operation with the builder and Stromaufwärts Photovoltaik GmbH, the photovoltaic contractor.

Decision process

The decision process was short as the builder knew exactly what was required and the materials and modules that were to be used. Given that the modules had several years of market exposure, the planning process requirements were made much simpler than if dealing with a new product and design strategy.

Architecture, project organisation

The architects were given a flexible brief but had to ensure an aesthetically pleasing building outcome. They decided to combine functionality and aesthetics in line with the corporate identities of the companies. This resulted in a compact and clear building shape with high PV visibility. The central staircase connects the offices with the production area and allows clients to watch the manufacturing process. The transparent atmosphere also has benefits for the employees. The workers in the production hall stay in visual contact with the office. The energy concept includes zones with different temperature levels: the offices are kept at 20 °C to 22 °C, the production hall at approximately 16 °C and the storage area is not heated.

COMPONENT CHARACTERISTICS

PV system power	South façade: 17.16 kWp
	West façade: 4.14 kWp
	Roof: 45.5 kWp
Type of building integration	Façade-integrated
Type of cell technology	Kyocera (south façade), Solarfabrik (west façade, roof)
Modular dimensions	142.5 x 65 cm
Array dimensions	133 m², 143 modules on the south façade, 450 m² on the roof (SAG Solarstrom AG)
Inverter	Fronius
Monitoring equipment	Thermal and electric monitoring; voltage, wind, energy production, air temperature, radiation at the south and west façades and horizontal surface insolation are monitored.
Other BiPV system elements	On the roof, photovoltaic modules by SAG are mounted in a conventional way. They produce 'clean and green' energy, 'Ökostrom', which is fed into the grid and can be bought by other ecologically minded people who want to support renewables.

Installation design
Integration design and mounting strategy

Pre-heated water collectors and photovoltaic modules were mounted so that the surfaces of both are on the same level. The builder had previously developed the mounting technology, where common elements of façade construction had been adapted for this special use. This worked well and at a reasonable price.

Lessons learnt from the project

The most innovative aspect can be found in the overall design of the project. Different strategies of ecological building were carefully harmonised in order to guarantee minimal impact on the environment. Due to the high quality of the building envelope (minimisation of heat losses), the heating demand is so low that active and passive solar energy gains (maximisation of gains) can cover it most of the time. Also, the backup system works with biodiesel, thus meeting the remaining energy demand.

The building was well received by the community and serves as an example for working ecological strategies. There is a permanent exhibition within the office section of the building and the production processes can be watched from the public area. The design process was assisted by building simulation tools that predicted the achievement of good building performance.

As the owner and builder was an experienced and environmentally conscious player in the solar business, it was possible to assemble and direct a strong team of planners and specialists. After the final completion of the construction, and the successful use of the building for several cold months in the Alpine region of Austria, the planning team released news of the building to the public. A further advantage was that the active solar implementation could be constructed, controlled and revised by the future users themselves. Doma, the owner, builder and user of Energiepark West, planned, produced and mounted the thermal water collectors and Stromaufwärts, the second user, planned, produced and mounted the BiPV installation.

Post-installation feedback

The design and the planning of the project are still seen as the best solution for the desired aims. There have been very few problems since completion. Three Wechselrichter Sunrise Midi (Fa. Fronius) inverters broke down, but were repaired within a few days. Six modules were destroyed at the end of 1999 as a result of the 'Lothar' storm. The project won the European Eurosolar 2000 Award, presented by Hermann Scheer.

Fig 2 Installation of mounted PV roof
Source: Stromaufwärts

Project cost breakdown

PV modules: EUR109,000

Construction of combined PV modules and glass: EUR39,000

Costs of monitoring system: EUR14,500

Performance characteristics

Energy performance for the first year:	South-facing PV façade produced 11,280 kWh
BHKW biodiesel:	17,728 kWh
PV electricity exported to the grid:	17,760 kWh
Electricity imported from the grid:	7,017 kWh
Consumption of biodiesel:	3,900 litres
Thermal façade:	82m² produced 22,400 kWh

More detailed monitoring data is not available.

CANADA: **TORONTO HIGHRISE ROOF**

BIODATA

PROJECT:	Ontario Power Generation Inc. head office: 5 kWp PV system
LOCATION/CITY:	Toronto, Ontario
COUNTRY:	Canada
TYPE OF PV BUILDING:	Flat roof
BUILDING TYPE:	High-rise office
NEW/RETROFIT:	Retrofit

CLIMATIC CHARACTERISTICS

LATITUDE:	43° 60'N
LONGITUDE:	79° 38'W
ALTITUDE:	400 metres above sea level
CLIMATIC TYPE:	Temperate
SUNSHINE HOURS:	Yearly average = 3.5 hours per day (horizontal surface)

Fig 1 Ontario Power Generation Inc. head office, Toronto, Canada
Source: Brian Dougherty, OPG

General project background

The photovoltaic array is located on the roof of the Ontario Power Generation Inc. (OPG) head office building in downtown Toronto, Canada. OPG, with an installed capacity of 30,873 megawatts, is the largest electric generating utility in the Province of Ontario. OPG was formed in 1999 when Ontario Hydro, a publicly owned utility, was restructured by the provincial government, as part of an electricity deregulation process that commenced in 2001.

This new electricity market encouraged OPG to examine the potential of green energy. The original utility, Ontario Hydro, was in fact founded on a renewable source of energy and the word 'hydro' has become synonymous with 'electric utility' in the Canadian setting. In the present mix of energy sources, hydro represents about 25 per cent of Ontario's total capacity. OPG Evergreen Energy is now looking to additional renewable sources such as small hydro, biomass, wind, and solar. PV represents the smallest potential source. However, PV has many positive features.

Photovoltaic systems are small and modular and therefore require minimal time to implement. It is also an environmentally friendly technology, eliminating lengthy formal hearings and assessments. Perhaps the greatest advantage over other green technologies is visibility. Wishing to make a green statement, a company can architecturally incorporate photovoltaic arrays into its buildings. The challenge for architects is to make this integration as creative and attractive as possible.

Fig 2 Split-level roof and PV installation
Source: Brian Dougherty, OPG

COMPONENT CHARACTERISTICS

PV system power	4.8 kW – 48 photowatt PW1000 modules
Type of building integration	Ballasted flat roof mounting
Type of cell technology	Poly-crystalline silicon
Modular dimensions	1,380 x 710 x 70 mm
Array dimensions	Approximately 12 x 12 m
Weight	10.5 kg per module x 48 plus 54.7 kg x 55 for ballast curbs
Visual details	The array is located on the roof of a 20-storey building and is not visible from the ground
Inverter	5 kW Arise GX 5000
Balance of system components	Photovoltaic system connected to standard building electrical panel
Monitoring equipment	Schott Applied Power – data accessible via OPG's website, www.opg.com
Other BiPV system elements	All metering and project description presented in a display in the main lobby of the office building

BiPV design process
Planning process, alternative designs

The OPG building was inherited from Ontario Hydro, the previous tenant. The office building was constructed in 1975 and includes features such as annual thermal storage for heating and cooling. While it was energy-efficient in its time, its location, shape and orientation made it a poor candidate for a building-integrated photovoltaic system. These factors limited the design options to a rooftop installation. Further, the roof is actually split into two levels, the top portion being filled with communications equipment. The top floor also shades much of the second level, narrowing the site to a small area at the south end. Having selected the site for this photovoltaic array, it was learned that the perimeter had to remain clear for the window washers' equipment. These physical constraints limited the design to about 5 kilowatts peak capacity.

Decision process

As the photovoltaic system designer/installer overcame each of the above barriers, another barrier appeared. While OPG is a long-term tenant of the building, the company does not own it. The photovoltaic array could therefore not be fastened permanently to the roof. As with all other equipment, such as OPG's computers, the photovoltaic system required the capacity to be unplugged and removed from the building. Approval for the proposed OPG system design therefore had to be obtained from the building owner before installation could start. The final photovoltaic design is a reflection of all the above factors. It also meets another requirement, namely low cost.

BiPV design

The photovoltaic array consists of seven rows of modules in varying lengths from five to nine modules to conform to the diamond-shaped area available. The modules are mounted horizontally at a 15° slope to maintain a very low wind profile. Electrically, the modules are wired in four strings of 12 modules. Individually, the modules are configured for a nominal 12-volt output, resulting in an array DC operating voltage of approximately 200 volts. AC output voltage is 208 volts. The support structure is made from standard 'off-the-shelf' aluminium lengths. All components are bolted together; there are no expensive welds. Since the structure could not be fastened to the roof, it is held in place by concrete ballasts. These ballasts are standard concrete curbing blocks used in parking lots. There are 55 blocks, each weighing 54.7 kilograms.

Installation design
Integration design and mounting strategy

Due to limited space, the photovoltaic array is located on the roof of the second top floor of the building. This small area was covered with patio stones. There is no permanent attachment to the roof. The array concrete ballast curbs were arranged on top of the patio stones and are interlocked with aluminium lengths that also serve as trays for the inter-row wiring. The photovoltaic modules are mounted lengthwise in single rows to maintain a low profile. The mounting angle was 30° in the initial design. While the Toronto area is not exceptionally windy (net 30-year return period wind pressure is 0.48 kPa), the curved form of the building and the array location make it difficult to predict wind exposure. The mounting angle was therefore lowered to 15° to minimise wind loading while still maintaining enough slope to clear snow.

Fig 3 Close-up of modules, showing mounting angle
Source: Brian Dougherty, OPG

Electrical configuration including grid integration

The array is configured in four strings of 12 modules connected together in a combiner box mounted externally on the building wall. This combiner box contains lightning spike protection, blocking diodes and isolation switches for the individual strings. The DC wire is fed through the wall to the inverter, which is mounted inside, about 20 metres from the combiner box, close to a building electrical panel. This is the point of connection for the photovoltaic system.

Planning approval and institutional processes

The installation of the photovoltaic system required the approval of the building owner. The main concerns were that the array would be secure on the roof and that the weather seal of the roof was not affected in any way. The installed system also required electrical inspection and approval by the Electrical Safety Authority. Both the photovoltaic and the utility interconnection aspects of this system represent fairly recent technological developments; there are few applicable electrical standards in place in Canada. Nevertheless, the system received full approval from the ESA using existing codes and draft photovoltaic guidelines from international sources.

Installation
Installation procedures and experiences

All components were transported to the roof via a single small freight elevator. This required some co-ordination with other building contractors since the top floor was undergoing extensive renovation during the installation period. The array was fairly simple to install. There were no tight tolerances and once the concrete curbs were arranged, it became a repetitive task to install the aluminium support structure and modules.

Successful approaches

Due to miscommunication, the concrete curbs were delivered on pallets too large for the elevator. Realising the manual labour required to transfer these heavy curbs to the roof, several OPG staff from the shipping department volunteered their services. It was just one example of the co-operation that helped make this a successful project.

Problems during realisation

The main problem was related to weather. The first day of installation coincided with the onset of the third-coldest winter ever recorded in Toronto. This made working on the roof of a 20-storey building quite unpleasant at times. Scheduled work was cancelled on several occasions due to cold temperatures and heavy snowfall, thus extending the installation time by many weeks. While the sudden onset of winter was unexpected and there can be no control over weather, there should be some consideration given to seasonal timing when planning this type of a project, at least in the Canadian environment.

Performance characteristics

Typically, a photovoltaic system installed in the Toronto area will produce about 950 kilowatt-hours per kilowatt peak per year. This value is based on systems installed with a slope of 30°. Because the slope of the array was lowered to 15° to reduce wind loading, the system is expected to produce 8–10 per cent less than this value. Based on the assumption that photovoltaic power would be displacing fossil generation, the Carbon Dioxide Emission Reduction Credit is 850 kg/MWh. Between start-up in early 2001 and the end of October 2002, 656 kWh had been produced by the system.

The power performance of the system was tested using a Spire PV Array tester. An I-V curve was produced for each string to assess module matching. The power tracking capability of the inverter and its efficiency were also evaluated.

Fig 4 Inspecting the photovoltaic system on the roof of the Ontario Power Generation head office
Source: Brian Dougherty, OPG

Fig 5 Photovoltaic array located on roof of 22-storey building in downtown Toronto
Source: Brian Dougherty, OPG

Project cost breakdown

The project was entirely funded by Ontario Power Generation Inc. Approximate costs (CAD$) were:

	CAD$
Design, engineering and management	10,000
Photovoltaic modules	30,000
Inverter	7,500
Array support structure	5,000
Installation including BOS	10,000

There are no maintenance tasks anticipated. Major components are warranted. Present electricity rates in the Toronto area are 5.65 cents per kilowatt-hour.

Post-installation feedback

The project demonstrated the flexibility of photovoltaics. OPG Evergreen Energy wanted to demonstrate green technology to its customers. While small hydro, biomass and wind all contribute to its energy mix, only photovoltaics could be installed in a downtown environment. Designing and installing the array on this particular building proved to be a challenge. Given the roof area available in most large cities and the fact that a photovoltaic system could be installed on this building indicates the enormous potential for photovoltaic generation in an urban setting.

The photovoltaic system 'officially' began on Earth Day, 2001. While 5 kW for a very large building only represents a small step in terms of electrical contribution, it is nevertheless a step in the right direction. The photovoltaic system represents an environmentally friendly way of generating electricity and has received much support from the employees in the building both during construction and following start-up. The lobby display keeps people informed on its performance. While people are interested, the most repeated question is 'What lights, computers and machines are connected to the photovoltaic system?' The concept of utility interconnection, that is the photovoltaic power mixes with the utility power, can be difficult for people to grasp.

Project organisation

The project was initiated and funded by OPG Evergreen Energy. This department is responsible for marketing green energy. The photovoltaic system was designed and installed by Sol Source Engineering under a fixed contract with OPG Evergreen Energy, which was responsible for the overall management of the project including performance monitoring.

Sol Source Engineering designed the photovoltaic system, built the support structure and assisted with the installation. The photovoltaic modules and the inverter were obtained from Canadian manufacturers and a local Toronto electrical contractor installed the system. The electrical contractor had previous photovoltaic experience, having installed an 80-kilowatt photovoltaic system on the roof of the Bloorview MacMillan Rehabilitation Centre.

CANADA: WILLIAM FARRELL BUILDING

BIODATA

PROJECT:	2.2 kWp (20 x 108 Wp) BiPV ventilation system
LOCATION/CITY:	Vancouver, British Columbia
COUNTRY:	Canada
TYPE OF PV BUILDING:	Façade-integrated
BUILDING TYPE:	Commercial
NEW/RETROFIT:	Retrofit

CLIMATIC CHARACTERISTICS

LATITUDE:	49° 15'N
LONGITUDE:	123° 15'W
ALTITUDE:	10 metres above sea level
CLIMATIC TYPE:	Coastal
SUNSHINE HOURS:	Yearly average = 4 hours per day

Fig 1 Telus BiPV ventilation system
Source: BCIT

Fig 2 Telus building in downtown Vancouver
Source: BCIT

General project background

The project is an extensive interior and exterior renovation to the William Farrell building in downtown Vancouver, British Columbia, Canada. This eight-storey office building was built in 1940 to serve the city's telephone system and housed communications equipment as well as offices for technical and administrative staff. In 1999, Telus, the owner of the building, wished to rehabilitate both the built form and internal occupancies of approximately 11,800 square metres into office, retail/commercial and presentation space with the requirement that the existing building be recycled and reused and that green strategies be incorporated.

By revitalising an existing building in a high profile location, the building establishes both a strong visual presence for Telus in downtown Vancouver and demonstrates a commitment to minimising environmental impact. The most visible aspect of the building renovation is the new double-glazed, fritted and frameless glazing system suspended 900 mm from the existing building face.

The structure of the existing building was left essentially unchanged with the exception of removing the original brick veneer. The building's new external cavity operates as a thermal buffer and natural ventilation intake and significantly enhances the thermal performance of the building envelope. The curtain wall around the cavity has a ceramic fritted pattern that, in combination with light-shelves, allows daylight into the interior. Electronic temperature sensors control the entire ventilation system to maintain the optimum temperature within the cavity. The air space is ventilated with fans powered by building-integrated photovoltaic (BiPV) solar modules incorporated directly into the glazed façade. All perimeter windows are user-operable for natural ventilation.

Considerable reductions in operating energy and attendant greenhouse gas emissions can be attributed to the tempering effect of the new glazed façade, resulting in reduced winter heat loss, useful winter solar gain and reduced summer cooling load.

A variety of other green strategies are incorporated in the renovation. Indoor air quality is maximised by the use of low volatile organic compound (VOC) paint, linoleum, water-based adhesives, low pile and tight weave construction carpet with low-emission backing. Light shelves and whitewashed concrete ceilings maximise daylight within the building and recycled and recyclable materials were used throughout the renovation. About 75 per cent of the material from the previous structure was retained for reuse or recycling. This includes the existing Andersite and granite stone that was re-cut and reused on the ground floor exterior walls. Windows, handrails, stairs, doors and fittings were also reused, while much of the new material in the building was selected to be easily recycled. The new glazing system is unitised and can be dismantled with framing and glazing intact. All new concrete contains 25 per cent recycled fly ash and recycled steel rebar.

BiPV design process

Using a renewable energy source to enhance the performance of the façade fits very well with the mandate to use green strategies in the building revitalisation. The architects Busby + Associates, and engineers Keen Engineering, recognised the potential to use BiPV in the renovation of the building and willingly adopted the choice of photovoltaic technology as an integral part of the building's energy strategy. An awareness of BiPV was provided through presentations by the BCIT Technology Centre.

Although the entire curtain wall could have been adapted to BiPV, this would have exceeded the energy requirements of the ventilation system. In order to achieve sufficient airflow within the façade, forced ventilation is required, as stack-effect flow was deemed inadequate. Keen Engineering determined the power requirements of the forced ventilation system and the BCIT Technology Centre designed the two 1 kWp photovoltaic arrays that supply the power. The integration of a PV array within the ventilated façade is an attractive source of energy because the ventilation fans are operated directly by the PV array during sunny periods when ventilation is required. In addition, the airflow behind the façade cools the PV array and enhances its performance.

The parameters for the system were clearly defined early in the design process, limiting the possible configurations for the system. The power requirements for the ventilation system were defined by the amount of air to be moved. Keen Engineering performed thermal energy modelling of the façade to determine the minimum airflow required to maintain the correct temperature within the wall cavity under various climatic conditions. The design of the BiPV array takes into account motor, fan and controller efficiency as well as building orientation and climate.

The BCIT Technology Centre was contracted to design the PV system. Consultation with the architect led to the selection of blue poly-crystalline cells as the most attractive and compatible choice of cell technology. The colour and texture of the cells integrated effectively with the other building elements. Crystalline silicon solar cells allowed the modules to be designed with appropriate space between the cells to allow daylight to enter the upper floors of the building. Amorphous silicon was deemed inappropriate due to its colour and the difficulty of obtaining modules of the required colour and transparency. Custom-size solar modules were required for this project.

The number of cells and the cell spacing within each module was determined by the daylight requirements of the building and the curtain wall dimensions. Thus, the electrical characteristics of the modules were fixed early in the design process and the remaining design choices involved the electrical configuration of the photovoltaic array, fan and motor selection, and motor controller selection.

DC fans were chosen due to the higher efficiency of the motors and simplicity of the design. A customised maximum powerpoint tracking (MPPT) controller, built by the BCIT Technology Centre, is used to optimise the performance of the array, to allow for a soft start of the fan motors and to control the voltage delivered to the DC fans.

   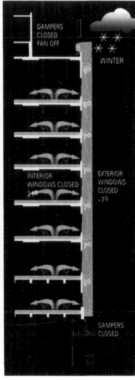

Fig 3 Seasonal operation for spring–autumn day; summer day; summer night; winter day
Source: Busby + Associates Architects

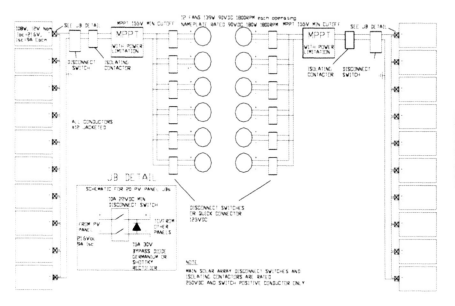

Fig 4 BiPV electrical schematic
Source: BCIT

Fig 5 BiPV module installation into the curtain wall system of the Telus building façade.
Source: BCIT

COMPONENT CHARACTERISTICS

PV system power	2.2 kWp
Type of building integration	Façade-integrated curtain wall
Type of cell technology	Poly-crystalline silicon (photowatt spider grid cells)
Modular dimensions	2,370 x 800 mm
Array dimensions	47.4 x 0.8 m
Weight	53 kg each (1.06 t total)
Visual details	Semi-transparent modules matched with the rest of the façade elements (shape and colour)
Balance of system components	MPP Controller (input 216 V DC, output 90 V DC)
Other BiPV system elements	Curtain wall mullions used as wire ways. Load: 12 1/6 hp DC exhaust fans

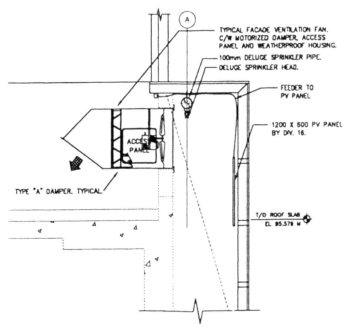

Fig 6 Architectural detail of fan/BiPV module layout
Source: BCIT

Fig 7 Maximum powerpoint tracking (MPPT) controllers
Source: BCIT

BiPV design

The BiPV electrical design was performed by the BCIT Technology Centre in compliance with the Canadian Electrical Code; figure 4 shows the electrical wiring.

Installation design

Two sub-arrays of custom-designed, semi-transparent, poly-crystalline solar modules are incorporated into the northwest and southwest walls of the office tower's new glazed curtain wall façade. Each module produces 108 watts at STC and is configured as a nominal 12-volt module with an open circuit voltage of 21.6 volts DC. The modules in each sub-array are connected in series to produce a maximum open circuit voltage of 216 volts DC.

The maximum DC power generated by this optimised photovoltaic array is 2.16 kilowatts, dedicated to powering 12 high-efficiency DC ventilation fans. A unique MPP controller, developed at the BCIT Technology Centre, regulates the output of each sub-array to 90 volts DC for the ventilation fans. Each sub-array drives six 1/6 hp exhaust fans. The fans are rated for an air volume of 2,800 cfm (cubic feet per minute) at 1,800 rpm (revs per minute).

Installation

From the glazing contractor's point of view, the installation of the complete façade segments with integrated BiPV modules was a relatively standard curtain wall installation. The BiPV modules were assembled as pre-manufactured sealed glazing units in the AGS factory and the mullions were pre-drilled to accommodate the electrical wiring. Glaziers and electricians were both present during installation of the modules. Western Pacific electricians carried out some of the electrical work as the modules were being installed. The curtain wall mullions are used as wire raceways.

Performance characteristics

The main purpose of the ventilated façade is to improve the thermal energy performance of the building and the main parameter affecting this is the temperature of the air behind the façade. Energy production of the solar array is not monitored; instead, measuring the temperature behind the façade monitors the overall performance of the ventilated façade.

The ventilation system is performing to the design specification. During winter, the fans in the façade are turned off and the air space is used to reduce heat loss from the building while trapping incident sunlight and warming the concrete mass of the building. During the summer, when the building must be cooled, the fans draw heat from within the façade and away from the building. The ventilation system maintains a maximum air temperature of 27 °C at the top of the façade.

Fig 8 Pre-assembled BiPV glazing elements
Source: BCIT

Project cost breakdown

Total project cost of the BiPV component was approximately CAD$90,000. One half of the cost was offset by a government grant towards reducing greenhouse gas emissions.

Post-installation feedback

The project is an excellent demonstration of building-integrated photovoltaics applied in an urban Canadian environment. The system operates efficiently and seamlessly integrates with the other building systems. This BiPV ventilation system demonstrates the simplicity of integrating BiPV with standard commercial construction methods. Additionally, this system was very appropriate as a retrofit to an existing building as it remained separate from the building's existing electrical system.

The experience gained by the design team in the effort and co-operation required between the various trades and engineering disciplines has already encouraged others to consider BiPV as a valid design option.

Project team

Dominion Construction of Vancouver, BC managed the renovation of the Telus building and Busby + Associates was the architect for the project. BiPV system design and co-ordination of the mechanical and electrical engineering as well as architectural integration of the BiPV system was carried out by the BCIT Technology Centre group. Keen Engineering and Reid Crowther and Partners were responsible for the mechanical and electrical engineering, respectively. The glazing contractor, Advanced Glazing Systems, and the electrical contractor, Western Pacific Enterprises, worked together on the BiPV module installation.

DENMARK: **BRUNDTLAND CENTRE**

BIODATA

PROJECT:	14.25 kWp
LOCATION/CITY:	Toftlund
COUNTRY:	Denmark
TYPE OF PV BUILDING:	Roof and façade system
BUILDING TYPE:	Commercial
NEW/RETROFIT:	New

CLIMATIC CHARACTERISTICS

LATITUDE:	55° 20'N
LONGITUDE:	9° 10'E
ALTITUDE:	20 metres above sea level
CLIMATIC TYPE:	Northern maritime
SUNSHINE HOURS:	1,922 hours per year (where direct radiation is greater than 120 W/m²)

Fig 1 Brundtland Centre
Source: BEAR Architecten

General project background

As a result of the recommendations in the UN report, *Our Common Future*, Denmark invited municipalities to suggest how a local municipality could become a 'Brundtland Town'. The municipality of Nørre Rangstrup was selected and Toftlund was chosen as the city to demonstrate the possibility of achieving an overall energy saving of 50 per cent. The purpose of the Brundtland Centre was to collect all information and lessons learnt from the Brundtland Town project and disseminate these to the public. The Centre was used for exhibitions and educational activities related to energy topics, as well as for disseminating results from energy-saving activities from the Brundtland Town project. After a few years, the building was sold to an IT company and it now functions as a regular, but energy-efficient, office. The building itself still serves as a demonstration project for energy-efficient design, and incorporates components related to photovoltaics, including daylighting systems, utilisation of passive solar energy and passive cooling.

The original site had a number of restrictions regarding orientation, footprint, height of the building and some general architectural concerns. During the political negotiation phase of the project, it was decided to move the project to another site, on the outskirts of Toftlund, overlooking surrounding open landscape. This decision allowed the design team to fully control all parameters of the shape of the building shell. This was the optimal starting point for developing an architectural idea while following the general design concept of the project, allowing careful landscaping of the site to reflect the design of the actual building, and integrating the building into the open landscape.

Project brief

The scope of the building was to demonstrate how a 50 per cent reduction in energy use, compared to normal Danish building standard, could be achieved. The aims of the project included:

- Low energy consumption: 50 per cent reduction compared to ordinary buildings;
- High indoor comfort levels and use of environmentally sound materials;
- Demonstration of daylighting systems integrated in sealed glazing;
- Demonstration of building-integrated translucent photovoltaics providing solar shading;
- Demonstration of an atrium integrated into the building for utilisation of passive solar energy.

Planning process

Close co-operation was achieved between the engineers and the architects, resulting in a positive experience for all parties. The design work of the project was initiated by a two-day design workshop, in which all members of the design team participated. The agenda for this workshop included topics such as building performance criteria, planning constraints and building design options; technical issues such as daylighting design, PV systems, ventilation and indoor climate; a review of strategies relating to the overall energy optimisation of building, and a review of planning issues.

The discussions and decisions made during these two days moved the project forward dramatically compared to more traditional working methods. Following the workshop, weekly meetings were held, following normal procedures in building project design. The initial decisions ranged from general overall problems regarding construction principles to specifics, such as selecting concrete slats supported by pillars on both floors and investigations regarding the design of the atrium roof. There were a large number of important parameters, including shading effects on PV cells; solar shading needs of the atrium; sizes and shapes that could be integrated during construction; mounting techniques; the effects of snow; and cleaning and maintenance. Looking at the whole design process, the atrium roofing consumed the most time, due to the large number of parameters to be taken into consideration.

The design team was structured so that architects and engineers shared the responsibility for quality, economy and durability, as required by standard design contracts for Danish building projects. Normally the architect leads the design team, but given the technical complexity of the building, the design team agreed to recommend to the client that the engineering company responsible for the construction should share the lead role. This was a sensible option as the engineer had experience in BiPV and building operation dynamics and was able to present options from which design strategies could be formulated.

A section of office space was located with a façade facing southeast, in order to demonstrate that even with a southeast orientation, a good thermal indoor climate can be provided without the installation of active cooling. The southeast façade of the office section, the 'energy façade', was developed to optimise solar energy use and diffuse daylight in the rooms behind the façade, in order to achieve the best possible indoor climate and a low energy consumption.

BiPV design process

Early consideration of PV in the design process resulted in a well-integrated, aesthetically pleasing and energy-efficient PV system. Two types of PV system were used in the building envelope. A PV array was integrated into the roof of the atrium, a central space connecting the adjacent two-storey buildings. Another array of PV modules was mounted on the southeast façade of the office section. In the overall scheme of the integration of PV into the building, the design team focused on more than just achieving optimal electricity production from the panels. This demonstrated that the use of photovoltaic elements must serve more than one purpose in order to make the technology, with the current state of development and economics, attractive to building designers.

Round, opaque PV cells were contained within the sealed, doubled-glazed transparent units that form the saw-tooth trussed atrium roof. The atrium roof, incorporating transparent PV modules, stretches out above the entrance of the building, creating a large canopy. The PV system mounted on the atrium roof is visible from the inside of the atrium as well as from the outside of the building. Mounting the PV system onto the roof had a great impact on the architectural expression, the building layout and internal atmosphere of the Centre. To achieve the optimum orientation and tilt angle of the PV array (60° south), the atrium roof was constructed as a saw-tooth form that runs diagonally across the space. The steel truss roof, combined with the alternating pattern of dark, round cells against the transparent glazing, gives the atrium a high-tech atmosphere.

Special attention was paid to providing a soft diffuse quality of daylight in the interior of the atrium. A thin diffusing glass fabric is integrated into the modules so that sharp edges from the circular solar cells are softened. The vivid blue colour of the PV system integrated in the façade has an even greater impact on the building's image. Other materials and colours used in the curtain wall are anodised aluminium and dark blue painted aluminium windows that blend well with the PV.

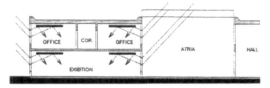

Fig 2 Cross-section view
Source: Henrik Sørensen, Esbensen Consulting Engineers

Fig 3 External structure
of PV atrium design
Source: BEAR Architecten

Fig 4 Detail of PV atrium
Source: BEAR Architecten

Energy strategy

The total expected energy consumption of the Brundtland Centre is approximately 50 kWh/m² per year (not including the photovoltaic power) while an average standard office building in Denmark uses 170 kWh/m². This reduction in energy consumption has been achieved through the combination of daylighting techniques, a compact building layout, the integration of PV elements, utilisation of the building mass for thermal storage and the use of various optimisation methods. In energy terms, the integrated PV system in the atrium roof was designed to both produce electricity and provide shading to prevent the atrium from overheating. The transparent modules allow 20 per cent of daylight to enter the atrium. The concrete floors and façades accumulate surplus heat during the day. At night, the atrium is cooled by natural ventilation, while the mechanically assisted ventilation is used to ventilate the adjacent rooms. Automatic controls stop the ventilation when a sufficiently low temperature is reached.

BiPV design process considerations

The photovoltaic elements in the façade were subject to several changes during the design process. Originally, it was planned to use the heat that built up on the back of the panels as a preheating source for fresh air to the offices facing this façade. During the design process, simulations were carried out to calculate the potential heat gain. The result was that the potential would be far too little, compared to the effort and cost to establish this possibility as an integrated part of the photovoltaic panels. It was therefore decided not to use the panels as preheating elements and instead a local fan and heat exchanger were installed for the offices facing the façade. The tilting of the photovoltaic panels was not a result of optimised angles to yield the maximum solar power, but was a geometrical solution taking into account the maximum shading of the daylighting windows, architectural aspects of the façade and the need to avoid shading between the solar panels.

Looking only at the net electricity production of the PV system of the building compared to the investment, a profitable investment is unlikely to be realised. When looking into all the other functions of the solar panels, especially on the atrium roof, more benefits than just electricity production can be identified, providing better profitability from the PV systems. For example, the panels provide sufficient solar shading for the atrium to avoid overheating, but also allow for sufficient daylight entering the atrium and the adjacent rooms during overcast sky conditions.

The façade modules are a more standardised design, compared to the atrium modules, but in the initial design phase, the modules were connected to a decentralised ventilation system for the office wing. The idea of the original design was to combine the need for efficient cooling of the back of the solar panels with the preheating of ventilation fresh air for the offices, but as described above, the energy potential of this solution unfortunately was too low compared to the investment.

The design of the façade panels changed instead to a more traditional design, but with the panels carefully designed in shape and framing to match the framing of the other parts of the energy façade. It was also important in the design of the modules to take care not to shade the daylight windows, as the efficiency of these windows is very dependent on the diffuse daylight from the parts of the sky almost in zenith. The PV façade modules were laminated with an almost fully transparent laminate to let some direct sunlight pass through the panels, giving the façade contrast and depth.

Fig 5 Interior atrium space of Brundtland Centre
Source: Henrik Sørensen, Esbensen Consulting Engineers

Lessons learnt from the design process

During the detailed design phase, the budget was not sufficient to include all elements. Additional funding had to be raised, which was only possible through direct subsidy from the Danish Energy Agency. In the project, a number of completely new elements were designed and included in the building. The main obstacle to realising these was to find manufacturers able to produce these elements with high quality. To some extent, completely new products were invented and the design team had largely underestimated the effort needed for all the different phases of design, feasibility analysis, lab-testing, production start, test production and final delivery. Another lesson learnt was the importance of very close collaboration between the members of the design team. Even though this design team had a good level of communication, a number of problems were unsolved at the time of construction commencement and had to be resolved quickly, and the best solution was perhaps not found every time. Especially with the façade, with a large number of restrictions regarding shading and utilisation of PV elements, daylighting elements and natural ventilation, a variety of details had to be defined for this project compared to an ordinary building project.

Performance characteristics

PV power generation: 13,500 kWh per annum (DC)
PV exported to the grid: 11,000 kWh per annum (AC)
Detailed monitoring data is not publicly available.

Provision and control of daylight, combined with high-efficiency artificial lighting and movement sensors saved approximately 70 per cent of electricity for lighting compared to traditional office lighting designs.

COMPONENT CHARACTERISTICS

PV system power	Translucent PV glass atrium: 17.16 kWp Façade: 4.14 kWp
Type of building integration	PV atrium and façade
Type of cell technology	Solel Ltd., Denmark

Project cost breakdown

	US$
Total building costs	2,337,000
Cost without property per m² (gross)	1,290
Typical costs for office building per m² (gross)	1,110
Design costs	821,000
EU funding for low-energy components	342,000
EU funding for design phase and measurements	323,000
Danish Ministry of Energy subsidy for building and components	367,000

GERMANY: FRAUNHOFER ISE

BIODATA

PROJECT:	20 kWp Fraunhofer Institut fuer Solare Energiesysteme ISE
LOCATION/CITY:	Freiburg
COUNTRY:	Germany
TYPE OF PV BUILDING:	Roof-integrated, façade-integrated, flat roof-integrated
BUILDING TYPE:	Research facility
NEW/RETROFIT:	New

CLIMATIC CHARACTERISTICS

LATITUDE:	48° 00'N
LONGITUDE:	7° 5'E
ALTITUDE:	260 metres above sea level
CLIMATIC TYPE:	Moderate (Temperature: January average = 1.8 °C; July average = 19.5 °C)
SUNSHINE HOURS:	Yearly average = 4.8 hours per day

Fig 1 Aerial view of Fraunhofer ISE
Source: Fraunhofer ISE

General project background

Fraunhofer ISE is a solar energy research institute. For many years the facilities were spread over some six buildings in an area of about 0.5 square kilometres. The new building brings these parts together and provides a modern research facility with up-to-date energy efficiency features and a healthy and pleasant atmosphere for its employees. The building site is located on the northern edge of the central area of the City of Freiburg. It is a narrow lot with a north–south orientation, and is subject to a little shading in winter. Its boundaries are an abandoned railway track to the east, another research institute building to the north, and streets to the west and south. The site was used for storing and shipping ammunition during World War II and had been vacant since. Its soil was heavily contaminated with organic chemicals from the war period.

Total floor area of the building is 12,000 square metres. One third of this area is used for offices, the remaining two thirds are for laboratories and workshops. The total building budget was about DM50 million (EUR25 million) including new laboratory facilities and equipment.

Position and rated power of the PV systems	
Southern facade	2.5 kWp
Saw-toothed roof	4.9 kWp
Spandrel south wing	2.9 kWp
Roof southern wing	4.5 kWp
Roof central wing	5.3 kWp

Building concept

The building features a sophisticated concept. Following the motto, 'Exemplary Building with the Sun', the aim was to combine a high-quality working environment and high functionality with a low-energy consumption and high design quality. Most of the building complex is three stories high. It consists of three parallel building wings connected by an access area, which is adjacent to a technical prototype laboratory. The comb structure and the interior zoning were deliberately chosen to achieve south-oriented offices for maximum daylighting. Thermal insulation, solar control, lighting and ventilation technology were designed for minimal energy demand.

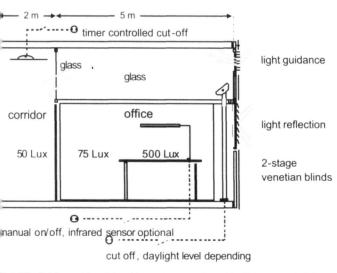

Fig 2 Office lighting combines light guidance, glare control and high-efficiency electric lighting
Source: Fraunhofer ISE

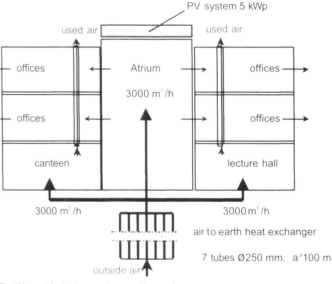

Fig 3 Air conditioning by an underground heat exchanger
Source: Fraunhofer ISE

Solar design features

The comb-like ground plan and the separation between the wings were chosen on the basis of minimal shading, good daylighting conditions, comfortable summer temperatures, passive use of solar energy and a pleasant indoor atmosphere (pronounced horizontal transparency). The main entrance foyer is dominated by an atrium with a saw-toothed roof and integrated PV modules. External venetian blinds with a light-redirecting function provide solar control. Beyond this, switchable glazing is installed in some areas (gasochromic systems). An underground heat exchanger cools or pre-heats the inlet air for ventilation of the entrance block. The building substance is cooled in summer by active nocturnal ventilation. Figures 2 and 3 show the office lighting and ventilation concepts.

Energy concept

The energy supply is based on a gas-fuelled combined heat and power unit. The waste heat serves to heat the building where needed and to cool the laboratories via an absorption cooling machine (power/heat/cooling combination). To avoid the need for air-conditioning units in the offices, the planners deliberately minimised the internal and external loads with energy-efficient lighting and solar control measures respectively. Thus, only the cooling load for laboratory and special-purpose rooms remained. This will be supplied via absorption cooling machines driven by the waste heat from a gas-fuelled combined heating and power unit. The electricity generated by the co-generation plant reduces the expensive peak load drawn from the public grid and provides the backup to the grid supply. The absorptive dehumidification of the inlet air to the clean-room laboratory shifts cooling to heating loads. The heating load in winter is reduced by using above-standard insulation (16 cm thermal insulation, optimised insulated double glazing units) and efficient heat recovery for the laboratories and many of the offices. Figure 4 shows an overview of the energy concept.

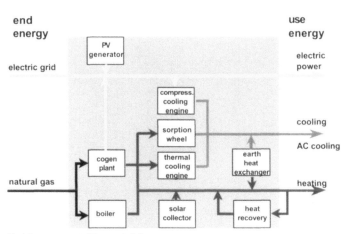

Fig 4 Energy concept of the new building
Source: Fraunhofer ISE

Building design process

In the early design process, three building concepts were pursued and evaluated: the 'block', the 'campus' and the 'wings' (figure 5). A thorough evaluation, guided by the aims of energy-efficient building and high indoor comfort, led to the selection of the 'wings' approach. The evaluation scheme is shown in table 2.

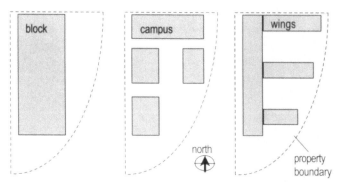

Fig 5 The three competing design concepts
Source: Fraunhofer ISE

The decision for the 'wing' concept was strongly influenced by the institute's Solar Building Design Group, which convinced the decision-makers that this approach was the most appropriate. The chosen concept has proved its value in similar applications and is appropriate to the pronounced north–south orientation of the site. The three building wings are oriented east–west and are widely separated to allow daylighting. The offices, which are not air-conditioned, are located on the sunny side, while the air-conditioned laboratories are on the shaded side. The flat roofs of the wings have been designed to function as outdoor test areas. The wing structure and the internal distribution of zones combine passive use of solar energy for heating in winter (sun low in the sky) with low (over)-heating loads in summer (sun high in the sky). A central access area extends more than 130 metres in the north–south direction and protects the inner courtyards and the wing façades from the summer afternoon sun. An entrance block at the southern end of the access area houses the administration and central services. The technical prototype laboratory, clean room and workshops adjoin the access area to the west.

Criteria	Block	Campus	Wings
Indoor climate	–	0	+
Daylighting	–	0	+
Energy consumption	+	–	0

Table 2 Matrix for evaluation of building concepts

BiPV design process

Research on photovoltaics is a major field of activity at Fraunhofer ISE. So is solar building design. Therefore, it was the aim of Fraunhofer ISE to demonstrate several approaches of building-integrated PV. Furthermore, it was of prime importance to demonstrate not only some 'mechanical' integration, but to also demonstrate functional integration as part of an energy concept. The PV modules should have another function besides electricity production.

Starting at a very early design stage of the project with integrating photovoltaics into the building offered an excellent opportunity for true integration. The PV elements are not add-ons, but integral parts of the building skin.

Planning process, alternative designs

The PV generators demonstrate different aspects of building integration. In the southern façade and the saw-toothed roof, the solar cells are encapsulated within heat insulating (double) glazing. They reduce the heat gain to the building and support the efforts to dispense largely with conventional air conditioning. The saw-toothed roof and south façade will employ the effects of translucent PV modules and use the shadow of the solar cells in the building interior as an architectural element.

The generator on the façade of the southern wing will demonstrate the application of photovoltaic modules as spandrel elements in vertical and tilted configurations.

The focus of this case study will be on the 'saw-toothed roof' array above the atrium. However, for comprehensiveness the other PV generators will be briefly discussed.

Cladding elements on the wings' roofs

The PV arrays on the roofs of the southern and central wings serve as cladding for the ventilation shafts behind them. They employ standard modules in a standard mounting structure. These modules replace an initially planned sheet metal wall. The cost for the support structure is completely recovered from the saved sheet metal wall.

This idea was developed during the construction process after the north wing had been nearly completed. It had not been included in the original call for tender. Therefore, only the later parts of the building, central and south wing, benefited from this idea.

Support structure

Concrete elements above the structural members of the building bear galvanised steel frames. These are inclined at 30°. On these frames an 'AluTec' mounting structure is screwed on to hold the Type ASE-100-GT-FT modules in place.

This module type was chosen because it employs the same cells (ASE EFG 100 x 100) as used in the access area façade. Of course, they also fit well into the available geometric boundaries.

Fig 6 The PV generator on the flat roof of building B
Source: Fraunhofer ISE

Façade of access area

This PV generator is an eye-catching structure for visitors coming from downtown. It employs EFG cells with their rather homogeneous appearance in a large area module. The cells reduce the heat gain of the south-facing glass façade. The cells were chosen as a compromise between the homogeneous, dark appearance of mono-crystalline cells and the lower cost of poly-crystalline cells. The wiring is run in a vertical duct at the eastern edge of the construction on the outside of the building. Inverters are mounted in the basement below the façade.

Decision process: atrium roof

The saw-toothed roof over the atrium offers some 70 square metres of area for PV modules and is a good example to illustrate the design process. Figure 9 illustrates the optimisation of the saw-toothed roof geometry as part of the building concept. Good daylighting conditions are essential for the use and the aesthetic effect of an atrium. However, enormous overheating would arise in summer if transparency is high and there were no shading elements. Simulations were undertaken for the indoor thermal and daylighting conditions to take into account both aspects. Electric power generation, daylighting and protection from overheating in summer were balanced. The first geometry of figure 9 was chosen yielding an inclination of the shed-skylights near the optimum of 35°.

The glazing is supported by a steel structure that holds a type of mullion/transom stick construction. Modules are placed on sealing rubber strips and fastened by a screwed cover profile.

Fig 7 South façade after completion
Source: Fraunhofer ISE

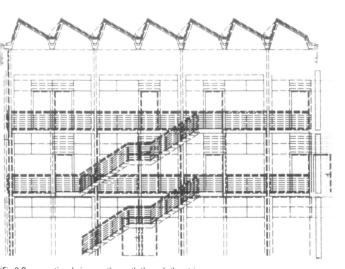

Fig 8 Cross sectional view north–south through the atrium
Source: Dissing + Weitling

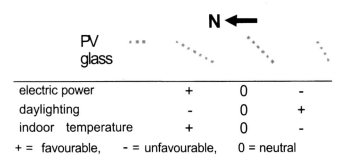

	PV glass		
electric power	+	0	-
daylighting	-	0	+
indoor temperature	+	0	-
+ = favourable, - = unfavourable, 0 = neutral			

Fig 9 Evaluation of different geometries of the shed roof structure
Source: Fraunhofer ISE

Module design

Module design was guided by:

- architectural design considerations;
- electric power production;
- heat gain reduction;
- daylighting.

The building design grid and the geometry of the 'saw teeth' determined the module dimensions. The basic shape of the structure had been fixed according to the optimisation illustrated in figure 9. From that, the basic dimensions were set to about 590 x 1900 mm. Further requirements for the modules were: cells from a partner company, a TSET (total solar energy transmittance) value below 30 per cent and light transmission above 15 per cent. The cell type was not important, but the rear view was, since the cells are visible from below but not from the front. After the basic dimensions had been defined, the modules were designed. Initially, three drafts were discussed (figure 10). Eventually the 'SHELL' cell (125 x 125 mm) was chosen and then the layout and the electrical circuit was designed. The architects opted for large gaps between cells for transparency and the design was fixed as shown in figure 11. A TSET value of about 30 per cent was achieved.

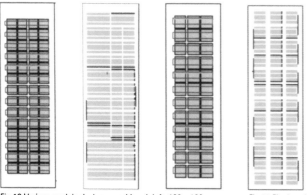

Fig 10 Various module designs considered; left: 100 x 100 cm cells; centre: 125 x 125 mm cells, densely packaged; right: 125 x 125 mm cells, 20 mm distance between cells
Source: St. Gobain Glass Solar

Fig 11 Final module design
Source: St. Gobain Glass Solar

Optimising the electrical design

After optimising the geometry of the PV modules, the design of the electrical circuits including the PV modules was optimised to take into account the partial shading by the southern sheds. Following a proposal by the module manufacturer, St. Gobain Glass Solar, the modules were split in three sub-modules (figure 12). The sub-modules were designed according to the rule of thumb for open field arrays: at winter solstice noon, the shadow of the front row should just fall below the rear module's lowest cells.

The module is constructed as heat insulating glazing with a gas-filled space between front and rear pane.

String wiring

The module layout, as well as the wiring, was designed to allow a flexible inverter concept (figure 14). String inverters as well as a central inverter can be used. Slight asymmetries in the data of the strings due to differing numbers of cells in series are negligible compared to the influence of turbulent winds and different natural ventilation.

Table 3 is a summary of the electrical data of the array. A newly developed central inverter, the SOLWEX 35300 E +, a transformerless inverter was chosen. This will yield an improvement in annual mean efficiency of 2–3 per cent.

	Number	V_{oc} V	V_{MPP} V	I_{sc} A	P_{MPP} W	Remark
Module	69	10.6	8.7	4.3	36	3 separate
		5.3	4.4		18	sub-modules
		5.3	4.4		18	
					Total: 72	
Strings	4	360	296	4.3	1220	Slightly
		371	305		1260	asymmetrical
		366	300		1240	
		366	300		1240	
Total		**371**	**298**	**17.2**	**4980**	

Table 3 Nominal array data at STC

4 mm PLANIDUR® Diamant
2 mm Zellzwischenraum
4 mm PLANIDUR®; PLT®-Futur
14 mm SZR (Krypton)
8 mm VSG mit 0,76 Folie

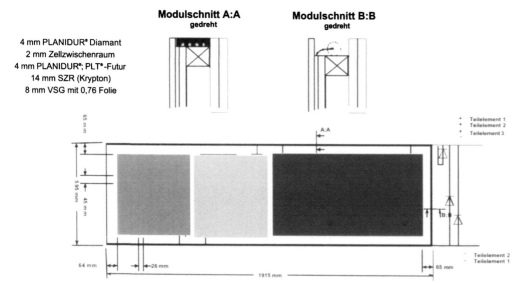

Fig 12 Construction of the modules. The sub-modules are marked by different colours. Bypass-diodes for each sub-module are mounted at the upper edge. The modules need a special two-layer back pane, because they are mounted 'overhead'.
Source: Fraunhofer ISE

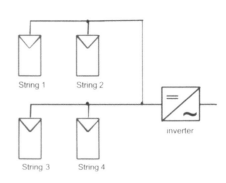

Fig 13 Block diagram of the array
Source: Fraunhofer ISE

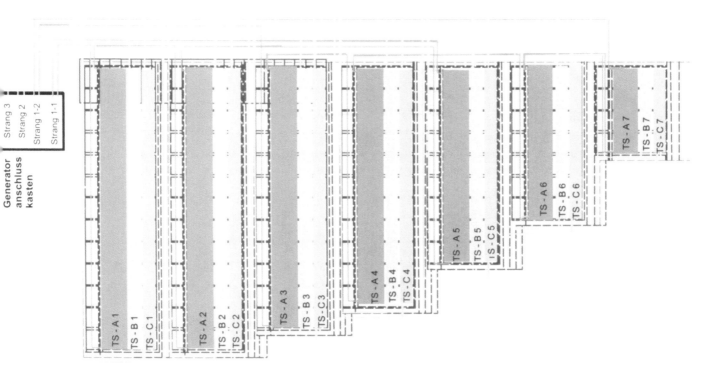

Fig 14 The wiring of the array. Despite the irregular structure of the shed roof, an even distribution of modules per string could be achieved.
Source: Fraunhofer ISE

Installation

The installation followed standard glazing procedures.

Fig 15 Modules were delivered in crates onto the roof by a crane. From there they were transported manually onto the shed structure.

Fig 16 Modules were then placed on the sealing frame. Each module weighs about 50 kg.

Fig 17 Close-up of a module after it had been placed onto the sealings. Connection wires of two sub-modules are clearly visible.

Fig 18 Cover profiles are mounted to secure the modules against up-lift winds

Fig 20 View from inside during mounting of the glazing

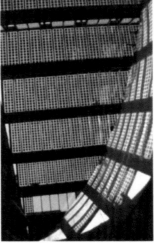

Fig 21 The roof is nearly complete

Fig 19 Modules have been secured by a cover profile. The sub-modules have been connected using crimp terminals and a protective shrinkable tube. The string wires are placed into a channel beside the top of the modules. The northern glazing has been mounted and secured (vertical cover profile). At the bottom of the picture the top of a ventilation flap is visible.

Figures 15 to 21
Source: Fraunhofer ISE

Problems during realisation

Some modules showed water condensation in the insulating air gap between panes. Presumably, this resulted from careless work by the glazing manufacturer. The modules affected were removed and resealed.

Performance characteristics

Performance data are not yet available. An annual yield of around 16 MWh is expected for all PV systems, which should meet the entire demand for office lighting in the new building.

Lessons learnt from the project

True integration of photovoltaics into buildings demands a holistic design approach. PV offers new opportunities for aesthetically attractive solutions to daylighting, and overheating prevention. A good design of the electrical systems minimises the effect of unavoidable shading.

Project cost breakdown

Total building cost was about 26 million Euro including laboratory facilities and equipment.

All electricity is fed into the utility grid at a rate of 0.50 Euro/kWh. This is according to the German law for electric power from Renewable Energies (EEG). Originally, it was planned to feed the electricity into the building grid using distributed feeders for the inverters. After the law for a higher buy-back rate had been passed these plans were changed to feed all electricity into the utility grid. The utility did not require distributed meters and requested a separate, additional grid connection for the feed-in meter. Regular electricity supply is connected on the medium voltage level of 20 kV.

	Façade of access area	Saw-toothed roof	Roof wings	Spandrel of south wing
Rated power [kWp]	2.5	4.9	10	2.9
Modules	20,000 [1]	50,000	42,000	35,000 [2]
Mounting structure	0 [3]	0 [3]	4,500 [4]	n/a
Mounting and array wiring labour	10,000	25,000		n/a
DC main cable and inverter wiring including labour	1,500	1,800	8,000 [5]	n/a
Inverter	1,600	2,800	7,000	2,200 [2]

Other costs:

Additional grid connection for feed into grid	3,500
Monitoring equipment including wiring and sensors	12,000

Table 4 Cost breakdown for the PV subsystems*. All costs in Euro.

*costs for the spandrel of the south wing are not yet known
[1] estimate
[2] initial offer
[3] included in building structure
[4] Alutec mounting profiles
[5] long cable runs to allow for flexible testing configurations

GERMANY: MONT-CENIS ACADEMY

IODATA

PROJECT:	1,000 kWp Mont-Cenis Academy
LOCATION/CITY:	Herne-Sodingen
COUNTRY:	Germany
YPE OF PV BUILDING:	Semi-transparent PV overhead glazing and PV glass façade
BUILDING TYPE:	Government training academy; multipurpose building
NEW/RETROFIT:	New

LIMATIC CHARACTERISTICS

LATITUDE:	51° 32'N
LONGITUDE:	7° 15'E
ALTITUDE:	153 metres above sea level
CLIMATIC TYPE:	Moderate humid climate (Temperature: winter average = 4.8 °C; summer average = 14.35 °C [30-year average, period 1961–1990])
SUNSHINE HOURS:	Yearly average = 3.98 hours per day

Fig 2 View from inside to the overhead glazing of glass envelope
Source: Flabeg Solar International GmbH

1 General view from southeast
ce: Flabeg Solar International GmbH

General project background

At the end of the 1980s the Minister of the Interior of the state of North Rhine Westphalia made the decision to move the Continuing Training Academy to Herne. It took ten years to complete this exciting architectural concept that included a 1 MW building-integrated PV system for the ecological and economic renewal of the region. The project began with a two-stage competition in 1991 for the Internationale Bauausstellung Emscher Park (IBA), won by French architects Jourda & Perraudin. In the second stage of this competition, the German architects Hegger, Hegger Schleif joined the design team and a fruitful German/French collaboration began.

The site is the former Mont-Cenis black coal mine at the centre of the Ruhr area and at the heart of the region dedicated to the Internationale Bauausstellung Emscher Park (International Architectural Exhibition). It is an important part of the IBA 'Emscher Landscaped Park' project, a series of green spaces developed in the last ten years to improve the quality of life in the Ruhr region.

Architectural concept

Situated slightly higher than the surrounding area, the Mont-Cenis Academy building consists of a huge micro-climate glass envelope that forms a shelter for several interior buildings. It is 176 metres in length, 72 metres wide, and 15 metres high. Originally planned as a training academy, the building today includes several other functions including seminar facilities, meeting rooms, accommodation facilities, a restaurant, a gymnasium, a library, a civic hall and leisure facilities. The building became one of the Ruhr region's new landmarks and serves, like the mines before, as the functional and urban centre of Herne-Sodingen.

The project uses a series of devices to preserve and improve the environment: decontamination of the existing polluted soil; collection of rainwater and re-use of groundwater; collection of gas escaping from former mine shafts used for urban heating; generation of passive solar energy by the use of the micro-climatic envelope of the greenhouse; generation of active solar energy for heating water and the production of PV electricity; and the utilisation of easily recyclable ecological building materials and construction techniques.

Passive solar energy use

The glass envelope of the Mont-Cenis Academy creates a climatic shift in the summer and winter. It keeps out wind and rain and creates a garden-like interior with a mild micro-climate similar to that of the Mediterranean. As a consequence the interior buildings no longer need to be absolutely weatherproofed against wind and rain. An air conditioning system is not installed. Instead, sophisticated ventilation and heating systems reduce the energy consumption considerably in comparison with conventional air-conditioning technologies. The ventilation of the glass envelope is controlled automatically from a central position. A meteorological station and sensor supply climate data. To prevent overheating in summer, the roof and façade elements can be opened variably. On hot days, doors in the lower façade can be opened as well. The shadows of the trees and the cooling effect of waterfalls and fountains are also used. Additional fresh air from cooler outside areas is supplied directly to the inside through ground ducts (diameter 1 metre). The air is naturally cooled or heated during very hot or very cold periods respectively, thanks to consistent below-grade temperatures.

The ventilation of the buildings inside the glass envelope is achieved by natural or mechanical means. In order to reduce energy consumption in winter and to cool naturally in summer, an air handling system unit with a heat exchange system was installed. The heating system will use less than 2 kWh/m²/year. In winter, the annual heat requirement is below 50 kWh, while the total energy requirement, with optimum control of the installation, is approximately 32 kWh/m² per year. This means that the buildings require about 23 per cent less energy than buildings with the same insulation standard (corresponds to about 18 per cent less CO_2 emissions). The direct air supply to the houses is warmed in winter to about 8 °C through the ground ducts.

Daylighting concept

Different daylighting technologies have been employed within the building. In addition to the special design of the PV roof (figure 2) light shelves were incorporated into certain façades of the buildings inside the glass envelope to reflect daylight deeper into their rooms.

Hologram films integrated into the roof micro-climate envelope redirect the sunlight down into the library and the entrance hall. In the library the hologram films act as a heliostat, which intensifies the light level. In the entrance hall they break up the light spectrum and create a rainbow effect (figure 3).

Fig 3 Detail of overhead glazing; clear glass pane with integrated hologram film; semi-transparent PV modules on the right and left sides
Source: Flabeg Solar International GmbH

Ecological building materials and construction

Preference was given to ecological building materials and construction techniques that allow easy maintenance and bear a high potential for recycling. This resulted in a limited range of building materials, mainly timber, glass and concrete.

The timber elements of the structure make use of local wood sources. The main structural support columns of the glass envelope consist of the trunks of 130-year-old pine trees which were felled in a nearby forest. Thanks to the protected climate, the wood did not have to be treated (figure 4). The uniform basic grid (12 x 12 metres) enabled cost-effective pre-finishing. On the outside, waxed larch wood and larch-laminated wood were used. The concrete structure acts as a heat sink, balancing out the temperature differences between day and night as well as periods of more extreme temperature differences.

Rainwater system

The rainwater falling on the large roof is collected by a special rainwater system, which minimises pipe diameters. It is collected in an underground storage tank, filtered and reused for cleaning purposes, and for the watering and maintenance of plants within the micro-climate glass envelope.

Fig 4 View from southwest to the south façade of the micro-climate envelope with vertical wood columns in front
Source: Flabeg Solar International GmbH

Fig 5 Town planning model of Mont-Cenis site
Source: Flabeg Solar International GmbH

Mine-gas driven co-generation plant

The mine gas of the disused Mont-Cenis mines allowed the installation of a co-generation plant on the site. Since November 1997, two co-generation plant modules have supplied 253 kW electricity and 378 kW heat. One of the plants is operated with mine gas and the other optionally with mine gas or natural gas. The heat is used for heating the Mont-Cenis Academy as well as a nearby hospital and 250 flats. The electricity generated is fed into the grid. The two co-generation plants worked so successfully that in the year 2000 a third co-generation plant module was installed, supplying 1000 kW electricity and 1200 kW heat. In total, this helps to reduce the CO_2 emissions by 60,000 t/year. Two peak-load boilers using natural gas and each providing 895 kW have also been installed to supply heat.

Consideration was also given to the requirements of people with various disabilities. To allow easy wheelchair access, barriers were avoided and all doors and lifts open automatically. To assist the blind, a feeler model is located at each entrance. A routing system and Braille information on railings, lifts, and doors make orientation easier.

BiPV design process
Planning process, alternative designs

The original competition brief in 1991 did not call for the use of PV (figure 5). Intake of solar light and heat was controlled only by sun screens and natural planting inside and outside the micro-climatic envelope. But in a later planning phase, when different shading systems were under discussion and the architects showed the clients how much energy the roof deck could produce, the installation of PV became a priority for the client. The idea of controlling incoming light and shade in this elegant manner was attractive to the client, particularly because there were no budgetary constraints on the additional costs of the PV system. The client's only requirement was to build the world's largest 1-megawatt building-integrated PV system.

Design process

A total of 20,640 square metres of glass were used for the micro-climate glass envelope, 10,533 square metres of which were fitted with PV cells. The roof itself provides space for only 9,744 square metres of PV modules. The rest was integrated in the west façade. The PV modules serve as truly multifunctional building elements. They provide shading, daylighting and electricity production at the same time.

The idea of integrating PV modules into the glass envelope was formulated when the design and structural concept of the striking glass cover had already been established. It was still possible to adapt the construction of the micro-climate glass envelope to the special requirements of the PV-system technology, but with certain limits. The roof-mounted PV modules are orientated south with an inclination of 5°. The façade-mounted PV modules are orientated west with an inclination of 90°. For this site location, the optimal inclination angle of 28° could not be achieved. The single glass panes of the original design of the overhead glazing were simply replaced by semitransparent PV modules of the same size.

In the first design, all the solar cells and modules were evenly distributed across the roof, but computer modelling demonstrated that the building interior would be too dark with this layout. The solution looks like a cloud-patterned sky (figure 6). The cells are concentrated over the internal buildings and left clear glass between the buildings and over the central thoroughfare. Six different types of PV modules with different densities of solar cells and glass of various degrees of transparency were used. The number of cells per module ranges between 128 and 260, with corresponding power outputs of between 250 and 419 Wp. Their densities varied from 86 per cent directly over the buildings to 58 per cent in transitional zones. The variation of the cell layout along with the passage of daylight creates exciting ever-changing patterns of light and shade inside the building. An automatic cleaning system employing recycled rainwater for the PV overhead glazing was envisaged, but not realised.

On the west façade, semi-transparent PV modules with mono-crystalline PV cells cover 30 per cent of the vertical glass façade. They help to avoid overheating (figures 7, 8, 9).

To achieve the highest system efficiency for the complicated PV modules and PV cell distribution on the roof and façade, an innovative string inverter concept was developed. All inverters are mounted on the rooftop. Originally it was planned to paint the 569 inverters in different colours and to locate them over the total roof area, symbolising 'flowers' in different colours. But for maintenance reasons, Stadtwerke Herne AG, the owner of the PV system, required easy access to all the inverters. All inverters are now located next to each other along the edges of the roof.

Fig 6 Aerial view of part of the roof deck of the micro-climatic envelope with integrated PV modules
Source: Flabeg Solar International GmbH

Fig 8 West façade with structural glazing and integrated semi-transparent PV modules
Source: Flabeg Solar International GmbH

Fig 7 Detail view of west façade with window openings for ventilation and the row of inverters on top
Source: Flabeg Solar International GmbH

Fig 9 Fixing detail of vertical structural glazing façade
Source: Ingo Hagemann

COMPONENT CHARACTERISTICS

PV system power   a) PV overhead glazing: 925 kWp; b) PV façade: 75 kWp; total: 1,000 kWp

Type of building integration   a) PV overhead glazing: semi-transparent PV modules with different cell area ratio; orientation: south; inclination 5°;
b) PV façade: semi-transparent PV modules; orientation: west; inclination 90°

Type of cell technology

Type of PV Cell	A	B	C
Size (mm)	125 x 125	100 x 100	100 x 100
Efficiency (%)	12.5	16	12.8
Material	poly-crystalline silicon	mono-crystalline silicon	poly-crystalline silicon
Colour	blue	blue	blue
Installed capacity (kWp)	500	400	100
Manufacturer	Solarex	ASE	ASE

Modular dimensions

Type of PV module	I	II	III	IV	V
Size (m²)	3.36	3.36	3.36	3.36	2.78
Size (cm)	120 x 280	120 x 280	120 x 280	120 x 280	116 x 240
Weight	114	114	114	114	70
PV cell area (%)	86	86	73	63	58
PV cell type	B	C	A	A	B
PV module capacity (Wp)	416	332	282	250	259
Voltage (V)	132	119	68	61	96
Manufacturer	Flabec Solar International GmbH; Pilkington Solar International GmbH				

Vertical section of the semi-transparent PV modules from outside to inside   a) PV overhead glazing: semi-transparent PV modules in glass-glass encapsulation technique: 6 mm extra white-heat strength glass; 2 mm cast resin with PV cells; 8 mm heat strength glass;
b) PV façade: semi-transparent PV modules in glass-glass encapsulation technique: vertical glass façade: 5 mm extra white-heat strength glass; 2 mm cast resin with PV cells; 5 mm heat strength glass

Array dimensions   a) PV overhead glazing: 2,900 PV modules, 3.36 m² each; total size 9,744 m²;
b) PV façade: 284 PV modules, 2.78 m² each; total size 789.52 m²;
total: 3,184 PV modules; 10,533.52 m²

Visual details   The PV cell area of the PV modules varies between 86% and 58%. The space between the single PV cells varies according to the translucent ratio.

Inverter Type   'Sunny Boy 1500'; maximum power: 1,650 W

Maximum efficiency   96%

Nominal Power   1,500 W

Size   w 434 x h 295 x d 214 mm

Weight   Approximately 24 kg

Battery   816 single batteries including a centrally controlled electrolyte circulator, a cell temperature stabiliser and an automatic water refiller; total energy output of 1.2 MW; total energy content of 1.2 MWh; expected lifetime: 20 years

PV plugs   Special development from Leopold Kostal GmbH & Ko KG for Flabeg Solar International GmbH.
The plug is not available on the market.

Data control unit   Sunny Boy Control

Mounting structure   a) PV overhead glazing: custom-made aluminium profiles with pressure plates manufactured by Wicona Bausysteme GmbH;
b) PV façade: custom-made structural glazing façade with aluminium profiles manufactured by Wicona Bausysteme GmbH

Installation design
Integration design and mounting strategy

PV modules and glass panes of the overhead glazing rest on aluminium profiles and are held in place with aluminium pressure plates. The vertical PV-and-glass façade is carried out as a structural glazing façade. The glass panes and PV modules are glued onto aluminium profiles. All aluminium profiles are mounted on the load-bearing wooden substructure and were designed specially for this project by Wicona Bausysteme GmbH, Ulm.

The interconnecting plugs and the required cabling are integrated in the aluminium profiles that hold the PV modules and glass panes into place. They are invisible and protected against weather conditions and ultraviolet light. This new mounting detail was possible due to the specially developed PV plug. The plug is not thicker than the 4 mm glass pane of the PV modules (figure 10) and therefore fits well in the rebate. The plug helps to cut down the installation time and costs.

Fig 10 Detail of new PV plug
Source: Ingo Hagemann

Electrical configuration including grid integration
Inverter

The inverter concept for this project is innovative. Special string inverters (Sunny Boy 1500), help to achieve the highest system efficiency for the complicated module distribution on the roof. Losses due to partial system shutdown, to temperature differences between the PV panels and to mismatching are reduced to a minimum. Individual management of each string optimises the energy output of the total system.

Fifty-five kilometres of cables were used to connect the 569 inverters with the 3,185 PV modules. However, the use of cost-intensive DC cabling was reduced to a minimum. The installation of the inverters took about three months; two weeks each were necessary to install the frames and to fix the inverters, which were built in a durable stainless steel case, on the roof. The cabling itself took a further two months.

The advantage of this electrical layout is that each string has its own MPP-tracking and is connected to a central computer, which shows not only the total power of the PV generator, but also the working condition and power of each string. It is possible to easily check the whole PV system with one glance at the computer screen. The whole inverter concept proved to be a good choice.

Monitoring system

With several Sunny Boy control units, system monitoring and remote diagnosis are possible. These central data acquisition and diagnosis units allows a flexible system management with remote control of the plant components and the transmission of all the relevant system data to a PC. The inverters transmit their weather data through the mains, using power-line communication, allowing access to the data of each single component at nearly every point of the building. This allows easy supervision of this large PV plant with its inverters.

Battery system

The battery systems refine the value and quality of the photovoltaic electricity and of the mine-gas driven co-generation plant, and transmit the current automatically to the network, exactly when it is needed. Part of the PV power is directly used within the building. Surplus power is stored in the battery and used at night or during periods of low light. The rest of the surplus energy is fed into the public grid.

With an output of 1.2 MW and an energy content of 1.2 MWh, the battery storage system fulfils three functions:

- it reduces peak demand loads;
- it compensates for system fluctuations, which can emanate from the PV installation; and
- it can supply emergency power for the Mont-Cenis Academy.

Automatic maintenance systems increase the service life of the battery up to 20 years. For this reason, a centrally controlled electrolyte circulator, a cell temperature stabiliser and an automatic water refiller were integrated into the system. The battery consists of 816 single batteries and weighs 90 tonnes. It is computer-controlled and works through three reversible converters parallel to the PV installation and to the low-voltage network of the local utility company.

This battery system was not necessary for this project as the PV generator is connected to the public grid. It serves more as a demonstration project for the export of such technologies. The idea was to show that energy storage on a large scale could be economically feasible.

Fig 11 Detail view of inverters at the roof edge
Source: Flabeg Solar International GmbH

Planning approval and institutional processes

During the ten years from vision to realisation, it took two years to solve the legal and financial problems and to resolve other reservations and doubts relating to the project.

With regard to the PV system, extra effort was required by the design team to be granted a building permit to use the glass-glass PV modules for the overhead glass construction. Special load tests were required and had to prove that the sandwich construction of the large PV modules (1.2 x 2.8 metres) was stable enough to walk on for maintenance and cleaning purposes, and that the PV modules would not collapse and fall after cracking.

During the mounting of the PV modules and cables, the installers walked without any problem on the roof and its PV modules. Today however, German industrial safety regulations demand additional rope safety devices for workers on the roof which can only be ensured by a crane.

Performance characteristics

The calculated energy output of approximately 700,000 kWh/year has not yet been reached. The measured energy output reached a maximum between 600,000 kWh/Year and 650,000 kWh/year. On the one hand, it was assumed that the calculated figure was too optimistic and did not take special site conditions into account, such as sub-optimal inclination and orientation. On the other hand, air temperatures below the PV modules were much higher than expected. This was likely due to the failure of the natural ventilation system of the micro-climate glass envelope. There were some problems with the opening mechanisms of the automatically driven ventilation openings of the roof. Although the PV modules were mounted only with a slope of 5°, the saw-tooth roof caused some shading on the PV modules in early morning and evening hours as well as in winter, when the sun is low on the horizon. The effect of this partial shading on the total energy output was underestimated.

Post-installation feedback

Building-integrated PV modules successfully allow control of the interior climate of this large micro-climatic glasshouse. The intake of solar heat and light in different zones of the building is controlled by the choice of different cell densities of the PV modules of the overhead glazing. For large projects such as this, the integration of PV in the glass envelope is a good choice to replace conventional shading systems as they not only provide shade, but also act as a weather skin and PV generator. It is also expected that they will have a longer lifetime than conventional shading systems. The multifunctionality of these PV modules makes them attractive. Access to and any work on the roof always requires the help of a crane. This is not cost-effective and is inappropriate for the required maintenance and cleaning of such a glass shelter.

Project organisation

Due to decisions on a political level, the local utility company, Stadtwerke Herne AG, owned by the City of Herne, took over the ownership and the client's responsibility for the PV system. Pilkington Solar International GmbH (now Flabeg Solar International GmbH) in Cologne served as the general contractor and manufacturer of the 3,184 PV modules, which were assembled in a nearby plant. The associated company SMA Regelsysteme GmbH, provided the 569 transformerless inverters, type 'Sunny Boy 1500'. Aabakus energiesysteme GmbH, Gelsenkirchen, developed, in close collaboration with the architects, engineers, manufacturer and the installers, the concept for the electrical interconnections of the 3,184 PV modules and prepared the working drawings. They also supervised the installation work on the building site and the commissioning of the total PV system.

Project cost breakdown

The overall cost of the PV system was DEM15.7 million (EUR8.0 million). This cost includes roofing, façade, shading and the solar generator. The PV system was financed 49 per cent by the state of North Rhine Westphalia and the European Commission. The Stadtwerke Herne, the local utility company, invested the remaining 51 per cent.

	DEM (million)	EUR (million)
Construction costs		
Site reclamation/new landscaping	28.4	14.5
Academy of the State Interior Ministry	35.0	17.9
District Town Hall, Herne Sodingen	11.4	5.8
Micro climate glass envelope	30.0	15.3
BiPV system	15.7	8.0
Total	120.5	61.5
BiPV system costs		
PV modules	11.1	5.67
Inverters	1.2	0.60
Switches, cabling, etc	0.6	0.31
Planning and engineering	1.1	0.56
Mounting	1.7	0.86
Total	15.7	8.00
Maintenance costs		
Total (estimated)	0.03	0.02

ITALY: THE CHILDREN'S MUSEUM OF ROME

BIODATA

PROJECT:	15 kWp roof and canopy
LOCATION/CITY:	Rome
COUNTRY:	Italy
BUILDING TYPE:	Commercial
NEW/RETROFIT:	Retrofit

CLIMATIC CHARACTERISTICS

LATITUDE:	42° 11'N
LONGITUDE:	12° 28'E
ALTITUDE:	15 metres above sea level
CLIMATIC TYPE:	Temperate (Temperature: yearly average = 15 °C)
SUNSHINE HOURS:	Yearly average = 4.3 hours per day

Fig 1 Interior view of skylight
Source: Abbate & Vigevano Design Studio

General project background

The project is located in the historic centre of Rome, on a site previously occupied by the public transportation warehouse, operated by the municipal government of Rome. The project is the transformation of this large complex of buildings into an exhibition space and multipurpose area for the new Children's Museum of Rome.

The structure of the main exhibition hall is the steel and cast-iron shelter built in 1920 from a design patented by the French engineer Polenceau. The total area of the project covers 2,500 square metres with 3,000 square metres of green outdoor spaces; the Children's Museum building has 1,500 square metres of exhibition space.

The overall approach to the project is based on respect for the environment, and all the material used for the construction is either recycled or recyclable and non-toxic.

It has been estimated that approximately 130,000 people, including children from infants to 12 years of age, schools and families visited the museum in its first year of operation.

Project brief

The Children's Museum of Rome project brief aimed at 'offering experience, play and sociality in a unique environment – an extension to learning which is beyond normal education. By bringing together the family, school and the world of work, children are encouraged to learn the skills that they must acquire in order to enrich the world around them.'

There was not a specific requirement to use photovoltaics, but the exposure of children to alternative energy was considered to be an effective illustration of the basic aim of this museum: to heighten awareness of the quality of urban life through a transparent guided itinerary' of everyday activities.

Fig 2 Site prior to renovation
Source: Abbate & Vigevano Design Studio

Fig 3 CAD design of PV system
Source: Abbate & Vigevano Design Studio

BiPV design process

The initial renovation project did not include any renewable source of energy due to a conventional approach to the design of technical systems, the requirement to limit the overall cost of the undertaking and the need to stay within the tight construction timetable.

The idea of integrating photovoltaic energy was suggested by Cinzia Abbate and Carlo Vigevano, the designers of the PV system, early in the working drawing phase of the project. The idea of including an alternative source of energy in the building, was enthusiastically accepted by the directors of the museum, since it conformed to its overall didactic purpose, and the goal of using only well-balanced, environmentally friendly materials.

The estimated cost of the PV installation was an obstacle until Abbate and Vigevano found the opportunity of joining a EU/THERMIE programme with two European partners, Cenergia from Denmark, and Ecofys from the Netherlands. The Children's Museum of Rome subsequently became not only the first, but also the largest building integration of photovoltaics in a historic city centre of Italy.

The first PV design approach proposed a moveable 3 kWp PV roof structure that would shelter the large skylight, and 12 kWp moveable canopies shaped like a 'Meccano' toy, placed along the south façade. Several systems were studied, and several comparative cost analyses were made, before deciding to limit the moveable parts to only those essential areas where the indoor climate and natural quality of the light would benefit substantially from the mechanical moveable elements. The cost analysis demonstrated that the solution of replacing conventional glass with double-glazed PV modules in the skylight reduced the cost of the PV installation and also eliminated the cost of maintenance of the mechanical roof structure. The PV skylight also contributes significantly to improving the overall indoor comfort and the aesthetic quality of the roof.

The architects proposed three different layouts of the skylight and the canopies, to be evaluated by Cenergia, in order to select the best solution to be built.

Cenergia used simulation programs to study the existing conditions and to verify the integration of the photovoltaic system in the roof and the façade as passive shading. Due to the large volume, and to the roof configuration of the building, the energy consumption required for heating with a conventional transparent skylight would have been approximately three to four times more than for cooling. Indeed, the results showed that the consumption would have been approximately 90 kWh/month for net heating and 32 kWh/month for cooling.

The results for the three solutions presented by the architects Abbate and Vigevano were almost identical, but obviously the one with a larger PV surface demonstrated the possibility of reducing the use of cooling systems during summer. Analysis of the indoor temperature and of the amount of time necessary for natural air exchange was used to estimate how much energy saving was generated by the use of photovoltaics and the specific design of the skylight surface. In order to decrease the cooling demand, the photovoltaic skylight was designed to integrate special solar reflecting glazing in the transparent area of the roof.

Other additional passive cooling devices were incorporated to improve indoor comfort, such as control of the ventilation rate during use and after-use of the building, maximised use of night cooling in the summer time and an evaporative cooling device. This was achieved by placing a large shallow fountain in the centre of the museum, underneath the photovoltaic skylight. The water sprinkled on the porous surface of the stone fountain evaporates and cools the air, particularly during the hot summer months.

The aesthetic value of the photovoltaic panels played an important role in convincing the architect of the museum to accept the presence of the poly-crystalline silicon cells, not only for the visual impact of the outdoor space, but especially for the indoor shading. Real examples of the modules, computer renderings, and pictures of other built projects with similar roofs were extensively used to persuade both the architect and the director of the added aesthetic value of the PV installation.

The design and realisation process of the photovoltaic installation took approximately a full year of man-power for the PV designers, over the three-year span of its completion. Before its completion, the project underwent at least four different variations of the roof design, and at least nine different solutions for the canopies. As an example, during the restoration of the existing cast-iron structure, the Monuments and Fine Arts Office prohibited the removal of several insignificant existing structural steel bars in the gutters, requiring a design modification of the support element of the canopies on the roof.

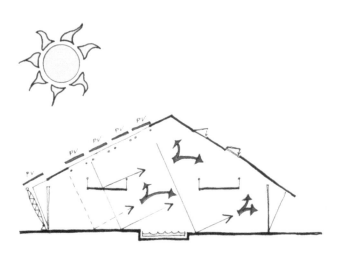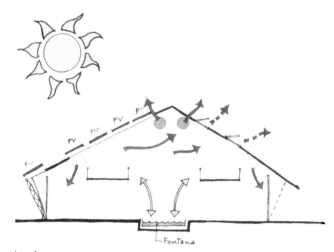

Fig 4 Diagrams (left winter, right summer) of the photovoltaic and passive solar system used for indoor thermal comfort
Source: Abbate & Vigevano Design Studio

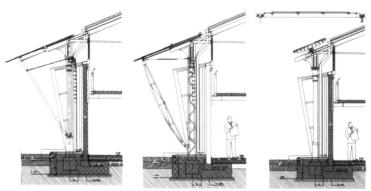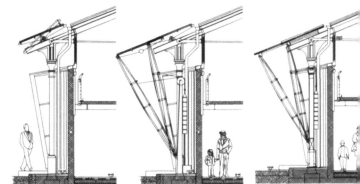

Fig 5 Different design proposals for the same canopy presented chronologically
Source: Abbate & Vigevano Design Studio

Installation design

The aim of the project was to encourage the retrofit application of PV systems as part of the restructuring and maintenance of older industrial buildings. In particular, the project aimed to improve the quality of the natural lighting and to decrease the heat load of the building through the innovative design of a grid-connected 15 kWp PV plant, located on the south pitched roof of the main building.

The PV installation is divided into two systems, with shading devices and skylights becoming an integral part of the industrial pavilion. The 7 kW PV canopy system works with alternating fixed and moveable sections connected to the lower part of the roof, which shades the southern façade. The 8 kW PV system replaced part of the old roof tiles with a specially designed skylight of transparent PV glass modules.

The photovoltaic poly-crystalline cells transform sunlight into a 15 kWp plant to supply 30 per cent of the energy required in operating the museum's exhibits, or 60 per cent of the total artificial lighting of the pavilion.

The motors, and all the mechanical parts of the PV building components, were designed to:

- simplify the mounting technology;
- reduce the costs of production, assembly and installation;
- reduce the maintenance costs and
- introduce a playful configuration, for a friendly introduction to PV technology.

The same principle of clarity and visual immediacy is encountered in many other components of the technical systems of the Children's Museum of Rome, thus stimulating the curiosity and interest of the visitors.

In particular, the detail of the mounting system of the roof adapts a typical PV Shuco window frame, but leaves all the junction boxes of the PV modules exposed, to show the electrical character of the installation. All the cables of the motors for the moveable sections of the canopies are exposed and accentuated by the bright colour of the cables. A child can easily follow the path of the cables to understand where the electricity is generated and where it goes.

A particularly playful sculptural installation of bells and water was designed to be mounted adjacent to the PV system, showing children how much energy is generated by the solar exposure of the modules. An electronic monitoring device and special graphic panels measure the amount of energy produced and show what could be done with the same amount of energy were it used in a different context.

To control the general cost of the PV installations, all the modules used at the Children's Museum of Rome were selected from the advertised standard catalogue of Eurosolare. In order to match the span of the roof's cast-iron structure to the sizes of the PV modules, the designers introduced special transparent glass panels over the existing roof structure, thus underlining its beauty with clear beams of natural light. The overall effect of light texture over the exhibition hall is quite spectacular.

The installation of this PV system, which is part of a larger EU THERMIE plan, has been accomplished in co-operation with the Danish partner Cenergia and the Dutch partner Ecofys. This collaboration establishes a conspicuous and prestigious European context for the Children's Museum of Rome. It also aligns the project with the Alborg Charter, and other conventions on climatic change, formulated in the 21st Agenda of Rome for sustainable development and outlined in the 1992 Rio de Janeiro Conference.

COMPONENT CHARACTERISTICS

PV system power	15.6 kWp
Type of building integration	Roof and canopy integration
Type of cell technology	Poly-crystalline silicon 125 x 125 mm
Modular dimensions	Canopy: 560 x 1200 mm Roof: 1200 x 1200 mm
Array dimensions	14.5 x 1.5 m (both sides of the skylight) 5 x 10.5 m (central part of skylight) 3.8 x 1.8 m (on canopies)
Weight	Modules: about 2930 kg
Visual details	All the components of the plant (modules, inverters, cables, gangways) are visible to the public
Inverter	6 inverters, type SMA Sunny Boy 2500, 2500 W max each, 15 kW max in total
Monitoring equipment	Remote plant monitoring through Datalogger Sunny Boy Control Plus
Other BiPV system elements	Public grid interface general switchboard

Fig 6 Interior view of skylight
Source: Abbate & Vigevano Design Studio

Fig 7 View of entry and south façade with PV canopies
Source: Abbate & Vigevano Design Studio

Fig 8 Module installation onto the left side of skylight
Source: Abbate & Vigevano Design Studio

Fig 9 Module installation and defective modules that were replaced
Source: Abbate & Vigevano Design Studio

Installation
Integration design and mounting strategy

As mentioned above, all the major construction components of the window frames were assembled from standard industrial elements. The only special parts that had to be produced, but again with standard components, were the joints fixing the PV skylight to the roof. The designers produced detailed drawings of the roof ridge and of the joints between the aluminium frame of the PV and the roof covering. The installation was performed using suction cups and a mobile platform.

For the moveable canopies, Abbate & Vigevano, in collaboration with engineer Bruno Masci, designed special structural posts to support the string of standard PV modules. These vertical lenticular shaped elements were placed next to the existing cast-iron columns, and echo the light 'spider web' design of Palenceau's original structure.

Since the structure of the Children's Museum of Rome is a registered historical landmark, the design for the museum and of the PV elements had to be submitted to the Monuments and Fine Arts Office for approval. For nearly six months the commission deliberated over all the elements of the building that could not be touched or removed. The specific details of the canopy are the result of the need to neither modify any existing component nor to screw any new part directly onto the original.

Electrical configuration including grid integration

The photovoltaic subarray on the canopy is composed of 6 rows of 12 modules each with 64 cells (125 mm x 125 mm) connected to 3 inverters of 2,500 W.

The photovoltaic subarray on the roof consists of 6 rows of 18 modules with 36 cells (125 mm x 125 mm) connected to 3 inverters of 2,500 W.

Lessons learnt from the project

The collaboration on the project produced an interesting exchange of ideas, not only among the designers but also among the different contractors and workers, with unexpected results of reciprocal training and accumulation of information. During the definition of the rules and responsibilities of each contractor, it emerged that some operations, especially related to the cabling of the PV installation, could be done under the direction of the PV expert with non-specialised workers. Gechelin, the PV installer and electrical designer, subcontracted Ciel, the general electrical contractor for the museum. Ciel workers were trained to mount a PV installation of this extent under Gechelin supervision. The result was a significant saving of operating time and costs, thereby simplifying building maintenance for the client. By using such a strategy, the number of potential personnel responsible for the maintenance of the electrical system of the entire museum was reduced.

Successful approaches

Using standard components for the aluminium window frames of the skylight resulted in rapid mounting time, requiring only four days of work for two people to mount all the PV modules on the roof, not including the time of wiring the modules.

Problems during realisation

An important lesson learnt during construction was that no one at the PV factory had verified the position of the junction box of the transparent roof modules during the electrical testing. The result was that the modules delivered to the site could not be mounted on the skylight, since the distance of the junction box to the glass border was different to that specified by the Eurosolare factory. Technicians from Eurosolare attended the site immediately and substituted all 72 boxes with new ones precisely positioned. A new electrical test for each module was then performed on site to guarantee the expected electrical performance.

Performance characteristics

At the time of the compilation of this case study the system was not yet in use, therefore performance and monitoring data were not available; all the data available below are predicted estimates.

The estimated power production shown in figures 10 and 11 was calculated using PV SYST software.

The estimated overall energy directly produced by the PV will be 18,000 kWh/year, but the PV used as a passive cooling and heating device will reduce the consumption of energy for heating by 11.3 per cent, and the consumption of energy otherwise needed for cooling by 52.8 per cent.

The energy produced will be directly used in the museum, so the predicted yearly saving for energy produced by the PV is 18,000 x ITL250 = ITL4,500,000/year (about US$2,200), plus the estimated energy saved with PV, (about US$4,000), will generate a significant total saving of US$6,200 per year.

A study by the German scientists T. Wetzel, E. Baake and A. Muelbauer [1] calculated that with a solar radiation of 1,100 kWh per square metre per year, and a 12 per cent energy-efficient poly-crystalline cell, it will take 5.6 years to pay back the cost of the energy used to produce the system. In comparison to the 570 grams of CO_2 emitted per kWh produced with conventional energy in Italy, the PV installation will produce only 50–60 g/kWh, saving 9.270 t of equivalent greenhouse emissions per year.

Simulation, version VC2: Pensiline – Balances and main results

	GlobHor kWh/m²	T Amb °C	GlobInc kWh/m²	GlobEff kWh/m²	EArray kWh	EOutInv kWh	EffArrR %	EffSyR %
Jan	54.2	6.20	61.6	59.3	427	385	9.38	8.46
Feb	66.9	7.10	72.4	69.8	502	459	9.40	8.59
Mar	92.6	9.70	97.1	93.7	660	603	9.21	8.41
Apr	129.6	12.30	131.4	127.0	881	810	9.08	8.35
May	173.5	16.50	169.0	163.4	1097	1011	8.80	8.11
Jun	190.2	20.60	179.9	174.1	1105	1018	8.32	7.66
Jul	201.3	23.70	192.7	186.4	1149	1061	8.08	7.46
Aug	173.9	23.30	173.1	167.5	1054	973	8.25	7.61
Sep	126.5	19.60	132.8	128.5	849	781	8.66	7.97
Oct	94.5	15.40	104.6	101.1	691	634	8.95	8.21
Nov	53.3	11.50	60.7	58.6	408	368	9.10	8.21
Dec	42.4	7.60	46.8	45.0	318	284	9.21	8.22
	1399.0	14.50	1422.2	1374.4	9142	8384	8.71	7.99

Key:

GlobHor	Horizontal global irradiation
T Amb	Ambient temperature
GlobInc	Global irradiation on inclined plane
GlobEff	Effective global, corrected for array incidence loss and shadings simultaneously
EArray	Effective energy at the array output (taking inverter behaviour into account)
EOutInv	Available energy at inverter output
EffArrR	Array efficiency: EArray/rough PV area
EffSyR	System efficiency: EOutInv/rough area

Fig 10 Estimated power production of the roof skylight
Source: Gechelin Group

Simulation, version VC1: Lucernaio – Balances and main results

	GlobHor kWh/m²	T Amb °C	GlobInc kWh/m²	GlobEff kWh/m²	EArray kWh	EOutInv kWh	EffArrR %	EffSyR %
Jan	54.2	6.20	61.6	59.3	435	394	6.92	6.27
Feb	66.9	7.10	72.4	69.8	534	488	7.24	6.61
Mar	92.6	9.70	97.1	93.7	726	664	7.33	6.70
Apr	129.6	12.30	131.4	127.0	1013	928	7.56	6.93
May	173.5	16.50	169.0	163.4	1309	1202	7.59	6.97
Jun	190.2	20.60	179.9	174.1	1370	1257	7.46	6.85
Jul	201.3	23.70	192.7	186.4	1460	1341	7.43	6.82
Aug	173.9	23.30	173.1	167.5	1318	1210	7.46	6.86
Sep	126.5	19.60	132.8	128.5	1009	927	7.45	6.84
Oct	94.5	15.40	104.6	101.1	789	723	7.39	6.78
Nov	53.3	11.50	60.7	58.6	428	386	6.91	6.24
Dec	42.4	7.60	46.8	45.0	311	280	6.52	5.88
	1399.0	14.50	1422.2	1374.4	10702	9801	7.38	6.76

Fig 11 Estimated power production of the canopies
Source: Gechelin Group

Fig 12 Aerial view of Children's Museum of Rome
Source: Abbate & Vigevano Design Studio

Project cost breakdown

	US$	
PV modules	53,630	
Electrical components and installation	47,500	
Structural support and installation for canopies	35,000	
Design and coordination with European partners	30,000	
TOTAL (no vat included)	**166,130**	**(11.07/Wp)**

The Children's Museum of Rome paid for 60 per cent of the cost while 40 per cent was financed by the EU.

Project group details

Patrizia Tomasich, President of the Children's Museum of Rome and the client of the overall project was an enterprising promoter of the museum from the beginning. She took the original initiative of proposing the project to the mayor 1994 and has since led the fundraising campaign. The Municipality of Rome and the Mayor offered the space, but the cost of the renovation and the entire organisation and co-ordination of the project is due to the vision of Mrs. Tomasich.

Studio Italplan of the Pagani architects, the designers of the renovation project and of the museum exhibits, were extremely open and enthusiastic advocates.

All the PV contractors understood the importance of the educational initiative of the project, and all agreed to work at competitive cost: Eurosolare as the producer of the photovoltaic modules, Gechelin Group as the installer and designer of the electrical system, and Italcarrelli as the builder of the special canopies. Obviously the high profile of the project, and the immense publicity and attention gathered by the new museum is an important indirect commercial benefit for the contractors. On the other hand, it is important to analyse the indirect commercial value that the project gathered with the introduction of passive solar strategies and the use of photovoltaic energy, for its entire fundraising campaign. The Children's Museum of Rome is the first of its kind to be constructed with the building integration of these technologies and under the aegis of the European Union.

All the contractors mentioned gave long-term guarantees for the components used, and they are liable for the function of the technologies. This was established under EU regulations, which required a ten-year guarantee of the PV components, and by the Italian standard construction contract which requires a minimum three-year guarantee of the work.

Note

1 Wetzel, Th, Baake, E, Mühlbauer, A, 'Energy requirements and ways of saving energy in the production process of photovoltaic modules', *International Journal of Solar Energy,* 2000 Vol. 20, pp.185–196.

JAPAN: **NTT DoCoMo BUILDING**

I O D A T A

PROJECT:	NTT DoCoMo building
LOCATION/CITY:	Yoyogi, Tokyo
COUNTRY:	Japan
TYPE OF PV BUILDING:	Façade integrated
BUILDING TYPE:	Office building and mobile telephone electric-wave relay base
NEW/RETROFIT:	New

CLIMATIC CHARACTERISTICS

LATITUDE:	35° 7'N
LONGITUDE:	139° 7'E
ALTITUDE:	40 metres above sea level
CLIMATIC TYPE:	Temperate (Temperature: January average = 5.2 °C; August average = 27.1 °C)
SUNSHINE HOURS:	Yearly average = 4.96 hours per day

Fig 1 South façade detail
Source: Kajima Corporation

Fig 2 DoCoMo building
Source: Kajima Corporation

General project background

The DoCoMo building is located in a downtown area of Tokyo. It is the prominent headquarters for the communication and information company NTT Mobile Communications Network Inc., which decided to integrate PV in the façade of the building to reflect its environmental credentials. DoCoMo was the first Japanese PV installation on a building over 200 metres in height.

The client was advised on options to harmonise the PV into the building fabric including installation costs and power generation outputs. Major challenges in the pre-design stage included avoiding an efficiency drop from partial shading of the systems and keeping the wiring as short as possible to minimise a decrease in voltage. The system provides a backup power supply for essential telecommunications equipment in an emergency and feeds directly into the grid.

The construction and design of the BiPV element was co-ordinated and implemented by the Kajima Corporation for the client NTT Power and Building Facilities Inc.

BiPV design process

The PV system was not applied in the original design. However, the policy to use BiPV for the exterior was incorporated over six months into the construction phase. The south façade presented a large surface area and ample solar gain. The voltage of the system is assumed to be the same for each unit and is linked to a central 10 kW inverter. Electrical connection boards are installed every three to four floors to reduce the overall length of thick wiring run (120 metres) from the modules to the inverter. Shading was an obvious problem for the proposed façade installation. Typically, partial shading of the modules greatly affects potential generation capacity. This impact was reduced by devising an internal parallel circuit design. As a result, the adverse shading effect on power generation was decreased, and the PV was installed without an adverse effect on the design of the building. The result was about a 70 per cent improvement in the amount of power generation, compared with conventional wiring strategies. The PV powers a large electrical display on an advertising tower during peak daytime electricity demand. It switches to supplying emergency power in the advent of a blackout or earthquake, keeping essential telephone communication systems running in the building.

BiPV design

Grey poly-crystalline silicon cells were adopted to promote harmony between the stone of the wall outside the building and the city view of the building. Dummy modules were also applied for both an aesthetic outcome and positioned in high shade areas to minimise the impact. Careful construction processes and mounting techniques were applied to ensure durability under the very high wind loads associated with tall buildings. Glass etching patterns on the front cover of the PVs nullified any undesirable reflection from the building. Parallel wiring was achieved by dividing modules into three east–west groups, thus decreasing the influence of the shade.

Fig 3 The back of a module
Source: Kajima Corporation

COMPONENT CHARACTERISTICS

PV system power	15.2 kWp
DC operating voltage	301 V
Type of cell technology	Poly-crystalline silicon cell (cell colour: grey)
Modular dimensions	H 888 x W 1940 x 48 x 123 Wp (for 2nd–25th floors) H 1348 x W 1940 x 28 x 184 Wp (for 26th–32nd floors)
Array dimensions	Approximately 156 cm²
Balance of system components	Grid connection
Inverter	Central 10,000 watts
Monitoring equipment	Original system by NTT Power & Building Facilities Inc.
Circuit composition	Three parallel connections, by which power generation losses due to shade are prevented
Power generation usage	The electricity obtained by PV is used for the 7th floor as a power supply on a large-scale display for an advertising billboard

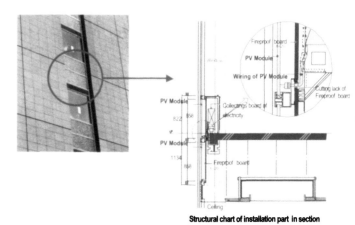

Fig 4 PV integration details
Source: Kajima Corporation

Structural chart of installation part in section

Installation

In order to restrict environmental impact and unnecessary waste, the project designed a reusable transportation container to deliver modules to the building site. As a result, the typical waste associated with packing material was eliminated. Thus, thought was given to conservation of energy and mitigating environmental problems during the construction stage.

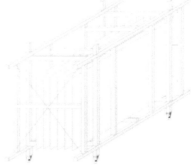

Fig 5 Transportation of modules
Source: Kajima Corporation

Sketch of transportation container

Performance characteristics

PV electricity generated totals approximately 4,600 kWh/year. Because the total extension of wiring was long, it was important to reduce the voltage drops in the vertical direction. Bundling a wiring box installed on every third or fourth floor alleviated unwarranted losses.

Project cost breakdown

Because the integration was applied to the façade of the high-rise building, PV module installation, construction and wiring work were positioned every two to three floors, increasing in cost with height. The PV system cost approximately 40 million Japanese yen.

JAPAN: J-HOUSE

IODATA

PROJECT:	J-House
LOCATION/CITY:	Tokyo
COUNTRY:	Japan
YPE OF PV BUILDING:	Tile-integrated roofing, double-glazed glass window-integrated
BUILDING TYPE:	Private residence
NEW/RETROFIT:	New

LIMATIC CHARACTERISTICS

LATITUDE:	35° 7'N
LONGITUDE:	139° 7'E
ALTITUDE:	40 metres above sea level
CLIMATIC TYPE:	Temperate (Temperature: January average = 5.2 °C; August average = 27.1 °C)
SUNSHINE HOURS:	Yearly average = 4.96 hours per day

General project background

J-House is located in a dense residential district in Shinjuku Ward, Tokyo. The plan was to build an environmentally friendly model house that utilises natural energy and solar power generation, and blends in with the surrounding area.

An underground cooling system and natural ventilation were introduced via large openings, and a wellhole in the living room, ensuring comfort in summer. In winter, the living room becomes a winter-garden with sunlight from the glass-covered openings; warm air is taken in from the roof, thus creating a passive solar house.

Hot-water radiant floor heating and a wood stove are installed in the wellhole in the living room; the floor plan of the house was designed so that hot air can reach individual rooms. The use of natural construction materials such as concrete, wood, and diatom earth ensures a healthy indoor environment.

Fig 1 South elevation installed with semi-transparent double-glazed PV glass
Source: Jiro Ohno

BiPV design process
Planning process, alternative designs

As the site is surrounded by dense low-rise houses, the available sunshine is relatively good. Originally, the development of a hybrid-type BiPV module using a combined PV and hot water supply system was discussed to effectively use solar energy from the roof, but the idea was abandoned because of problems of heat-collecting efficiency, cost and design. Though moveable louvre-type PVs were originally discussed for the glass PV on the south side, this idea was also abandoned, because the PVs interfere with each other in Japan when the sun is high in the sky, making effective installation difficult. Japan has more residences with PV installations than any other country, but there are only a few examples of building-integrated design. Accordingly, it was decided to install a transparent PV module for the glass surface to achieve an effective design.

Fig 2 Semi-transparent double glazing – exterior view
Source: Jiro Ohno

Decision process

The shade temperature is close to the annual average air temperature of the region, and the temperature of the groundwater in Tokyo is stable at an average of about 17 °C. A panel-type air-conditioning system using groundwater was planned, but Tokyo's hot and humid summers necessitated countermeasures for condensation. It was difficult to install both a dehumidification system and natural ventilation. It was also found that digging of the well and the panel-type air-conditioning system would be expensive, and so finally the idea was abandoned. A simulation of the solar power generation showed that the annual power generated could be equivalent to approximately two thirds of the required electricity for this residence, and the owner, who is an architect, decided to proceed with the installation.

Fig 3 Semi-transparent double glazing – interior view
Source: Jiro Ohno

COMPONENT CHARACTERISTICS

PV system power	4.94 kW
Type of building integration	Roofing tile PV integration and double-glazed PV glass integration
Type of cell technology	Mono-crystalline silicon cell (roof) and poly-crystalline silicon cells (windows)
Array dimensions	34.2 m²
Weight	120 kg (double-glazed PV glass)
Inverter	Line Back FX (Nihon Denchi Co. Ltd.) 4.5 kVA
Monitoring equipment	Horizontal pyrheliometer, inclined pyrheliometer, air temperature meter, module temperature measurement device, PC for measuring various sensor signals, PC for information processing and .telecommunications

Fig 4 Rooftop PV array
Source: Jiro Ohno

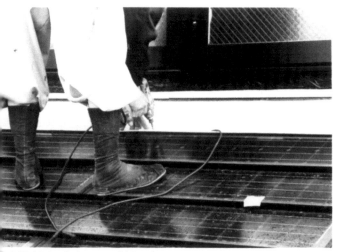

Fig 5 Module installation onto prefabricated tray
Source: Jiro Ohno

Fig 6 Module installation onto prefabricated tray alongside
a roof-integrated solar hot water system
Source: Jiro Ohno

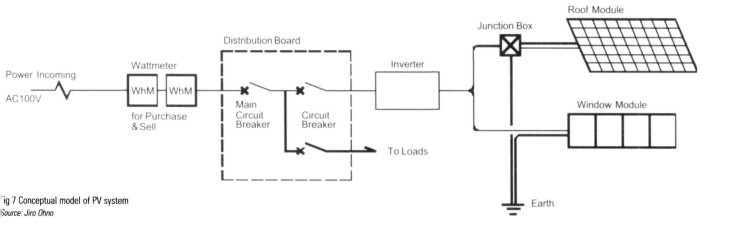

Fig 7 Conceptual model of PV system
Source: Jiro Ohno

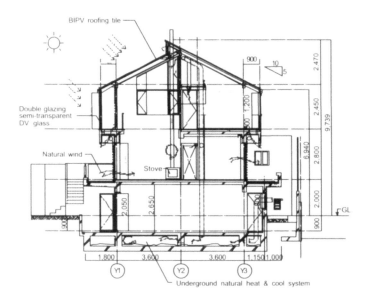

Fig 8 Cross-sectional diagram of the J-House
Source: Jiro Ohno

Installation design

The black sash and blue cells of the double-glazed PV glass provide an interesting design effect. The semi-transparent cell arrangement protects privacy, as well as generating power and providing shade. The roof and window arrays are connected to one inverter after adjusting the voltage, and are interconnected to the distribution network of the electricity provider. An electrical company took control of all negotiations with the interconnected electricity provider.

Installation

No problems were encountered during the installation of the solar power related equipment, as all work was performed by experienced contractors. However, a drain board covering the BiPV cell surface on the roof was reconstructed and the BiPV roof module was moved upward by one step to install the snow-guard attachment. It was found that when the top skylight was fully opened, it created a shadow over the module, so attention must be paid when opening and closing it. The window module is divided into four parts, however, in view of the shadow of a pole transformer, the electric circuit was designed so that three of the parts can be connected to the inverter, thus isolating any shaded module.

BiPV design

To harmonise with the roofs of surrounding residences, a grounding aluminium long-plate PV module, which is very common in Japan, was adopted for the roof module, and semi-transparent double glazing was adopted for the glass part on the south side. The conceptual model and the cross-sectional diagram are shown in figures 7 and 8.

Electrical system

The PV performance is being monitored by the Solar Technology Centre of JQA, which is commissioned by NEDO under site code ST082 (Tokyo–4). The Centre collects and organises the information by computer and telephone line, sends a report to clients every month, and also compiles data for the whole nation.

Project cost breakdown

The total construction cost of the J-House (total area 314 square metres) was approximately 80 million yen. The system cost relating to BiPV was 5 million yen, including a grant of 1,732,500 yen from NEDO. (The subsidy was 350,000 yen/kW in 1999.) The selling price of electricity through the interconnection was the same as the buying price: 23.85 yen/kWh. Under a time-of-use power contract, the price becomes 34.55 yen/kWh during the day, and 6 yen/kWh during the night.

Performance characteristics

Performance of the J-House system is mostly good and is summarized in figure 9. The system provided 2,607 kWh electricity over eight months following a detailed evaluation in 2000; system availability average was 8.4 per cent. Power conditioning efficiency was over 90 per cent. The performance ratio of the system was around 60 per cent, which was about 10 per cent less than the annual average performance ratio of 65 sampled Japanese residential PV systems in 1999. Because of lower reliability of irradiation data, caused by occasional shadow on its pyranometers, irradiation data measured by the Japanese Meteorological Agency at Tokyo weather station, which is 3 kilometres from the system, was used for calculating the performance ratio. The conditions of irradiation measurement at the weather station are better than in general urban areas, so that the substituted irradiation could be slightly higher than actual irradiation on the PV arrays, causing a slight decrease in performance ratio estimates.

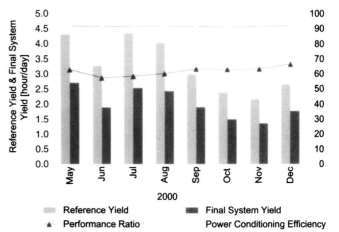

Fig 9 Power generation performance
Source: Jiro Ohno

Post-installation feedback

The semi-transparent double-glazed PV glass module for the window surface used in the residence is an excellent design. Light and shadow in a lattice pattern enter the room, protect privacy, and also generate electric power. As the voltage of the roof module is 215 V and that of the glass module is 268 V, inverter control is an issue.

In Japan, trees can reduce the external heat load during the summer by their screening effect. The reduction in power generation from the cherry tree in the garden is estimated to be rather small, although it creates some shadow effects. This cherry tree was planted by the owner's parents, and is considered symbolic of the region.

JAPAN: **SBIC EAST BUILDING**

Fig 1 West elevation with semi-transparent vertical louvre PV modules
Source: Jiro Ohno

General project background

The SBIC East building is located in Shibuya, where a newly developed city centre is being expanded. The site was vacant, having previously housed a freight car terminal of the former Japan National Railway. The site was sold and converted to an integrated business district through regional redevelopment conducted by seven private companies. The SBIC East Building incorporates large conference rooms, training rooms and exhibition rooms, and an anti-seismic structure. It has a feeling of transparency with its design themes of tiles and glass. In order to conserve energy, excellent weatherproofing and insulation are installed, all windows are double-glazed, and small windows are installed underneath the glass curtain walls to ensure natural ventilation during the appropriate seasons. The three-storey atrium allows natural light to enter, acts as a thermal buffer to individual offices and is designed to serve as a communication area for employees through glass stairs.

Fig 2 Looking up at semi-transparent eave array
Source: Jiro Ohno

BiPV design process
Planning process, alternative designs

In the original plan, introduction of BiPV was not considered. However, effective adoption of vertical louvres was considered in order to decrease the air conditioning load during afternoon sun. Ambient temperature is typically around 30 °C in summer. Though the building is surrounded by the newly developed city centre area of Shibuya, the sunshine condition is relatively good, with a railway line along the west side of the site. Both the estimation of annual power generation and the conceptual system design were performed based on the results of studying the shadow conditions, and the adoption of BiPV was finally decided as an integrated element of the outside building design. Environmentally friendly buildings are now one of the most important considerations for every company, and SBIC East Inc. decided to adopt BiPV for the outside design with venture capital to help promote small and medium businesses.

Decision process

From the beginning of this design, the adoption of vertical louvres was planned in order to decrease the air conditioning load during afternoon sun. The architect thought that BiPV integrated vertical louvres could add a power generation function, and also presented an opportunity to create a new design for the outside of the building utilising light and shadow with semi-transparent modules. As it is located in the centre of a city, the building is affected by the shadows of the surrounding buildings. However, with the owner's approval, BiPV was introduced to create a building that has great environment protection appeal. During the design process, annual power generation was estimated using a simulation program that takes into account the shadow effects of the surrounding buildings. Based on the calculated results and past experience of system design, four types of BiPV were installed.

Project organisation

The owner was an organisation related to MITI, Japan's Ministry of International Trade and Industry (now METI, Ministry of Economy, Trade and Industry), which considered the global environment, especially energy issues, to be of great importance. The design process used information from discussions with engineers working for related solar power generation makers. The project was performed as a field test of NEDO with 50 per cent financing, and was designated as a collaborative study.

BiPV design

The SBIC East building has four types of BiPV: eave-type array on pergola; inclined-type array on roof; furring-type array on parapet; and shade louvre-type array.

Fig 3 Semi-transparent vertical PV louvre view from outer deck
Source: Jiro Ohno

Fig 4 Eave-type array on pergola
Source: Jiro Ohno

Fig 5 Shade louvre-type array
Source: Jiro Ohno

PONENT CHARACTERISTICS

PV system power	Total power is 30.5 kWp (dummy cell is 985 W)
of building integration	a) Eave-type semi-transparent array (0.9 kWp, dummy cell capacity is 240 W); b) Inclined-type module (5.1 kWp, no dummy cell); c) Furring-type module (4.4 kWp, no dummy cell); d) Shade louvre-type semi-transparent module (20.1 kWp, dummy cell capacity is 745 W)
ype of cell technology	Mono-crystalline silicon (eave-type semi-transparent module, shade louvre type semi-transparent module) Grey coloured poly-crystalline silicon (furring-type module) Poly-crystalline silicon (inclined-type module)
Modular dimensions	a) Eave-type module (480 x 1528 x 14 mm), 15 modules b) Inclined-type module (530 x 1200 x 35 mm), 60 modules c) Furring-type module (190 x 1750 x 23 mm), 160 modules d) Shade louvre-type (480 x 1825 x 14 mm), 240 modules
Array dimensions	a) Eave-type array (926 W, 11 m²) b) Inclined-type array (5,130 W, 38.2 m²) c) Furring-type array (4,436 W, 48.4 m²) d) Shade louvre-type array (20,112 W, 38.2 m²)
Weight	a) Eave-type module (28.5 kg) b) Inclined-type module (8.5 kg) c) Furring-type module (6.1 kg) d) Shade louvre-type module (31.3 kg)
Inverter	Capacity 30 kW (Nisshin Denki Ltd.)
Monitoring equipment	Horizontal pyrheliometer, inclined pyrheliometer, air temperature meter, anemometer, module temperature measurement device, PC for measurement, PC for telecommunication
BiPV system elements	In order to create a new design of light metal fittings which support the semi-transparent vertical PV louvres, the DPG method was used to fix glass by wings which are extracted from round pipes. The cantilever DPG method was also used to support the semi-transparent PV eaves.

Installation

The semi-transparent shade louvre installed on the west side of the building has both energy-saving and power generation functions, and is an important element of the exterior design of the structure. The semi-transparent louvres constitute a new design for the interior of the building.

Problems during realisation

Installation did not progress smoothly due to a failure in the module manufacture, so some of the modules were remanufactured and the installation procedure was changed.

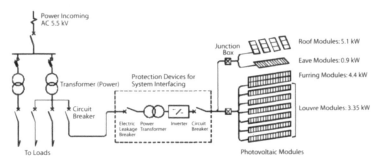

Fig 6 Electrical configuration
Source: Jiro Ohno

Performance characteristics

The SBIC East building system has operated well since April 1998, and its performance characteristics, following detailed evaluation, are summarised in figure 7. The system provided 30,400 kWh electricity during the 30 months from April 1998 to December 2000, and the average of system availability was 4.9 per cent. In 2000, the final PV system yield, that is, electricity per PV rated capacity, was 421 kWh/kW. This value is less than half the typical value for the Japanese Field Test Program's PV systems, (1000 kWh/kW) and can be accounted for by the large vertical array. Power conditioning efficiency varied between about 70–90 per cent, and the average power conditioning efficiency was 82 per cent. The performance ratio of the system varied between 40–60 per cent, and the average performance ratio was 53 per cent, which was about 20 per cent less than the annual average performance ratio of 150 Japanese Field Test Program's PV systems in 1998. Because of the lower reliability of the building's own irradiation data, caused by frequent shadow on its pyranometers, irradiation data measured by the Japanese Meteorological Agency at Tokyo weather station, which is 6 kilometres from the system, was used to estimate the performance ratio. The conditions of irradiation measurement at the weather station are better in sky factor than in general urban areas, so the substituted irradiation could be slightly higher than actual irradiation on the PV arrays, and cause a marginal decrease in performance ratio estimates.

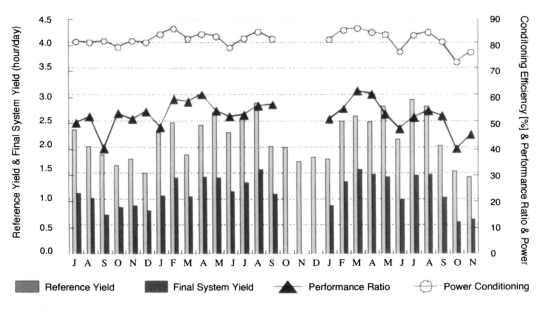

Figure 7 Power generation performance
Source: Jiro Ohno

Project cost breakdown

Total cost of SBIC East building	3.131 million yen
Solar energy plant	108 million yen
Installation costs	20 million yen
Maintenance, monitoring, other (5 years)	17 million yen
Project finance	50% SBIC East; 50% NEDO

Post-installation feedback

The SBIC East building was reported in building magazines as Japan's first case of total integration of BiPV in building design. Many visitors flocked to see this key BiPV building, and it was used as the location for a TV drama. The building is reported to be instrumental in the hiring of outstanding new employees. Benefits have been great for SBIC East Inc., which promotes advanced small and medium companies and venture businesses, as an owner of an environmentally friendly, sophisticated building.

As 30 kW inverters were installed for four 30 kW power generation arrays in total, the conversion efficiency sometimes decreases during operation, so the use of smaller capacity inverters should have been considered.

Fig 8 Shade louvre-type array
Source: Jiro Ohno

THE NETHERLANDS: **ENERGY RESEARCH FOUNDATION (ECN) – BUILDING 31**

IODATA

PROJECT:	Energy Research Foundation (ECN) – Building 31
LOCATION/CITY:	Petten
COUNTRY:	Netherlands
ᵀPE OF (PV) BUILDING:	PV lamella system, canopy and curved roof integration
BUILDING TYPE:	Office building and research laboratories
NEW/RETROFIT:	Retrofit

CLIMATIC CHARACTERISTICS

LATITUDE:	52° 47′N
LONGITUDE:	4° 40′E
ALTITUDE:	5 metres above sea level
CLIMATIC TYPE:	Moderate west-European maritime (Temperature: January average = –1.6 °C; July average = 22 °C)
SUNSHINE HOURS:	Yearly average = 4.05 hours per day

Fig 1 Canopy roof above the lamella shading system of Building 31; custom-made semi-transparent solar modules produced by Shell as glass–glass elements
Source: BEAR Architecten

General project background

The Netherlands Energy Research Foundation (ECN), is the leading institute for energy research in the Netherlands. The ECN mission is to contribute to a clean and reliable energy supply for a viable world, by research and development in the fields of greater energy efficiency, the implementation of renewable energy and the reduction of environmentally harmful emissions from fossil fuels with optimal cost-effectiveness. Research at ECN is carried out under contract from the government and from national and foreign organisations and industries. ECN's activities are concentrated in seven areas: solar energy, wind energy, biomass, clean fossil fuels, energy efficiency, policy studies, and renewable energy in the built environment. As ECN aims to strengthen the synergy between market and sustainability as a technology developer, the decision was made to demonstrate an ecological transformation of its own buildings.

ECN Building 31

The General Laboratory of the Netherlands Energy Research Foundation ECN, known as Building 31, was built in 1963 as a laboratory building and has a total floor area of 3530 square metres. Prior to the renovation, most of the rooms of Building 31 were in use as offices.

Fig 2 Building 31 laboratory before renovation
Source: BEAR Architecten

Fig 3 Building 31 during façade renovation
Source: BEAR Architecten

In 1997, the ECN unit for Renewable Energy in the Built Environment conducted a study to evaluate the building condition and energy performance of Building 31. The study found that the building had several technical and thermal problems:

- poor building insulation and thermal bridges;
- overheating by sunshine;
- inefficient lighting system;
- high rate ventilation system for the laboratories with low efficiency and comfort;
- high heating and electricity demand;
- deteriorating façade leading to cold ingress due to thermal bridges;
- draught, due to ventilation system and badly distributed heat.

The core aims of the renovation were to enhance the indoor climate and comfort and reduce the energy and greenhouse gas emissions by using solar energy effectively.

The electricity consumption before renovation was 80 kWh/m² with 140 kWh/m² of space heating consumed. The study emphasised the need to reduce the energy consumption for space heating by 75 per cent and electricity demand by 35 per cent, targeting a total primary energy demand of less than 80 kWh/m².

The ambitious renovation included:

- Façade renovation, with high insulation values, and total renewal of the complete façade construction on the south and north elevations;
- A PV system for the roof (72 kWp) and façade (42 kWp) to provide approximately 30 per cent of the electricity demand and to assist with sun shading and daylight optimisation;
- Balanced ventilation with heat recovery and night ventilation for summer cooling;
- New heating system;
- Better artificial lighting system;
- Low energy consuming installations and products (including computers).

Further, there was already a small utility building with a CHP plant, serving Building 31 and several other buildings. The plant had recently been renewed and a heating system for Building 31 and the new Building 42 was planned with:

- CHP-combined heat and power generation;
- heat and cold storage using a heat pump.

BiPV design process
Planning process, alternative designs

Building 31 was not only to be renewed, but a change in function was also planned. The former laboratories along the south façade were to be transformed into office spaces and the rooms on the north side converted to laboratories.

Hours of Overheating ECN 31 per Month

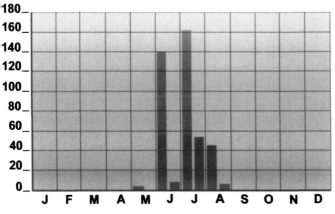

Left: before (measured), right: after renovation (calculated)

Fig 4 Graphic showing the hours of overheating measured on the third floor in the south-oriented rooms of Building 31 (left bar graph) and the projected values after renovation (right bar graph).
Source: ECN; Graphic: BEAR Architecten

The overheating problems on the south façade required a solution. Either an air-conditioning system or a good shading device was required. Preference was given to a PV-lamella design that reduced the heat load on the building. Computer simulations showed that air-conditioning would not be necessary. Figure 4 depicts the situation before renovation, where most working hours during June and July show overheating and the projected situation after renovation with a PV shading system. The calculated values are within an acceptable number of overheating hours for the users and are compliant with Dutch building regulations. The money saved by not purchasing an air-conditioning system was put towards installing the PV shading system. Further, a PV system was planned for the roof and for the staircase façade, which had to be renovated with insulated glazing. The PV system for the staircase turned out to be prohibitively expensive and hence greater effort was focused on installing increased PV capacity on the roof. The alterations involved a decision by ECN to provide more laboratory spaces, which required highly technically defined indoor conditions and a cooling plant for the laboratories.

Decision process

After the decision for a PV lamella system was made in 1996, a THERMIE project contract was successfully agreed by the European Commission to commence in 1997 (THERMIE SE/115/97/NL/DK). The THERMIE project group included:

- Netherlands Energy Research Foundation ECN, (Netherlands);
- Dasolas International Production (Denmark)/ALCO, (Netherlands);
- ENW/NUON utility, (Netherlands);
- BEAR Architecten, (Netherlands);
- Studio di Architectura di Cinzia Abbate, (Italy).

Design of PV lamella façade:
Step 1 – computer animation

The design aimed for both architectural quality and views from the interior to the exterior. After developing the first design proposals for the PV systems, a computer simulation was made in order to calculate daylighting effects of the proposed lamella designs.

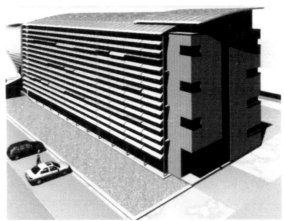

Fig 5 Computer animation of Building 31 with PV lamella system, canopy and curved roof PV integration; in the background, newly planned Building 42 with a PV atrium
Source: BEAR Architecten
Artist impression: Mart van der Laan

A choice had to be made between mounting the shading device close to the façade and mounting it within a certain distance. Further, the size of the lamellas had to be discussed: should a few, wide lamellas be chosen or a larger number of slim ones? What should the length of the lamellas be? From the point of view of maintenance, accessibility and window cleaning, it was decided to have the shading/PV device constructed as a separate façade, about 80 centimetres from the building, but connected to the main structure of the building. The length of the lamellas was guided by the width of the rooms behind. A general decision was taken to work with standard PV modules to save costs. Hence, a metal lamella system had to be developed, which would be capable of holding the standard modules and would fit with the modularity of the building.

Design of PV lamella façade: Step 2 – 1:10 scale model and experiments in the artificial sky

In a second design step, a 1:10 scale model of a laboratory room was built, to find the optimal design solution. Research was carried out in a daylight chamber and on a solar table at the Technical University, Delft. In order to choose the width of the lamellas, various solutions for an integrated system were examined:

- two large lamellas with modules, at a vertical distance of 1.5 metres in a fixed position;
- as above, but with a moveable tracking system;
- seven small lamellas with modules at a vertical distance of 0.5 metre in a fixed position;
- as above but with a moveable tracking system.

Fig 6 Computer animation of PV lamella system: view from office space to exterior with lamella at eye height in horizontal position
Source: BEAR Architecten

TNO–TUE, the Centre for Building Research at the Technical University of Eindhoven, conducted a computer simulation study with Radiant software. The study focused on the questions of:

- optimal solar gain for the PV lamella system;
- heat load of the building and passive solar gains in the winter;
- shading of the building;
- self-shading of the PV lamella modules;
- outside view from the interior;
- optimised daylighting conditions.

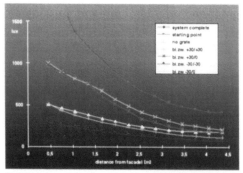

Fig 7 Computer calculations of light intensity (in lux) for different distances
Source: Measurements: TNO–TUE; Graph: BEAR Architecten

The study showed that the best results for solar gain, shading and daylighting were obtained with a model using four fixed lamellas per floor. Considering the solar ratio between fixed and moveable systems, the solar gain is only approximately 10 per cent higher with a moveable system. Considering the high costs of a moveable system compared to a fixed structure and the small difference of solar gain, it was decided to select a system that is fixed in the optimal position for the Netherlands with an inclination of 37° to the horizon. However, the occupant of the room can move one lamella at eye level in a horizontal position, in order to have a good outside view. After a defined space of time, for instance 20 minutes or so, the lamella will automatically take its position of 37° again. Thus, a continuously varying architectural view will be created.

Design of PV lamella façade: Step 3 – mock-up and measurements of light intensity in one of the laboratory rooms

As a third design step, a mock-up of the PV lamella system was developed at a 1:1 scale and placed in front of a laboratory room on the first floor of Building 31. Monitoring by TNO–TUE took place between September and November 1998. In the test room, ten light measurement cells (Hagner type SD1 with a sensibility of 10 pA/lux) were placed, according to the rules defined in IEA Task 21 for daylighting measurements. Two light cells were placed at a 1.5-metre height on the inside of the window glazing, one in the test room, the second one in a reference room without shading. The other eight cells were placed horizontally at a height of 70 cm above the floor to measure light intensity in the working environment.

The following questions were to be clarified with the 1:1 testing:

- effects of design and dimension of the PV lamellas on daylight intensity at the work station compared to a room without sunshades;
- tests of electrical and mechanical components;
- minimisation of damaging side effects such as noise produced by wind on the metal structure;
- optimisation of construction and detail design of the metal lamella structure;
- best strategies for production and mounting of the metal structure with PV.

The participants in the projects concluded that the mock-up was very helpful in avoiding problems and especially for gaining experience with:

- dimensions of the components;
- required and acceptable tolerances of dimensions;
- mounting process for metal lamellas and PV modules;
- acceptance of the PV lamella system by future users.

As a result of the monitoring, it was found that the lamellas shade the building during the summer period with an efficiency of about 85 per cent (which means that 15 per cent of the sunlight that falls on the façade is not reflected by the lamellas). For fine-tuning the daylighting aspects, especially in winter, a second very simple interior shading system was subsequently planned, as light intensity was still bright on sunny test days. To avoid glare, the light intensity was to be distributed with further measures such as horizontal white lamellas or light distributing measures on the inside. Additionally, it was proposed to give the window frames a white or light coloured finish, to reduce the contrast between the frame and the sky. The users had various opinions about the PV lamella system. In general, people were quite curious about the final result. Some did not like the lamellas because they hindered views when the users are in a sitting position, some did not like the view of the rear of the lamella. It should be mentioned that, after the façade was completed, acceptance by users has increased markedly. In general, the testing was successful and system components were optimised according to the results.

Fig 8 Mock-up: 1:1 testing of the lamella shading system of Building 31
Source: BEAR Architecten

PV roof design

The PV roofing system was originally meant as a kind of parasol, a passive cooling device for the roof. The roof construction underneath would provide water tightness. As the design of the interior of the building developed, it became clear that the space between the parasol and the existing roof could be used for technical devices such as ventilators and air ducts. Hence it was decided to construct the parasol as a watertight part of the building.

The roof is constructed from curved IPE 240 steel profiles, which carry corrugated sheet. The sheet is covered by a layer of rock-wool insulation, above which EPDM foil is placed as a rain-tight layer. Above this, standard PV modules are mounted with the BP Sunflower system. The profiles to hold the Sunflower module mounting points are fixed on pieces of laminated wood, which are fixed onto the IPE 240 profiles with bolts. As it is not possible to walk on this construction, a roof trolley in bridge form was installed, to enable easy maintenance.

BiPV design

The BiPV design process was conducted co-operatively between BEAR Architecten, Studio di Architettura di Cinzia Abbate, ECN and DASOLAS/ALCO. Each metal lamella was to be about 840 mm wide, 3000 mm long and covered by three standard poly-crystalline PV modules. The lamella at eye-height for a sitting person working in the interior is moveable, to allow exterior views. The lamellas are made from folded aluminium sheet, enamelled for a higher durability, and are mounted on vertical IPE 120 steel profiles, which are interconnected with horizontal IPE 120 profiles. These carry the metal grid for façade maintenance and are fixed onto the concrete floors of the existing building. The rear of the metal lamellas have holes for ventilation of the PV panels. The electric wiring is led in the hollow cores of the lamellas. The vertical steel profiles were designed with enamel aluminium sheet covers clipped on front and back. The vertical wiring is placed under these clipped aluminium covers. Along the steel elements, the PV façade modules are vertically interconnected to form 13 strings. Each string consists of 42 standard modules from the lamellas and 12 transparent modules from the canopy. All inverters are installed underneath the upper roof, which covers an installation space. The BP modules of the upper roof are interconnected into 19 strings.

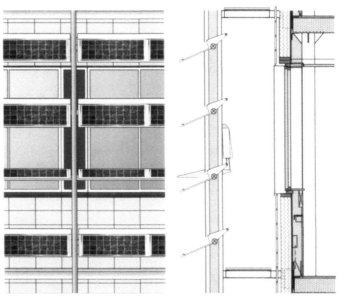

Fig 9 Lamella shading system of Building 31: south façade elevation and section
Source: BEAR Architecten

The façade system is made from a wooden construction, which holds the insulation. Ceramic façade cladding panels were chosen as exterior weather protection. Metal elements interconnect the façade layers. Galvanised light metal grids between PV lamellas and façade allow cleaning and maintenance of the PV system and façade.

Fig 10 Lamella shading system of Building 31: south façade
Source: BEAR Architecten
Photographer: Marcel van Kerckhoven

COMPONENT CHARACTERISTICS

Type of building integration	Total PV systems	PV lamella shading system	PV canopy	Roof-integrated
PV system power	71.88 kWp	26.21 kWp	6.91 kWp	38.76 kWp
PV module area	Approximately 700m^2			
Type of cell technology		Poly-crystalline cells	Poly-crystalline cells	Mono-crystalline solar
Modular dimensions		478 x 1006 mm Shell Solar type RSM 50 with 48 Wp	550 x 1100 mm custom made Shell Solar IRD 50 with 44.3 Wp	525 x 1183 mm type BP-585-L with 85 Wp
Number of PV-modules		546	156	456
PV-module construction		Standard framed PV laminates produced by Shell Solar	Custom-made framed glass–glass PV modules produced by Shell Solar	Frameless standard PV laminates made by BP Solar
Weight		Approximately 30 kg/m^2	Approximately 50 kg/m^2	Approximately 30 kg/m^2
Visual details		Opaque laminates	Semitransparent glass–glass modules	Opaque laminates
Inverters		13 string inverters 'Sunmaster 2500-150' with 2500 watt capacity	13 string inverters 'Sunmaster 2500-150' with 2500 watt capacity	19 string inverters SMA Sunny Boy 2400
Monitoring equipment	The PV system is permanently monitored by ECN			
	Calibrated solar cells are included in single PV modules			
Other BiPV system elements	A special metal substructure has been developed by ALCO and Dasolas for the shading lamella system			
	The canopy roof profiles were manufactured by Allicon			
	The curved roof-integration was made with the BP Sunflower system			

Installation design and installation

A special metal construction was designed and produced to fit standard modules, as this seemed to be more cost-efficient than to integrate custom-sized PV. The 1:1 testing ensured that the metal construction details were proofed and the installation did not have any problems. In the beginning of the planning process, several parties thought it would be easier and safer to pre-install the metal lamellas with the PV in the fabrication hall or on the ground. But as a result of the testing it turned out that the PV modules would be exposed to damage when mounted together with the huge metal lamellas. Responsibility for broken glass would have to be agreed between the parties. So it was decided to first install the ALCO metal lamellas with a huge installation bridge and to then install the Shell Solar PV modules with the same installation bridge.

Fig 11 Integration of PV modules into the metal shading system
Source: BEAR Architecten
Photographer: Marcel van Kerckhoven

Performance characteristics

Measurements are permanently performed on the system by the 'wind and sun' unit of ECN with specific scientific aims: in particular, the PV modules include single cells which have been exactly calibrated before being incorporated into the modules. This allows for exact calculation of the efficiency of the PV system.

Project cost breakdown
Support

The PV system has been supported with

- 40 per cent contribution by the European Commission within the THERMIE programme;
- 9 per cent support by the Dutch government via NOVEM, the Netherlands Energy Agency;
- EUR97,853 by NUON utility.

The local utility NUON/ENW supports PV plants with EUR1.36 per Wp. For the 71.9 kWp installed on ECN Building 31, this represents a sponsorship of EUR97,853.

Costs of PV system

- PV system for shading lamellas south facade: EUR181,512 for 26.21 kWp turnkey PV-system with standard modules by Shell Solar (costs include PV system installation by Shell in pre-installed metal shading lamellas by ALCO, PV-engineering, inverters, electric installation and components);
- PV modules canopy roof: EUR79,865 for 6.91 kWp glass–glass PV module canopy roof by Shell Solar: (price only for PV modules, without mounting and electric components);
- PV roof integration curved roof: EUR237,972 for 38.76 kWp turnkey delivery of installed PV system by BP Solar (with Sunflower profiles including engineering, inverters, mounting and all components).

Post-installation feedback

The project team is convinced that 1:1 testing is always a good idea, if innovative systems such as the PV shading system are to be realised on a large scale for the first time.

Fig 12 Building 31 during façade construction
Source: BEAR Architecten

THE NETHERLANDS: ENERGY RESEARCH FOUNDATION (ECN) – BUILDING 42

BIODATA

PROJECT:	Energy Research Foundation (ECN) – Building 42
LOCATION/CITY:	Petten
COUNTRY:	Netherlands
TYPE OF (PV) BUILDING:	PV integration in conservatory glazing
BUILDING TYPE:	Office building and research laboratories
NEW/RETROFIT:	New

CLIMATIC CHARACTERISTICS

LATITUDE:	52° 47'N
LONGITUDE:	4° 40'E
ALTITUDE:	5 metres above sea level
CLIMATIC TYPE:	Moderate west-European maritime (Temperature: January average = –1.6 °C; July average = 22 °C)
SUNSHINE HOURS:	Yearly average = 4.05 hours per day

General project background

As Building 31 was being renovated (see ECN Building 31 case study), a new building complex was planned adjacent to it to serve the growing demand for space at the ECN. Building 42 was designed as a modular extension of the existing solar-renovated Building 31, with a conservatory connecting the two buildings and acting as a common entrance for both buildings. Building 42 consists of three building blocks, which are grouped around a glazed conservatory space covering the east–west circulation axis and will be built in three steps. Building 42 unit 1 was finished in March 2001. Construction of unit 2 began in Spring 2002 and unit 3 is foreseen as a further extension possibility to be built in the future.

Fig 1 Aerial view of Building 31 and Building 42, units 1 and 2 from the north side
Source: BEAR Architecten
Artist impression by Mart van der Laan

Fig 2 Conservatory roof of Building 42 with PV modules and connection to Building 31
Source: Het Houtblad
Photographer: John Lewis Marshall

Fig 3 Conservatory roof as shading device for the office blocks
Source: Het Houtblad
Photographer: John Lewis Marshall

ig 4 Entrance area for Building 42 (left) and Building 31 (right)
ource: Het Houtblad
Photographer: John Lewis Marshall

Ecological approach for Building 42

When the planning began, it was decided to build Building 42 as a demonstration project for renewable energies in the built environment and for low energy consumption standards, contributing to the research and consulting work carried out by ECN in this field. ECN's aim is to promote Building 42 as the 'most energy efficient office building in the Netherlands'.

Energy consumption can be directly related to the building shell and installations, such as lighting needs and elevator power. But even elevator and the lighting energy consumption are dependent on how intensively a building is used and for what purpose. Laboratories, for example, sometimes require a 10-times air exchange rate per hour. This means a lot of mechanical venting is needed and occasionally, energy intensive air-conditioning. Heating and cooling demand for such spaces cannot be compared with the energy demand for other uses, such as offices. The electricity demand is also dependent on equipment used and the work carried out in the space. Even in the same type of room in Building 42, one floor higher for example, the energy consumption could vary by 100 per cent, depending on the use.

In the Netherlands, the energy efficiency of buildings is benchmarked against an EPC (energy performance coefficient measure), resulting from a calculation method laid out in the building codes. Energy consumption is reflected in the EPC. The lower the coefficient, the better the building performs. The use of renewables will further reduce the energy performance coefficient. The EPC value foreseen for office buildings is 1.6. Building 42 reaches an EPC value of 0.86. If combined heat and power generation and the PV installation is taken into account, the EPC value is 0.43. A small utility building with CHP plant, which served Building 31 and several other buildings until recently has been renewed to supply energy to both Buildings 31 and 42. The actual planning stage foresaw combined heat and power generation and heat and cold storage with a heat pump.

Research was carried out to determine whether the CHP plant could run on bio-fuel. If bio-gas or bio-oil is used as fuel, the EPC value of Building 42 will be zero and the building will run completely on renewable energy. Another aspect of sustainability for Building 42 is the flexibility of use; to avoid cost and energy intensive future building processes, the building has been designed to serve as an office as well as laboratories. Huge ventilation channels, adaptable installations and installation channels, non-load-bearing walls and good daylighting, suited to different functions, are required. As a first step in the design process, five relevant buildings throughout Europe were visited to study the technologies, building strategies and practicalities for Building 42.

The final options chosen were:

- daylight-controlled artificial lighting system, as users are not always conscious of the importance of lighting for electricity consumption;
- central switches for electricity to avoid standby losses;
- ventilation concept with heat recovery system;
- summer night ventilation through centrally openable windows;
- optimised daylighting through the conservatories and atriums;
- high insulation values for windows and façades;
- compact building form;
- unheated conservatory space as a climatic buffer;
- reduction of cooling load by the conservatory glazing (umbrella idea);
- air-heating system to cover the low demand.

PV plays an important role in the daylighting and in sunshading, and hence in conditioning the indoor climate in the conservatory.

BiPV design process
Planning process, alternative designs

In the beginning, plans were made to cover not only the conservatory area with PV glazing, but to span the PV screen above the building blocks as well. Above the blocks, opaque PV roofing with standard modules was foreseen. The plans were revised due to cost restrictions.

In the final design, the roofs on the left and the right side covering the building blocks were changed. The blocks of units 1–3 are now attached as cubic building volumes to the central conservatory, where the PV is placed. The conservatory is divided into two parts: the curved and south oriented roof, interconnecting units 1 and 2, was built in two steps together with units 1 and 2; the straight and north oriented central conservatory without PV will be attached in a later stage, when unit 3 is built.

Decision process and project organisation

All decisions in the building project were taken by the building directorate of ECN. The ECN unit for Renewable Energy in the Built Environment was involved in the overall planning process from beginning to end as an energy consultant.

The utility NUON is the investor and owner of the PV system. ECN and NUON have a contract that ensures the ECN will buy back the green solar electricity over a contractual period of 15 years from NUON. The payment for the green electricity allows financing of the PV system by NUON. After 15 years the PV system is given to the ECN for free. The ECN was responsible for the architectural design, engineering, installation and maintenance of the PV system. This allows overall control and integral planning of the project by ECN. The PV system was delivered turnkey to NUON.

BiPV design

As a choice for renewable material with low embodied energy, it was decided to build the conservatory with a laminated wood construction. Two curved glued timber beams are interconnected with small round steel profiles to form strong beams. The wood beams are horizontally interconnected by 80 mm-high IPE steel profiles. On top of the IPE steel profiles, aluminium glazing profiles are placed to hold the PV modules horizontally and vertically. To be able to follow the curved wooden roof beams, the glazing profiles have been cut out and bent every 120 mm to form straight segments of the curved roof shelter. The greenhouse construction company Allicon was chosen as a partner to develop the PV glazing and glass façades. Glass profiles developed for greenhouses are used for the conservatory roof.

The conservatory of unit 1 is single glazed, as the conservatory space functions as a buffer zone and is not heated. The PV modules were produced and installed by BP Solar. The laminated glass–glass modules with PV cells are encapsulated with EVA. Each module is 575 x 1175 mm. BP Solar mono-crystalline LGBG (Laser Grooved Buried Grid) cells have been used. The cells are spaced with a distance of 1 and 2 centimetres to allow light transmittance of 30 per cent to the interior conservatory space to avoid overheating but allow daylighting.

COMPONENT CHARACTERISTICS

PV system power	Total (units one and two) 43 kWp Step 1: Building 42 Unit 1: 26.73 kWp Step 2: Building 42 Unit 2: 16.31 kWp Step 3: Building 42 Unit 3: no further PV
PV module area	Total approximately 400 m²
Type of building integration	Conservatory glazing
Type of cell technology	BP Solar mono-crystalline LGBG cells (125 x 125 mm, efficiency approximately 16.5%)
Modular dimensions	575 x 1175 mm with 32 PV cells per module
Number of PV modules	Total: 570 PV modules (354 for Building 42.1; 216 for Building 42.2)
PV module construction	Frameless custom-sized PV glass–glass laminates made by BP Solar Front: 3 mm hardened white glass Back: 4 mm glass
Weight	PV modules: approximately 40 kg/m²
Visual details	PV modules for the conservatory roof are semitransparent glass–glass modules
Inverters	Total: 24 string inverters Building 42.1: 14 x type BP Sunny Boy 2400 and 1 x type BP Sunny Boy 1100-E Building 42.2: 9 x BP Sunny Boy 2400
Monitoring equipment	The PV system is permanently monitored with modems included in the inverters, which transfer data to a datalogger ECN calibrated solar cells and temperature sensors are included in nine of the PV modules
Other BiPV system elements	The conservatory PV glass roofing was made by Allicon with standard profiles for greenhouse glass

Fig 5 View to conservatory roof of Building 42 unit 1 with PV integration; laminated wood beams hold the conservatory glazing with its elegantly small profiles
Source: Het Houtblad
Photographer: John Lewis Marshall

Fig 6 Conservatory glazing of Building 42 with PV integration. The conservatory connects Building 42 with the renovated Building 31
Source: Het Houtblad
Photographer: Herman Van Doorn

Installation

The following images show various stages of the construction process.

Fig 7 Left: Building 42, unit 1 under construction; right: Building 31
Source: BEAR Architecten

Fig 8 Conservatory construction of Building 42: the laminated timber beams are installed
Source: BEAR Architecten

Fig 9 Conservatory entrance of Building 42
Source: Het Houtblad
Photographer: John Lewis Marshall

Fig 10 Left: Building 42; right: Building 31
Source: Het Houtblad
Photographer: Herman Van Doorn

Performance characteristics

Monitoring will be permanently performed by the Solar Energy Group of ECN. The Sunny Boy string inverters are equipped with modems, which permanently transfer data to a datalogger, which has been developed by SMA–Regelsysteme (producer of Sunny Boy inverters) to store the data of several string inverters. The data is transmitted over the electricity network. ECN calibrated solar cells and temperature sensors are included in nine of the PV modules, to scientifically evaluate the exact performance efficiency of the PV system with measured data.

Project cost breakdown
Costs of PV system

PV system curved conservatory roof for Building 42 unit 1 and 2: EUR384,352 for 43 kWp complete PV system by BP Solar.

Price includes PV engineering, PV modules, inverters with modems and datalogger, electric cabling and components, and electric installation. Price does not include mechanical installation of PV modules in the roof or glazing profiles to hold the PV modules.

Financing

The utility NUON normally supports PV plants with EUR1.36 per Watt-peak. In this case, NUON owns the PV system. ECN and NUON have a contract, which foresees that ECN buys the green electricity over a contractual period of 15 years back from NUON. The payment for the green electricity allows financing of the PV system. After 15 years the PV system will be owned by ECN.

Support

The PV system has been supported with support by the Dutch government via NOVEM the Netherlands Energy Agency.

THE NETHERLANDS: **LE DONJON**

BIODATA

PROJECT:	Le Donjon
LOCATION/CITY:	Gouda
COUNTRY:	Netherlands
TYPE OF PV BUILDING:	PV canopy above façades
BUILDING TYPE:	Office building
NEW/RETROFIT:	New

CLIMATIC CHARACTERISTICS

LATITUDE:	52° 00′N
LONGITUDE:	4° 43′E
ALTITUDE:	Sea level
CLIMATIC TYPE:	Moderate West-European maritime (Temperature: January average = –1.6 °C; July average = 22 °C)
SUNSHINE HOURS:	Yearly average = 4.05 hours per day

General project background

The 'Le Donjon' office building is situated in Gouda, a small town in the green heart of the Netherlands. Gouda is about 30 kilometres east of the Dutch capital The Hague and enjoys a mild but humid maritime climate. The building site is located between a residential neighbourhood with two-storey brick homes from the 1930s on one side, and a train line on the other side, and is the site of a former elementary school. Town planning regulations and building codes strictly required integration of the new building design into its built environment context. Therefore, the choice was a flat roof and dark red bricks. The name 'Le Donjon' (The Castle) was chosen for the compact two-storey building consisting of three building blocks, which are grouped around an inner courtyard. The units are owned by three investors.

The same low-energy design is used for all three units: ground floor and roof are made from concrete with a U-value of 0.27 W/m²K, the walls from sand-lime bricks with red-brown bricks outside and an U-value of 0.24 W/m²K. The windows are larger on the south and smaller on the north side. The window frames are environmentally friendly and made from pinewood, covered with an aluminium metal sheeting on the outside as weather protection. All south and west windows are equipped with integrated metal louvres (triple glazing) with a resulting U-value of 0.98 W/m²K. The windows to the north and east have inner shutters and a U-value of 1.1 W/m²K.

Fig 1 Le Donjon office building, west elevation
Source: BEAR Architecten
Photographer: Marcel van Kerckhoven

Further environmental measurements have been included in the building, which is part-owned and used by BEAR Architecten:

Energy:

- PV system 6.2 kWp;
- small thermal collector (approximately 2.5 m²) for cleaning, kitchen tap water and dishwasher;
- high-efficiency gas boiler;
- floor heating system for office spaces and low-temperature radiators for the meeting rooms;
- balanced ventilation with heat recovery system (winter);
- summer-night ventilation for cooling;
- building management system for outdoor climate regulation of heating and ventilation;
- daylight controlled and energy saving HF (high frequency) lighting and daylight spectrum;
- electricity saving equipment such as flat screens, high energy-efficient fridge and dishwasher;
- electricity switches for each workstation, to avoid standby losses of equipment such as computers at night;
- EPc – Energy Performance Ratio of 0.6 – building code requires EPc of 1.6.

Ecology:

- natural materials such as Forest Stewardship Council (FSC) certified wood, natural paint and linoleum;
- water saving installations and rainwater use for toilet flushing;
- rainwater pond to slow down rainwater discharge and roof-garden;
- nesting boxes for bats and swifts.

The BEAR Architecten office is listed as a 'Panda Office' by the World Wildlife Fund (WWF) for its energy saving and renewable aspects.

BiPV design process
Planning process, alternative designs

BEAR Architecten has successfully completed more than 15 different projects with PV integration and was named the most well-known architectural firm for building-integrated PV in the Netherlands in a study carried out by Novem in 2000. Therefore, it was a must to integrate PV in the firm's own new office building. The design had to fit into the neighbourhood, so a very futuristic high-tech building with a huge south oriented PV-roof or glass façade was not considered. Instead, a very basic building design was developed, based on classical brick buildings, but including many renewable features. A flat roof was chosen as the most efficient roof form for office buildings. A classical flat roof PV installation with independent support structures was not desirable, as the PV system should be visible and enhance the architecture. Due to financial limits only the unit used by BEAR Architecten includes PV.

Fig 2 North elevation with WWF panda during opening, November 2000
Source: BEAR Architecten

Fig 3 North façade
Source: BEAR Architecten
Photographer: Marcel van Kerckhoven

For the BEAR Architecten unit, a flat roof PV system, developed by 'TNO-Axys', with an installed capacity of 3.5 kWp was constructed in 2001. The overall potential is about 1,000 square metres flat roof area. Depending on the inclination of possible PV installations the potential could be up to 80 kWp and at least 35 kWp.

Decision process

For architectural as well as for economic reasons, it was felt that the PV system should be a multifunctional, integral building element. This led to the combination of solar energy gain and rain protection for the walls in a PV-element as a canopy roof on top of the walls. The roofing function for the walls meets the sustainable building code 'DuBo maatregel S061' in the Netherlands: to protect the walls by adequate façade design. A further advantage of the chosen design is an aesthetic but relatively cost-efficient mounting construction for the PV system, compared with other façade or roof-integrated systems. In the beginning it was planned to integrate sun-protection functions as well, but a fixed sunshade would have had the problem of taking away too much light when it is needed.

Unfortunately, it was not possible to convince the two other investors in the building complex to integrate PV. So the other two blocks have a glass canopy roof with the same substructure and hardened glass, but without PV. Taking into account that subsidies for the PV system have been given and the system earns some money back by producing green electricity, the pure glass installation is not really more cost efficient than the PV system.

Project organisation

The building market in the Netherlands is quite tight and for private as well as for business purposes, it is very common to own the building used. When BEAR Architecten wanted to move, it decided to build its own sustainable office as a reference project as well.

To find the most economical way of integrating a PV system, a feasibility study was completed by Ekomation financial consultants. The study was to provide an overview of the economic parameters, so as to find a solution for optimal financial PV project organisation. The following parameters were evaluated:

- What kind of public sponsorship is available and optimal for the PV installation?
- Should the PV system be self-owned or leased?
- What are the tax advantages of different investor packages?
- Is amorphous silicon more cost-efficient than crystalline silicon for the site?

As a result of the study and the design approach of BEAR Architecten, it was found that around 6 kWp of semi-transparent PV modules integrated as a canopy roof was the best option. The Netherlands tax-system allows 'EIA – Energy Investment Deduction' and the 'VAMIL – Accelerated Depreciation for Environmental Investments'. The Netherlands Energy Agency NOVEM paid subsidies for the feasibility study, for the construction and the PV installation.

Responsibilities

BEAR Architecten completed the building and PV integration design. Construction drawings as well as construction co-ordination and management were done by the project manager ARCOM, one of the three investors. The construction company Nees de Jong, a general building contractor, carried out the building. Blonk Staal was contracted by Nees de Jong to deliver and place the metal construction for the PV, and the glass

Fig 4 Southeast elevation
Source: BEAR Architecten
Photographer: Marcel van Kerckhoven

Fig 5 Southeast elevation from courtyard
Source: BEAR Architecten
Photographer: Marcel van Kerckhoven

ompany Verloop was contracted for mechanical mounting of
the PV modules. The PV modules were directly supplied by BST,
as a general contractor, and Siemens supervised the mounting
process. BST itself worked together with Siemens as subcontractor
or the electric installations, including the placing of inverters.
The connection to the grid was made by Jos Struik, the company
responsible for all the building's electrical installations.

BiPV design

The PV modules are produced as custom-sized frameless
laminates. The transparent back foil allows a semi-transparent
visual appearance. The use of PV modules as little roofs above
walls for buildings with flat roofs was a good strategy, especially
as the roofs are small and the attic walls will shade part of the
roof. The PV modules therefore:

- function as rain protection for the attic walls and
  substitute a horizontal metal covering;
- protect the façade from rainwater;
- could function as sunscreen if they were larger.

The more or less horizontal installation allows optimal use of
space, especially in an inner-city neighbourhood with small roof
areas and a large proportion of façade shading. For design
reasons (rainwater flush) the PV canopy was designed to run
around the roof with a uniform inclination of 5° to the horizon.
For that reason, the PV elements are all inclined to the centre of
the roof, so the PV modules on the south side are 5° inclined to
the north and so forth.

COMPONENT CHARACTERISTICS – CANOPY SYSTEM

PV system power	6.23 kWp
PV module area	72.5 m²
Type of building integration	Canopy roof over façade walls
Type of cell technology	AP-106 mono-crystalline silicon solar cells made by Astro Power
Modular dimensions	1150 x 1150 x 8 mm
Number of PV modules	51 laminates of 111 Wp each and 8 triangular modules of 567 Wp in total
PV module construction	Custom-sized frameless PV laminates with hardened glass for the front, treated with PV-Guard and a back sheet of tedlar
Array dimensions	PV is placed around the roof border with a length of 25 x 12 metres
Weight	The PV modules have about the same weight as glazing, approximately 17 kg/m²
Visual details	The laminates' rear tedlar foil is transparent
Inverter	3 x type 'Fronius Sunrise Midi' with 1200–2400 Wp nominal power
Balance of system components	The PV system is divided into three strings
Monitoring equipment	The program 'EnergieMonitor' developed by Econergy is installed on a separate computer for monitoring
Other BiPV system elements	A special metal substructure has been developed for the canopy roof above the walls

Figs 6 and 7: An installation wall incorporates the inverters and meters (right) in the entrance area (left)
Source: BEAR Architecten
Photographer: Marcel van Kerckhoven

121 •

Installation design

A special metal construction was designed and produced to fit the Le Donjon office building. Triangular steel profiles, to hold the modules, are welded to a horizontal steel tube, held by vertical tubes, which are fixed onto the concrete mass of the flat roof. The frameless PV modules are fixed with simple metal strips and screws to the triangular profiles.

The system is divided into three strings, connected to three inverters. This electrical design was made without differentiation between the module orientations, as the inclination is only 5° (always to the centre of the roof) and the three strings could be interconnected with short wiring runs to the cable inlet. The inverters are placed in the entrance area, directly beside the three electricity meters. One of the electricity meters registers electricity fed back to the utility's grid, the others register electricity used during the day and during the night. However, electricity is generally fed directly to the BEAR Architecten office. Only the PV electricity that is not needed immediately is fed back to the grid – for example on sunny weekends, when the office is closed.

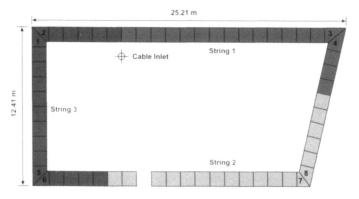

Fig 8 Roof plan showing PV design and strings
Source: B.S.T-Group and BEAR Architecten

Fig 9 Nearly horizontally installed PV laminates
Source: BEAR Architecten
Photographer: Marcel van Kerckhoven

Installation

Problems in the construction process and between the partner companies only occurred where the PV modules and the metal construction did not fit perfectly in certain spots. The shape and measurements of the PV modules were relatively complicated because the modules are inclined by 5° and had to be cut exactly to size to fit together as a canopy running all around the building's roof. Because of the characteristics of the site, the building has two corners that are not rectangular. This required eight differently shaped PV modules for the corners.

Fig 10 PV canopy during installation
Source: BEAR Architecten

Misunderstandings in the measurements led to size differences in the metal substructure and the laminates. Due to the precise building integration and high visibility only small tolerances were acceptable. Problems also occurred because the welded metal supporting structure could not be produced with the same precision as the bolted construction, which was originally proposed. Temperature changes during the process made the structure less precise. The lack of adjustment was a problem and made any changes after mounting the constructions very difficult. Therefore one of the substructures was reassembled and welding had to be carried out on the roof.

The inverters have been installed in acoustically separated stainless steel metal profiles to avoid noise emissions. Because the inverters are installed in the entrance area of the open-plan office, attached to an internal wall made from wood, any noise transferred to this wall had to be strictly avoided as it would have been heard in the whole office building.

Performance characteristics

Measurements are continuously made using the program 'EnergieMonitor', developed by Econergy. It is installed on a separate computer for monitoring, placed directly beside the inverters.

Figs 11 and 12 thermal collector for tap-water and dish-washer on roof (PV in background); right: boiler and hot water storage
Source: BEAR Architecten;
Photographer: Marcel van Kerckhoven

Project cost breakdown

The overall office complex has been equipped with a glass canopy on the outer façade of the three office blocks with PV over the BEAR Architecten office instead of hardened white glass. The steel construction, to hold the modules, is the same as for the other blocks and installed by the same glazing company. This provides a unique opportunity to compare the costs for the PV system and the simple glass canopy. The question 'What are the real additional costs for building-integrated PV?' can be answered exactly for this project on the basis of actual prices.

The price of the mounted steel structure was 219 Euro per linear metre of façade, including the work to place the glass panels/PV modules by the glazing company. The hardened glass was 157 Euro per linear metre (with a depth of 1150 mm) and the PV system costs were 828 Euro per linear metre. Thus the additional costs for the integrated PV system totaled 671 Euro per linear metre. Each running metre PV canopy of Le Donjon has an installed capacity of approximately 91 Wp. The cost of the addition of PV is therefore about 7.20 Euro per Watt peak. This is just as expensive as a completely separate flat roof PV system would have been.

The PV building integration has not proven to be a money saver. To install a PV laminate instead of a plate of hardened glass, and to pay all the necessary system costs, including electric installations costs, equates to 828 Euro per linear metre. This is more than five times the 157 Euro cost of installing the hardened glass.

Subsidies are very important to make investments in PV systems possible. In the Netherlands, an innovative PV installation such as the canopy can be sponsored up to 40 per cent, with 25 per cent expected for further installations. The subsidy is calculated over the full installation costs, including the steel support structure. The tax laws allow further savings. Taking into account that the glass canopy would have been installed anyway, the additional costs for the PV system reduce to only 100 Euro per metre. This is further reduced by the electricity produced by the system and hence the additional costs are almost gone (32 Euro per metre over a 10-year production period). Even with a lower subsidy, a PV system is relatively cost-effective.

Table 2 shows the effects of subsidies on the first calculation:

Type of Costs (Euro)	Glass Canopy	PV Canopy 6.23 kWp	Glass Canopy	PV Canopy	
	Total system (68 m)	Total system (68 m)	Per m/ façade	Per m/ façade (= 91 Wp)	per Wp
Engineering steel structure	749	749			
Anchoring steel elements	2216	2216			
Module supporting profiles	5150	5150			
Mounting steel anchorage	2591	2591			
Mounting supporting profiles	1917	1917			
Mechanical mounting PV modules/glass elements	1995	1995			
Subtotal	14,619	14,619	219	219	2.34
Hardened glass/PV laminates (price includes system components as inverters, cabling and electrical mounting of complete PV system)	10,512	55,353	157	828	8.89
Total costs	25,131	69,971	376	1,044	11.24
Less costs of standard glass canopy		−25,131		−376	−4.04
Additional costs for PV system		44,840		671	7.20

Table 1 Project cost breakdown for PV system

	Cost in Euro for PV canopy system per metre façade (approximately 91 Wp/m)	
	40% subsidy	25% subsidy
Investment – PV system	1,046	1,046
Subsidy NOVEM (Netherlands Energy Agency)	418	261
Net investment	628	785
EIA – energy investment deduction	121	151
VAMIL – accelerated depreciation for environmental investments	31	39
Standard glass canopy costs	376	376
Net costs	100	219
Income solar electricity sales over 10 years (approximately 680 kWh x 0.1 Euro x 10 years)	68	68
Extra costs – PV system	32	151

Table 2 Cost comparison of PV vs standard glass canopy under different PV subsidy levels

Post-installation feedback

The idea of canopy PV roofs placed on attic walls of flat building roofs is very viable, as it always is a problem that modern architecture, with its flat roof design, leads to façades that are not protected by rain. Water running down the façade will colour and pollute the façade. If the flat roofs are very small and an attic wall is envisaged, it will often be difficult to find enough unshaded flat roof area for PV use. Further, the canopy results in less weight on the roof, as no additional concrete weight is needed as for the independent support structures used in classical flat-roof PV systems. The PV canopy is also functional as an aesthetic solution.

THE NETHERLANDS: **NIEUWLAND**

B I O D A T A

PROJECT	1 MWp PV project
LOCATION/CITY	Amersfoort
COUNTRY	Netherlands
TYPE OF (PV) BUILDING	Various: roof and façade integrated
BUILDING TYPE	Residential
NEW/RETROFIT	New

CLIMATIC CHARACTERISTICS

LATITUDE	52° 09'N
LONGITUDE	5° 23'E
ALTITUDE	Sea level
CLIMATIC TYPE	Moderate coastal maritime (Temperature: winter average = 2.6 °C; summer average = 16.2 °C)
SUNSHINE HOURS	Yearly average = 4 hours per day

Fig 2 Nieuwland's urbanscape
Source: Ecofys

General project background

Nieuwland is a new housing area of the City of Amersfoort (population 123,000) in the centre of the Netherlands. It was developed by a public–private partnership called 'Overeem', in which the city authorities worked with property developers and construction companies to develop the area. The total settlement of Nieuwland comprises some 4,000 dwellings for 11,000 residents.

Nieuwland began before the Energy Performance Code (EPN) was established in the Netherlands. If translated to EPN criteria, all houses have an energy performance lower than 1.2, which is better than the required level in the building code at the time of the project. In addition, great attention was paid to passive solar (80 per cent of all houses are oriented between southeast and southwest) and solar thermal (900 solar water heaters were installed in the settlement). Condensing boilers are used for space heating.

Fig 1 Nieuwland's 'solar portal', bridging the central street that runs through the project
Source: Jan van Eijken

BiPV design process

The overall criteria of sustainability, set by the city authorities, provided an excellent opportunity to develop one section of Nieuwland into a true solar suburb. Also, the local regional utility, REMU, was looking for opportunities to develop a number of PV projects as part of its green energy strategy. Prior to the 1 MWp project, a few smaller PV projects had been carried out by REMU in the first sections of Nieuwland. These projects focused on familiarisation with the technology and on the process of working with architects, property developers, contractors, the PV industry and others involved in building-integrated PV projects. These first projects include the SCW project (figure 3), the Thomasson-Dura project (figure 4) and the De Border school project.

The SCW project comprises 50 row-houses, developed together with a social housing association and designed by architect Han van Zwieten (Van Straalen Architecten). The aim of the project was to technically integrate PV into roofs together with solar thermal and daylighting. Compared to the original design of the project, Han van Zwieten decided to 'lift' the PV section, not only to create a vertical window, but also to give the PV a well-defined and separated place within the building envelope.

The Thomasson-Dura project comprises 19 row-houses with individual PV roofs. The project was developed by a commercial property developer and the design by Artes Architects. The aim of the project was to technically develop the prefabrication of PV roofs. Prefabrication was not feasible at the time of the project (1996), given the state of the technology. The design of the project was, however, appreciated by architects worldwide. The strong points include the combination of colours and materials, and the use of PV as a protruding element of the building envelope.

Fig 3 The SCW-project: 50 row-houses, developed together with a social housing association
Source: Ecofys

Fig 4 The Thomasson-Dura project: 19 row-houses with individual PV roofs
Source: NOVEM

Fig 5 Nieuwland's K2 section with roof-integrated PV
Source: Source: Jan van Eijken

Fig 6 Nieuwland's K2 section roof-integrated PV solar porch
Source: Carl-Michael Johansson

Fig 7 Nieuwland's 'solar portal', bridging the central street
Source: Jan van Eijken

1 MWp Nieuwland project

The 1 MWp project was developed by REMU as a follow-up to these pilot projects. Together with the Overeem developer consortium and the urban designer working on Nieuwland, an appropriate section of Nieuwland was selected. The main criterion was that it should be large enough for one compact PV site with PV on all buildings. The urban design should allow sufficient south-facing roofs with limited shading.

The urban designer formulated a design that allowed for sufficient roof space facing south in order to install more than 1 MWp PV. In addition, an information package was prepared for the property developers and architects involved in this section of Nieuwland. The development of this information package was an important aspect of the project; it assisted in learning how to co-operate in a standardised way with large numbers of developers and architects on a large-scale project.

The most relevant section within the information package was a shortlist of technical requirements to take into consideration, set forward by the project partners Ecofys, REMU and ENEL. The information package also contained sections on aesthetics, project planning, technical and electrical details. A standard electric design scheme was developed as part of the preparation process. This design was based on the principle of one inverter per house. In order to limit the number of spare parts to be kept in stock for replacements, the same inverter was used throughout the whole project, resulting in a requirement to integrate 2 kWp to 3.5 kWp per building.

The objective of the project was to utilise, as much as possible, standardised, well-proven concepts provided by the PV industry during the design process, which started before the PV supplier was selected. This created some difficulties as the architects were unable to determine the size of the PV modules that would be used, or which mounting system would be provided by the PV supplier. Thus, some redesigns were required after selection of the PV supplier due to specified module sizes and mounting systems.

In March 1997, a call for tenders for delivery of the PV systems was publicised in the official journal of the European Union. Out of the eight companies that responded to the invitation, three were invited to submit a tender. Detailed terms of reference were formulated and sent to the selected companies. Contract negotiations were completed in 1998. Shell Solar Energy built the bulk of the systems, with BP Solar as backup. PV roof tiles from RBB and sun shades from Colt International were applied as special fixtures in small sections.

During this process, many designs were modified and the following issues were resolved:

- At sections K3 and N2, non-integrated, flat-roof-based PV systems were decided upon. Roof-integrated PV systems did not fit into the overall programme of requirements set for these houses. In N2, the building costs were limited (social houses), leading to the design of flat-roof houses (lower costs than sloped roofs). In section K3, the architect insisted on flat roofs for architectural reasons. Here the houses are among the most expensive in Nieuwland, regrettably without a visible and outstanding PV-flag.

- Sections K1 and N2 required a complete redesign of the houses. The property developers in these sections fully supported (financially) this redesign, showing their dedication to find attractive solutions for the integration of PV into houses. In section K1, however, the architect could not meet the PV requirements set by the PV project team without additional construction costs. The property developer in this case was not willing to fully bear these costs. As a result, REMU contributed to the additional costs incurred by the PV system requirements.

- In sections K2 and K4, the design of the houses was straightforward and without conflict from the design requirements for the PV. For these sections, the architect was highly experienced in building-integrated PV. Non-PV architects were involved in the other sections.

Fig 8 Close up of Nieuwland's 'solar portal' *Source: Jan van Eijken*

Architectural concept
Section K2
Architect: Han van Zwieten, Van Straalen Architecten

This section involves the smallest 'island' in the project: two rows of houses with a central street (figure 5).

In most other sections with sloped roofs, each row of houses has its street on the south side. However, in this section, the overall design brief asked for a symmetrical view from the street to both sides. This necessitated a symmetrical building design, with entrances facing south on one side of the street and north on the other side. If the houses were to be kept completely symmetrical (architecturally and functionally), this sloped section would need to face north on one side of the street and south on the other side. The architect therefore decided to create a rather narrow central section for the PVs, which could be 'turned around' on the roof without disturbing the symmetry.

A north window was incorporated behind the PV for daylight entrance to the central staircase underneath, maximising the solar gain of the building – PV not just as PV, but as a multifuctional solar element including daylighting. The architect was not completely satisfied with the standard profiles for the windows, isolated glass, see-through panels, northern panels, and so on. Despite being inexpensive, he felt that they were not very elegant and the problem of finding better ways to detail with PV without profiles or glue has yet to be solved.

As a finishing touch, the two sides of the island were connected with two 'solar portals' bridging the central street of the project. The overall appearance of the island now resembles an ancient racecourse with an intimate inner court (figure 10).

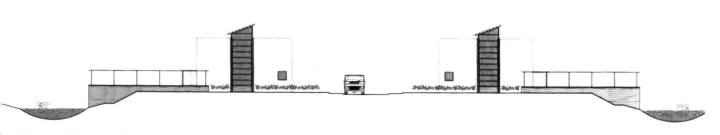
Fig 9 Side view of K2 housing profile

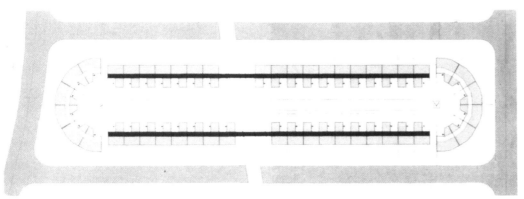
Fig 10 Plan view of development resembling an ancient amphitheatre
Drawing: Cinzia Abbate

Fig 11 The roof under construction, showing the smaller additional sections with sheet material to cover up gaps between the PV sections
Source: Ecofys

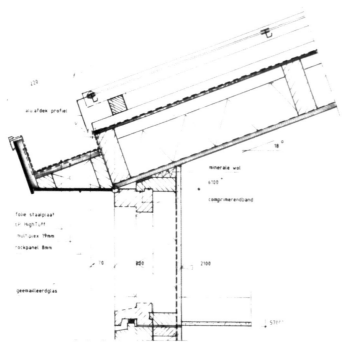

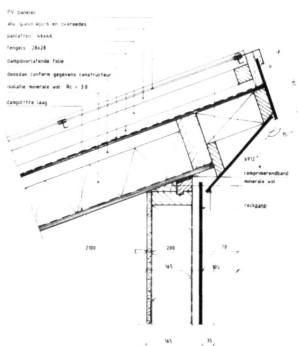

Fig 12 Cross sections of bottom and top end of the PV roof. Small ventilation gaps are designed to allow air to flow behind the modules. The gutter section at the lower end only is used for those sections of the roof where there is a vertical wall below.
Drawing: Han van Zwieten bna

BiPV design

The strategy for section K2 was to use standard laminates and standard mounting systems provided by the PV supplier (Shell Solar) to create a central, tilted section on the flat roofs of the houses. This vertical section runs over all houses, without interruption. To achieve this, the width of each house (7.2 m) was adapted to the modules' width (1.2 m), allowing for six modules per roof. The final module size is 545 x 1185 mm, with a centre-to-centre distance between the vertical profiles of 1.2 m.

In practice, some additional smaller sections with sheet material were required to fill up some gaps in the PV roofs as shown in figure 11. Mounting details were designs based on experiences from previous projects.

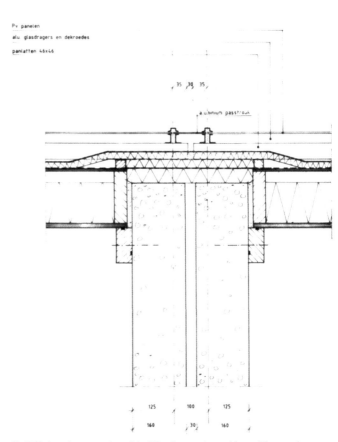

Fig 14 Horizontal cross sections of the PV roof across the partition wall between houses. Again, a narrow piece of sheet material is used to fill the gap. Also note additional measures to avoid transmission of noise from one house to the other.
Drawing: Han van Zwieten bna

Fig 13 Nieuwland's 'solar portal', bridging the central street
Source: Jan van Eijken

Installation design
Electrical configuration including grid integration

A standard configuration was developed for the whole 1 MW project, based on the principle of 'one house, one inverter'. Reasons for this approach were:

- To connect the PV to the energy system of the occupant. Even in the utility-owned sections, the utility wanted to establish a relationship with the building owner to ensure good and reliable operation of the system. In the parts of the project where the occupant co-financed the systems in return for ownership of the system, direct benefit by the occupant was a prerequisite;
- To achieve the lowest wiring costs. No extensive DC grid was required, with limited additional AC grid features;
- To limit requirements for switches, fuses and blocking diodes (which were omitted).

Contrary to the initial concept, no remote monitoring via the cable network was used (too expensive).

Planning approval and institutional processes

All designs were reviewed by a team consisting of the urban planner, the environmental supervisor, representatives of Overeem, and by REMU for PV-specific elements.

Installation
Installation procedures and experiences

Installation of the PVs was commissioned by REMU to Shell Solar and its team. The construction of the buildings was commissioned by the developer Van Zwol to its own contractor. This meant that there were two teams on site, each with their own principal, their own planning scheme, and so on. Prior to installation, agreements were made between Shell Solar and Van Zwol about the use of site huts, storage facilities, and scaffolds.

Successful approaches

For the installation of the PV system, a quality control strategy was developed by Ecofys and REMU to allow for a smooth realisation of large numbers of PV systems. This quality control strategy included pre-installation tests, on-site inspections, commissioning checks and performance contracts.

Problems during realisation

Mastervolt was not able to deliver its Sunmaster 2500 inverters in time for the K2 section. These were used in most of the project. Consequently while the PV roofs were being laid down, the inverter was being tested. This resulted in a situation where the PV roofs were finished long before an inverter had been placed. Due to this delivery problem, some of the houses were commissioned before the PV installation was complete, and in some cases, the houses were already inhabited when the inverter was installed. This was not a desirable situation for the PV suppliers and other parties involved because it limited access to the houses and made the procedure for finishing the PV systems more costly.

Fig 15 Sample of the Mastervolt inverter and connection box, used in K2 section
Source: Ecofys

COMPONENT CHARACTERISTICS

PV system power	2.25 kWp per house; 38 houses
Type of building integration	Roof and façade-integrated
Type of cell technology	Poly-crystalline silicon
Modular dimensions	Module type: RSM 75. Size: 545 x 1185 mm (frameless) – 560 x 1200 mm (including mounting system)
Array dimensions	6 x 5 modules – 2800 x 7200 mm
Weight	8.25 kg
Inverter	Mastervolt, Sunmaster 2500-150
Monitoring equipment	The Ecofys Eclipse is used for global monitoring
Other BiPV system elements	Connectors for module connection: Hirschmann B12 GDME; Module wiring: VUV / VYD 1 x 2.5 mm²

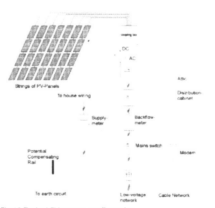

Fig 16 Typical PV system configuration

Project cost breakdown
Financing, ownership

The Nieuwland 1 MW PV project was financed by the local utility REMU, with financial support (49 per cent) from the Netherlands PV programme and the European THERMIE programme.

To allow investigation of the effects of the various forms of ownership and management, approximately half the installations remain the property of REMU. Agreements have been made with the developers, setting out the arrangements concerning accessibility of the installations and liability for any damage. These agreements are part of the purchase contracts that the building owners entered.

A right of superficies (building right) was established in respect of the plots. It has also been stipulated that the solar panels should remain unshaded. The residents are remunerated by REMU for the use of their roofs. Twenty per cent of the energy generated on their roofs will be paid for at the normal domestic consumer tariff. The other half of the solar power installations were transferred to the residents in return for a 25 per cent contribution to the system costs. Agreements regarding the legal and financial arrangements were concluded for this purpose with the developers concerned (Mabon, Achtgoed and Schoonderbeek).

The PV costs were approximately 7 Euros per Wp, though detailed costs varied per sub-section.

Feed-in tariffs

In the sections where REMU remains the owner, the occupant is financially compensated for the use of the roof. This compensation equals 20 per cent of the produced PV power, rated at consumer tariff. In the sections where the ownership was transferred to the occupant, a net-metering scheme with double counters is used. Residents receive the same rate for PV power as they pay, approximately EUR0.14 per kilowatt hour.

SOUTH KOREA: KIER SUPER LOW ENERGY BUILDING

IO DATA

PROJECT:	KIER (Korea Institute of Energy Research) SLEB project
LOCATION/CITY:	Taejon
COUNTRY:	South Korea
TYPE OF PV BUILDING:	Roof-integrated
BUILDING TYPE:	Office
NEW/RETROFIT:	New

CLIMATIC CHARACTERISTICS

LATITUDE:	36° 18'N
LONGITUDE:	127° 24'E
ALTITUDE:	77 metres above sea level
CLIMATIC TYPE:	Temperate (Temperature: winter average = 5 °C; summer average = 22 °C)
SUNSHINE HOURS:	Yearly average = 6.0 hours per day

General project background

The Korea Institute of Energy Research (KIER) is a national research agency located 100 miles south of Seoul, South Korea. In 1998, it initiated a national project to design and construct a very low-energy consumption building, the so-called 'Super Low Energy Building' (SLEB). The building has six major energy efficient systems including:

- double-skin façade;
- thermal storage water tank;
- ground coupled heat exchanger system;
- integrated system control;
- high-efficiency vacuum type solar thermal collector, and
- 30 kWp solar system.

The purpose of the project was to design and construct the SLEB and target an annual energy performance index (EPI) value of 74 Mcal/m²yr from the use of various Energy Conservation Opportunities (ECO) including the six major technologies described above.

Fig 1 Front view of KIER SLEB building
Source: KIER

BiPV design process
Planning process, alternative designs

The project was led by KIER and special committee members in various areas reviewed the proposed design. The reference model was the Obayashi Kumi Building in Japan and design modifications were proposed and implemented for the SLEB design. KIER considered three alternative BiPV designs:

Option 1: 25.4 kWp roof with 30° tilt;

Option 2: 30.8 kWp roof with 30° tilt;

Option 3: 34.0 kWp roof and wall systems with 30° and 90° angle tilt respectively.

Decision process

The SLEB BiPV design process was based on a combined system of independent and dependent evaluation. The PV was required to supply electricity for lighting as well as have the capacity to export to the grid in periods of excess supply. The interconnected inverter provided mains electricity back to the lighting in reduced supply conditions when cloudy and at night. This design strategy was chosen because it could reduce the initial, as well as the operating cost, and also provide relatively constant electricity on the demand side. The final design had an effective area of 265 square metres of poly-crystalline silicon solar cells, providing maximum power of 28.1 kWp and an annual energy generation of 24,856 kWh.

The ESP-r dynamic building energy simulation program was used to predict the energy consumption and to visualise the impact of some energy conservation opportunities (ECOs) on EPI. Estimated annual consumption was predicted as 73.9 Mcal/m^2 and also showed that annual heating demand amounted to 15.5 Mcal/m^2 (21 per cent), cooling demand was 10.4 Mcal/m^2 (14.1 per cent) and lighting and office equipment demand was 47.9 Mcal/m^2 (64.9 per cent). It indicated that the major portion of the energy was due to the internal load attributed to lighting and equipment needs.

COMPONENT CHARACTERISTICS

PV system power	30 kWp
Type of building integration	Roof-integrated
Type of cell technology	Poly-crystalline silicon
Inverter	30 kW 3-phase, 4-wire
Monitoring equipment	LG Honeywell
Other BiPV system elements	4 PV junction boxes

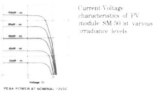

SAMSUNG HIGH-POWER PV MODULE

PRODUCT FEATURES

SAMSUNG's advanced production facilities have produced a high performance photovoltaic module with maximum power of 50 watts.
- 50mm heavy extruded aluminum frame designed to withstand hurricane force loads.
- High efficiency polycrystalline square silicon cells and design for long life and maximum reliability.
- Encapsulation is made by using tempered glass in the front side and humidity resistant plastic material at the back-side.

Fig 2 Catalogue view of SM-50 PV module
Source: KIER

SM-50

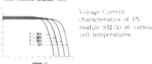

APPLICATIONS

- Microwave/Radio repeater stations
- Electrification of villages in remote areas
- Medical facilities in rural areas
- Telemetry
- Security system
  Electric fence chargers
- Navigation lighthouses, and ocean buoys
- Water pumping system
- Aviation obstruction lights
- Cathodic protection systems
- Desalination systems
- Recreational vehicles
  Railroad signals
- Sailboat charging systems

TECHNICAL SPECIFICATIONS

MODEL	SM-50
Typical Peak Power	50W
Guaranteed Min. Peak Power	47.5W
Voltage @ Peak Power	17.1V
Current @ Peak Power	2.92A
Current @ Operating Voltage	3.12A
Operating Voltage	15V
Weight	6.3kg

Note: The electrical specifications are under test conditions of irradiance of 100m W/m² spectrum of 1.5 air mass and cell temperature of 25°C.

PHYSICAL DESCRIPTION

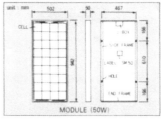

MODULE (50W)

For business and technical issues, please contact

ELECTRICAL CHARACTERISTICS

Current-Voltage characteristics of PV module SM-50 at various irradiance levels

PEAK POWER AT NOMINAL 12VDC

Voltage-Current characteristics of PV module SM-50 at various cell temperatures

POWER OUTPUT AT NOMINAL 12VDC

QUALITY ASSURANCE

SAMSUNG polycrystal photovoltaic modules exceed government specifications for the following tests.
- Thermal cycling test
- Thermal shock test
- Thermal/Freezing and high humidity cycling test
- Hail impact test
- Mechanical, wind and twist loading test
- Salt mist test
- Light and water exposure test
- Field exposure test
- Electrical isolation test

MATERIALS

- Polycrystalline silicon cell
- Tempered high transmission glass
- Anodized aluminum frame
- Silicon sealing
- Encapsulant: Ethylene Vinyl Acetate(EVA)
- Back: White colored Tedlar
- Junction Box: for external screw terminal connection

SAMSUNG ELECTRONICS CO., LTD.
- SEOUL OFFICE: Daehwa B/D, 169, Samsung-Dong, Kangnam-Ku, Seoul-City, Korea TEL 82-2-559-4094 7 FAX 82-2-559-4009
- SUWON FACTORY: 416, Maetan 3-Dong, Paldai-Ku, Suwon-City, Kyungki-Do, Korea TEL 82-331-200-4709 FAX 82-331-200-4722

Fig 3 Specification of SM-50 PV module
Source: KIER

Installation design

Installation design was fixed as follows:

- Poly-crystalline solar cell, made in Korea, was selected and installed with 580 modules of Samsung SM-50;
- Solar cell with one side open, rectangle type support was approved;
- Module connectors were used to connect all the output from each module;
- Inverter was installed to monitor the load and to provide it to internal lighting and other uses;
- Overall monitoring system was installed;
- Lightning rod was installed;
- Side support height on the top of the building was reduced from 90 to 45 cm;
- The installation was completed as per figure 4.

Installation

The installation process was as follows:

- All frames and supports were installed as in drawing and installation guide.
- The material for the frame was steel angle (75 x 50 x 2.3 t) and anchor, concrete angles were carefully installed to support them and also to prevent rain penetration into building.
- The flat portion of the top and treated support zinc film were joined with SUS bolt, nut and washer.
- The support column and concrete were joined with set anchors and treated with mortar.
- Electrical connection for solar modules was connected in series with 20 units and CV 5.5" (HI 22C) configured between array junction boxes.
- CV 8.0" (HI 22C) was used between array junction box and inverter
- Commissioning tests were done over the group of solar cells, configuration as noted in the installation guide.

Problems observed during realisation included:

- Shading effect of guide support over the cell.
- Fuse defect in the inverter and ABB converter defect was observed and resolved.

Fig 5 Framing mounting process
Source: KIER

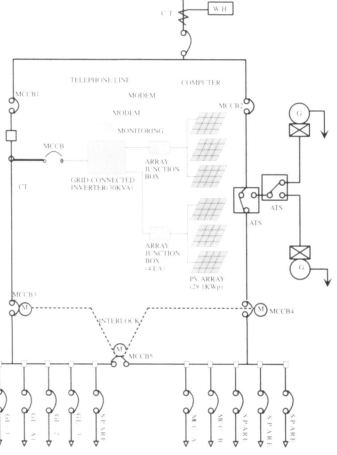

Fig 4 Block diagram
Source: KIER

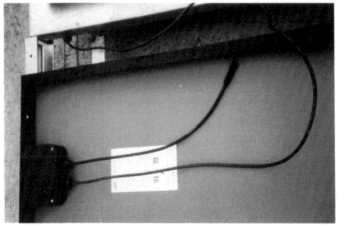

Fig 6 Panel rear connection
Source: KIER

Performance characteristics

Actual performance between 1 January and 31 December 1999 resulted in electricity production of 29,214 kWh during a detailed monitoring phase.

Project cost breakdown

The overall project cost was about US$170,000 (as at October 1998).

Post-installation feedback

Inverter operation could be improved through better design, installation and program testing. Pre- and post-testing of the inverter operation needs to be thoroughly considered.

Fig 7 PV array
Source: KIER

SPAIN: **UNIVER**

IODATA

PROJECT:	200 kWp
LOCATION/CITY:	Jaén
COUNTRY:	Spain
TYPE OF PV BUILDING:	Façade-integrated, pergola (building canopy), parking canopy
BUILDING TYPE:	Commercial
NEW/RETROFIT:	Retrofit and new (pergola)

CLIMATIC CHARACTERISTICS

LATITUDE:	37° 73'N
LONGITUDE:	3° 78'W
ALTITUDE:	578 metres above sea level
CLIMATIC TYPE:	Mediterranean with dry and warm summer
SUNSHINE HOURS:	Yearly average = 4.9 hours per day

Fig 1 General aerial view
Source: Universidad de Jaén

General project background

This project was carried out under the THERMIE programme, part of the fourth RDT framework programme of the European Union, and the PAEE programme of the MINER (Spanish Ministry of Industry and Energy).

The main objective of the project was the integration of a medium-scale PV plant into a university campus using different architectural solutions (such as parking canopies, pergola, façades) for research into, and dissemination of information about PV. The project focused on safety and protection solutions specially designed for crowded public areas. It is designed to produce about 15–20 per cent of the electricity consumed at the university, which is estimated to be around 280 MWh per year.

BiPV design process
Project organisation

The main organisations that collaborated on the Univer Project were:

- University of Jaén (project co-ordinator): the project was developed by Grupo Jaén de Técnica Aplicada, comprising architects and lecturers from the departments of electronics and electrical engineering. One of the main fields of research of the Group is photovoltaic solar energy and the integration of photovoltaic (PV) generators in buildings;
- Instituto de Energía Solar: photovoltaic solar energy R + D Centre at Madrid Polytechnic University. Responsible for quality control and technical advice for the engineering system;
- Newcastle Photovoltaic Applications Centre: R + D Centre of Newcastle University, UK. The centre has wide experience with these systems and was responsible for monitoring and evaluation of the project results;
- Isofotón, S.A., manufacturer of the photovoltaic modules. Supplied the modules and was the technical advisor for the photovoltaic generators;
- Solar Jiennense, installer of renewable energy systems. It was responsible for the electrical installation and the technical assessment of the civil works and the supporting structures;
- Compañía Sevillana de Electricidad (Grupo Endesa) (utility company). Collaborated in the grid connection.

BiPV design
Description of the installation

The installation is divided into four PV sub-generators, with different architectural solutions and configurations (PV generators and inverters). The intention was to analyse the performance of different PV modules (mono- and poly-crystalline), inverters (central inverter or string oriented inverters) and the potential for use on buildings such as pergolas, parking canopies and façades in the south of Spain, especially in Jaén. Figure 2 shows the layout of the PV plant.

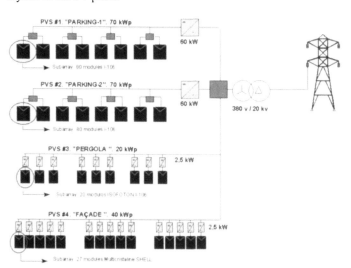

Fig 2 General layout of the PV plant
Source: Universidad de Jaén

Photovoltaic System 1

This PV system is integrated into one of the covered parking areas on the university campus. The system is composed of a 68 kWp generator, a 60 kW triphasic inverter made by Enertrón, and safety and monitoring components. The generator comprises 640 Isofotón I-106 modules (see figure 3). The total power of the system is 67,840 Wp under standard test conditions.

The modules are grouped as eight parallel arrays and every array comprises 80 modules, serial connected.

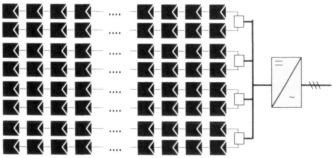

Fig 3 Layout of PV System 1 and System 2
Source: Universidad de Jaén

The existing parking canopies at the university were used for the integration of the photovoltaic generator. The canopies are almost totally free of shadow, with a 38° southeast orientation and tilted 5°.

PV System 1 modules

The values of the most interesting parameters of the generator in System 1 are shown in Table 1.

MODULES		PV SYSTEM 1	
Manufacturer and model		ISOFOTON I – 106	
PV technology		Monocrystalline silicon (square shaped cells)	
Number of serial cells	12	Number of serial modules	80
Number of cells in parallel	6	Number of modules in parallel	8
Total number of cells	72	Total number of modules	640
Area of modules	103.6 cm²	TONC (80 mW/cm², 20 °C):	47 °C
Electrical profiles under standard test conditions: 1,000 W/m², Tc = 25 °C			
Short-circuit current	19.6 A	Short-circuit current	156 A
Open-circuit voltage	7.26 V	Open-circuit voltage	580 V
Maximum power current	18.3 A	Maximum power current	146 A
Maximum power voltage	5.8 V	Maximum power voltage	464 V
Maximum Power	106 Wp	Maximum Power	67,840 Wp

Table 1 Operating parameters of Generator 1

PV System 1 inverter

The inverter is used for connecting the PV generator to the grid. It is an ACEF-SOLAR model from Enertrón, with 60 kW of nominal power. This inverter has been specially designed for photovoltaic solar energy applications.

Photovoltaic System 2

System 2 has the same design, modules and BOS as PV System 1 (figure 3). It is located in a parking canopy parallel to PV System 1, in the same parking area, as shown in figure 4. The original roofing was easily removed and the existing support structure was used to accommodate the PV modules (figure 5).

Fig 4 PV Systems 1 and 2
Source: Universidad de Jaén

Fig 5 Parking canopy structure after removing the conventional roof
Source: Universidad de Jaén

Photovoltaic System 3

This PV generator is integrated into a pergola close to the connection and control building of the project. The inverters, the data acquisition system and the safety and protection system are located in this building. The PV system consists of a photovoltaic generator with 20 kWp of nominal power and 9 string oriented inverters of 2,000 W, made by Fronius (Austria), as represented schematically in Figure 6. As for the two previous systems, the installation includes all the necessary elements to ensure safety protection to the public.

One of the aims of this integrated system was to use the PV to provide a shady area, very useful in this part of Spain, providing a comfortable outdoor space for students (figure 7).

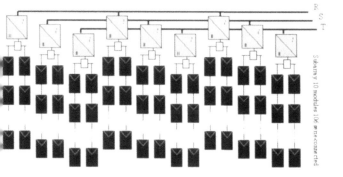

Fig 6 Layout of PV System 3 pergola
Source: Universidad de Jaén

Fig 7 PV System 3 pergola in front of the PV façade
Source: Universidad de Jaén

PV System 3 module

The PV generator consists of 180 semi-transparent (glass-glass) Isofotón I 106 modules, with a total power of 19,080 Wp in standard conditions, with a 52° southeast orientation and tilted 13°. This generator is divided into 9 sub-generators of 20 modules each, which are grouped in two parallel arrays with each array having 10 modules connected in series.

Table 2 shows the values of the most interesting parameters of each sub-generator of System 3

MODULES		PERGOLA SUBGENERATOR	
Manufacturer and model	ISOFOTON I – 106 transparent		
PV technology	Monocrystalline silicon (square shaped cells)		
Number of serial cells	36	Number of serial modules	10
Number of cells in parallel	2	Number of modules in parallel	2
Total number of cells	72	Total number of modules	20
Area of modules	103.6 cm²	TONC (80 mW/cm², 20°C):	47 °C
Electrical profiles under standard test conditions: 1000 W/m², Tc = 25 °C			
Short-circuit current	6.54 A	Short-circuit current	13 A
Open-circuit voltage	21.6 V	Open-circuit voltage	216 V
Maximum power current	6.1 A	Maximum power current	12 A
Maximum power voltage	17.4 V	Maximum power voltage	174 V
Maximum Power	106 Wp	Maximum Power	2,120 Wp

Table 2 Operating parameters of System 3

PV System 3 inverter

The inverter used for connecting the PV generator to the grid is a string inverter from Fronius (Sunrise-maxi model) with 2 kW of nominal power. It is specially designed for grid-connected photovoltaic solar energy applications.

Photovoltaic System 4

This PV generator is integrated into the south façade of a building located close to the connection and control building. It consists of 40 kWp PV poly-crystalline modules and 15 string-oriented inverters of 2,500 W made by Fronius (Austria). There were two main objectives of this system:

- to evaluate the potential of the façade as an integration element in the south of Spain; and
- to achieve a positive visual impact for all visitors coming to the University campus.

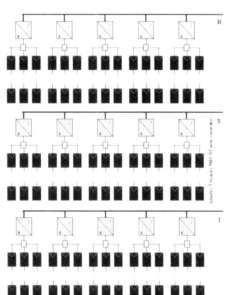

Fig 8 Layout of PV System 4 façade
Source: Universidad de Jaén

PV System 4 module

The PV generator consists of 405 Shell RSM 100s modules, with a total power of 40,905 Wp in standard conditions, with a 52° southeast orientation and tilted 90°. This generator is divided into 15 sub-generators of 27 modules each, which are grouped in 3 parallel arrays, with each array comprising 9 modules connected in series.

Table 3 shows the values of the most interesting parameters of each sub-generator of System 4.

MODULES		FAÇADE SUB-GENERATOR (X15)	
Manufacturer and model	SHELL RSM 100s		
PV technology	Polycrystalline silicon		
Number of serial cells	54	Number of serial modules	9
Number of cells in parallel	1	Number of modules in parallel	3
Total number of cells	54	Total number of modules	27
Area of modules		TONC (80 mW/cm², 20 °C):	47 °C
Measures in standard conditions: 1,000 W/m2, Tc = 25 °C			
Short-circuit current	4.4 A	Short-circuit current	13 A
Open-circuit voltage	31.9 V	Open-circuit voltage	287 V
Maximum power current	4.0 A	Maximum power current	12 A
Maximum power voltage	25.1 V	Maximum power voltage	226 V
Maximum Power	101 Wp	Maximum Power	2,727 Wp

Table 3 Characteristics and operating parameters of System 4

PV System 4 inverter

The inverter used for connecting the PV generator to the grid is the same as used in System 3, a string-oriented inverter from Fronius (Sunrise-maxi model) with 2 kW of nominal power.

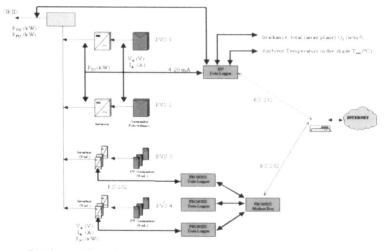

Fig 9 PV System 4 façade
Source: Universidad de Jaén

Fig 10 Layout of the monitoring system
Source: Universidad de Jaén

Description of the monitoring system

The monitoring parameters of the overall system are those recommended by JRC–Ispra:

- global radiation in the PV generator: GI (W/m²)
- ambient temperature: Tam (°C)
- generator voltage: VA (V)
- generator current: IA (A)
- power generated, measured at inverter output: PIO (kW)
- grid-injected power: PTU (kW)
- power obtained from the grid: PFU (kW)

The measurement of these parameters is performed by the corresponding sensors (such as calibrated cells and temperature sensors) connected to different dataloggers. Dataloggers take data every 10 minutes and generate daily files. A diagram of the monitoring system is presented in figure 10.

The data acquisition is based on an HP 34970 Data Acquisition Switch Unit, with a multiplexing card of 20 channels and two global purpose cards. The datalogger is connected to a computer which is also an internet service provider, allowing the remote display of the logged values.

Safety system performance

The safety and protection aspect is one of the most outstanding outcomes of the project, and has also been the most studied, due to the high number of students at this campus. The studies on safety and protection have been developed from two points of view: from the installation itself, and from the public safety viewpoint. In this sense, it is important to note the lack of legal regulation related to such installations in Spain. In general, the risks that can affect an electric installation are due to over-voltages and over-currents, although in this particular case, and because of the characteristics of the photovoltaic systems, the PV was only affected by over-voltages as a consequence of atmospheric discharges. In this case, the installation includes voltage limiters that reduce over-voltages to a value under the insulation level required by the equipment. These limiters are placed at the inverter input and output, at the general DC junction box, and at the junction boxes of the different arrays.

With the aim of maintaining public protection, the installation includes passive and active measures to avoid direct and indirect contact with the active parts of the system. The installation includes:

- a floating system configuration;
- a cover of wiring;
- a permanent insulation controller to detect the earth faults of the generators; and
- an earth grid.

Similarly, positive and negative wiring are completely separated in different connection boxes.

System	Parameter		Unit	N° Sensors	Range	Datalogger	Channel
	Voltage dc	V_{A1}	V	1	0 – 700	HP	3
PV System 1	Current dc	I_{A1}	A	1	0 – 200	HP	4
	Power ac	P_{IO1}	kW	1	0 – 70	HP	5
	Voltage dc	V_{A2}	V	1	0 – 700	HP	6
PV System 2	Current dc	I_{A2}	A	1	0 – 200	HP	7
	Power ac	P_{IO2}	kW	1	0 – 70	HP	8
PV Systems 1, 2	Radiation	G_{I1}	W/m²	1	0 – 2,000	HP	1
	T. cell	T_{c1}	°C	1	–15 – 100	HP	10
	Voltage dc	V_{A3}	V	9	0 – 350	Fronius	
	Current dc	I_{A3}	A	2	0 – 15	HP	19,21
PV System 3	Power ac	P_{IO3}	kW	9	0 – 3	Fronius	
	Radiation	G_{I2}	W/m²	1	0 – 2,000	HP	9
	T. cell	T_{c2}	°C	1	–15 – 100	HP	11
	Voltage dc	V_{A3}	V	9	0 – 350	Fronius	
	Current dc	I_{A3}	A	2	0 – 15	HP	20,22
PV System 4	Power ac	P_{IO3}	kW	9	0 – 3	Fronius	
	Radiation	$G_{I3.4}$	W/m²	2	0 – 2,000	HP	12,13
	T. cell	$T_{c3.4}$	°C	2	–15 – 100	HP	14,15
	Ambient Temp.	T_{am}	°C	1	–15 – 60	HP	2
All systems	Grid injected Power	P_{TU}	kW	1	1 – 3,600	HP	C1
	Power obtained from the grid	P_{FU}	kW	1	1 – 3,600	HP	C2

Table 4 Parameters monitored and their main characteristics

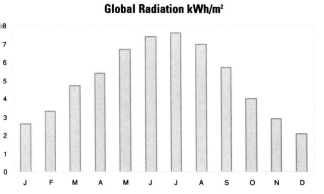

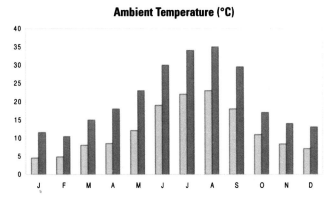

Fig 11 Values of radiation (left) and ambient temperature (right) for the Jaén site
Source: Universidad de Jaén

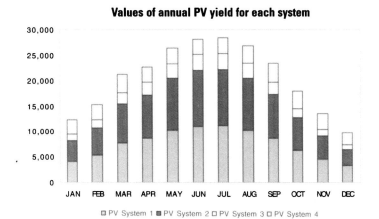

Fig 12 Values of annual PV yield for each system
Source: Universidad de Jaén

Energy and environmental benefits

Typical values of radiation and ambient temperature for the site in Jaén are shown in figure 11. Values of PV yields of each system are depicted in figure 12.

Since the UNIVER project annual PV yield is around 250 MWh, its related environmental benefits include an annual remission of 125 tonnes of carbon dioxide and 350 kg of sulphur oxides.

Lessons learnt from the project

After PV System 1 began in August 1999, some operating problems emerged. First, the inverter introduced a current harmonic (5 kHz) in the grid as a consequence of the poor fitting of the output filter, which had damaged the inverter capacitors. As a result, the protection system of the utility company disconnected the PV plant many times (two or three times per day). This current harmonic also produced interferences in a computer room in a nearby building and the PV system had to be disconnected. The Enertron company had to design a new capacitor filter, which eliminated this current harmonic and the grid interferences. Except for these disconnections, the PV plant operated without any further disruptions.

In the following months, some inverter disconnections were detected during cold and wet mornings, due to the decrease of the insulation resistance (about 5 kΩ) of the PV system. The public safety system of the Univer PV plant is based on the use of a Permanent Insulation Controller (PIC) device that is able to detect the loss of system insulation. When this fault appears, the PV field is short-circuited and connected to the ground and a manual re-connection is necessary. On some cold and wet mornings, the PIC device detected a decrease of the insulation resistance (the usual value of this resistance varies between 40 kΩ and 700 kΩ, depending on the environmental conditions) and disconnected the PV plant. This problem was due to drops of condensation in the connection boxes of the modules. A new control system, which is able to reconnect the PV system automatically, has been developed. This new control has been incorporated in PV Systems 1 and 2. As a consequence of the disconnections, some capacitors and a driver of Insulated Gate Bipolar Transistors (IGBTs) of the inverter were replaced.

It is worth pointing out that the excellent performance of the safety system of the installation was one of the most outstanding aspects of this project.

Project cost breakdown

The project cost breakdown is summarised in table 5.

TOTAL COST (excluding cost of demonstration phase)		
	ESP	Euro
Engineering	9,540,490	57,340
Architecture	326,699	1,964
PV modules	151,177,987	908,598
Inverter	31,319,502	188,234
Electrical material	4,364,970	26,234
Installation, cabling, civil work, support structure	54,638,999	328,387
Project visa	399,493	2,401
TOTAL	251,768,141	1,513,157

Table 5 Project cost breakdown

Operating cost

At present, all systems in the Univer project are under guarantee, so the operating cost is practically negligible. A real estimate of annual operating cost would be around 0.4–0.5 per cent of the installation cost. Currently, maintenance contracts are being prepared with the companies that have collaborated in the project, with the aim of ensuring its continuity and a stable energy production.

Economic viability

As a consequence of the publication of two decrees by the Spanish Government, which regulate assistance to energy produced by photovoltaic installations and guarantee a level of 0.22 Euro/grid-injected kWh, the economic situation has changed radically. The annual production calculated for the whole system is 250 MWh, which would mean an annual income of around 55,000 Euro.

SWITZERLAND: ABZ APARTMENT BUILDINGS

IODATA

PROJECT:	2 x 26.52 kWp roof installations
LOCATION/CITY:	Zurich
COUNTRY:	Switzerland
TYPE OF PV BUILDING:	Roof-integrated
BUILDING TYPE:	Residential
NEW/RETROFIT:	Retrofit

LIMATIC CHARACTERISTICS

LATITUDE:	47° 23'N
LONGITUDE:	8° 33'E
ALTITUDE:	440 metres above sea level
CLIMATIC TYPE:	Continental (Temperature: winter average = 5 °C; summer average = 25 °C)
SUNSHINE HOURS:	Yearly average = 3.6 hours per day

General project background

The apartment flats, owned by the Allgemeine Baugenossenschaft Zurich (ABZ), were built in the suburbs of Zurich in the 1970s.

In 1998, ABZ tendered for the power utility of Zurich's Solarstock exchange programme. Based on the client's wish to install an easily accessible and highly visible PV plant, ABZ decided that the 'Marchwartstrasse' building was an appropriate site. Both promotional opportunities and technical issues were favourable for this site. ABZ had set a target to cover 5 per cent of its annual general energy consumption from PV power.

The critical issue for ABZ was the not the cost of a turnkey installation, but to maximise installed PV power. In view of these aspects, BP Solarex and the Solrif integration system were selected as the suppliers. The goal was also to build a cost-effective and visually integrated PV installation. The power utility of Zurich accepts offers for solar energy of less than US$0.70/kWh for its Solarstock exchange.

As it was a retrofit, no sustainable traits influenced the building. Several environmental issues emerged on the PV side. All cabling was completed using PVC and halogen-free materials. Further, electro-magnetic resonance was an area of concern for the tenants. Due to an undercurrent of scepticism among the tenants, it was decided to implement some modifications on the AC installation (all cable grounded and earthed) and the inverters were placed on a copper metal sheet, which was also grounded.

Fig 1 Aerial view
Source: Enecolo AG

Thought was given to all installation work to minimise the CO_2 impact of the construction process. Workers used the public transportation system and were also instructed to use environmentally friendly construction materials. Reducing noise pollution during the installation phase was also an important aim and every effort was made to liaise with the tenants who often remained in their flats during the day.

BiPV design process
Planning process, alternative designs

Under the framework of the power utility of Zurich's Solarstock exchange program, the PV power plant was integrated into the roofs of two rows of apartments, using the Solrif system. In total, 624 BP Solarex model 585 panels, framed with the Solrif system were installed onto 410 square metres of roof, providing a total capacity of 53.04 kWp installed power. Maximum available roof area was 660 square metres. The roofs are covered with standard clay tile sloped roofs. The client required that not all of the available roofing area was to be covered with PV and that a minimum of 2–4 rows of tiles should remain on all sides. All tiles earmarked for the PV area were simply removed. A flush BiPV design was chosen with a pleasing appearance and was easy to mount (less construction work), in combination with the existing clay tiles. One row with flats consists of three blocks, two blocks are at the same level (roofs are joined) and the first block (facing west) is built 4.5 metres higher (figure 1).

The shading resulting from the house construction and the flanking trees had to be carefully evaluated. The energy loss due to the shading was calculated using Meteonorm software. The installation had to be adjusted in view of cost, effectiveness and shading influence. Some 2 metres above the roofing area was not covered with PV modules due to cost and effectiveness issues. The string wiring especially had to be seriously evaluated, planned and carried out. The design of the PV installation was in process for about four months before construction commenced and was accepted by all parties. The strategy was originated by the PV engineering company in discussion with the building owner and was carried out as planned.

Decision process

Specific client requests, such as the exact integration of the PV array (picture frame, at lower edge, centred) and the lightning protection, required several changes to the planning process. But most planning, like electrical installation AC side, location of the inverters, chosen products and the monitoring, were standard procedures and needed no additional effort. The DC cabling had to be carefully planned and carried out, because the shading influence of nearby trees. Other planning issues included:

- Proposal (permit) for the grid connection to the power utility;
- Building proposal to the city authority;
- Issuing of a tender. Comparison of all offers and presentation to the client;
- Negotiation with suitable installers and distributors; timetable defined.

Based on the offers and the client request to have a maximum installed PV capacity it was decided to give the contract to BP Solarex and Energieprojekte Kottmann as installer. The decision was based on factors such as the maximum installed PV capacity, the cost, the performance and the aesthetics.

The specific procedure was that one month before construction began, the project and the complete installation was announced to the occupants of the two affected house rows. All residents were asked to raise any questions or issues from the start, so that they could be taken into account during the design process.

COMPONENT CHARACTERISTICS

PV system power	53.04 kWp
Type of building integration	Sloped roof
Type of cell technology	Mono-crystalline
Modular dimensions	520 x 1118 mm
Array dimensions	2 x 66.3 m², 2 x 63.5 m² and 2 x 77.4 m² (total of 410 m²)
Weight	Approximately 12 kg per element (including new frame)
Inverter	Swiss products ASP TopClass 2.5 x 4 KVa
BOS components	SOLRIF integration frame system
Monitoring equipment	The fault relays are serial connected by all inverters. An error lamp indicates the failure and the responsible person will be contacted.

Installation design
Integration design and mounting strategy

The two rows of the apartment house roofs face southeast (between –25° and –30° azimuth), (figure 2). The 53.04 kWp installation is divided into two rows covering each house roof (approximately 200 square metres), producing 26.52 kWp of PV power.

The Solrif mounting system (figure 3) is used to integrate the PV modules. This system is suitable for almost any type of inclined roof in existing or new buildings and also meets high aesthetic demands. The unit consists of any type of solar laminate and is framed by four specially designed aluminium profiles. As an option, the profiles are available in various colours to assist with visual integration to the surroundings.

624 BP Solarex model 585 laminates were installed. Each module is rated 85 Wp and they are connected to 18 ASP 2500 and two ASP 4000 inverters. Figure 4 shows the electrical design for one house row (26.52 kWp). All AC cables were grounded to reduce the possibility of electro-magnetic resonance. No external switch is necessary in Switzerland, but a power switch is required to protect against a possible islanding situation. This is the case if the number of inverters is greater than one.

Fig 3 Solrif module
Source: Enecolo AG

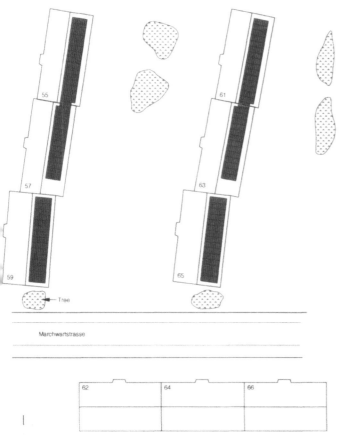
Fig 2 Site plan
Source: Enecolo AG

Planning approval and institutional processes

The standard building approval process was undertaken. No special building code was required for the installation. The proposal was submitted to the local (City of Zurich) building authority three months before construction started. Standard forms were used and all the necessary building drawings and blueprints were provided. The important issues for the building authority included the method of integration of the PV plant into the roof, the aesthetic appearance and the national safety guidelines for construction, (SUVA, Swiss Accident Insurance group and construction guidelines). Since the Solrif system can be easily introduced in existing roofs in combination with any standard clay tile, a very aesthetically pleasing BiPV installation was achieved. Consequently, the Zurich building authority approved the proposal to build the PV power plant. The building inspector suggested that the details of the project concerning the visual appearance and how the construction might affect them should be outlined in advance to all neighbours. With this approach, potential problems were avoided and no objections were submitted during the building approval process. The fire and construction safety issues were checked by the local inspector.

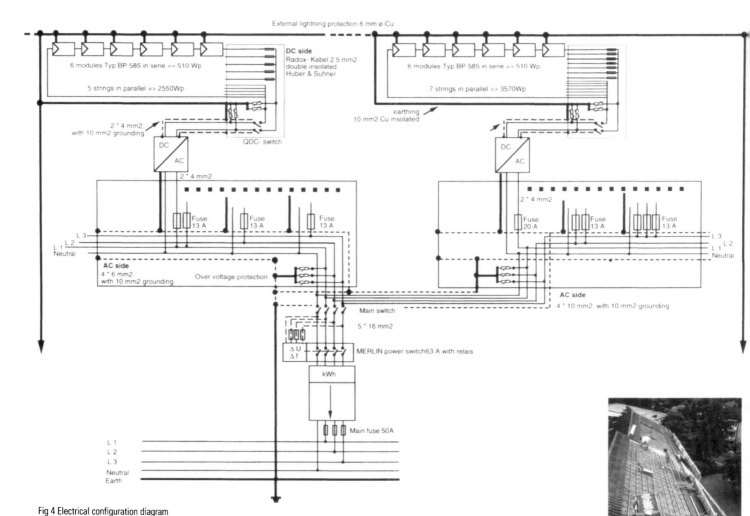

External lightning protection 6 mm ø Cu

6 modules Typ BP 585 in serie => 510 Wp

5 strings in parallel => 2550Wp

DC side
Radox- Kabel 2.5 mm2
double insolated
Huber & Suhner

QDC- switch

2 * 4 mm2,
with 10 mm2 grounding

DC / AC

2 * 4 mm2

6 modules Typ BP 585 in serie => 510 Wp

7 strings in parallel => 3570Wp

earthing
10 mm2 Cu insolated

DC / AC

2 * 4 mm2

Fuse 13 A   Fuse 13 A   Fuse 13 A

Fuse 20 A   Fuse 13 A   Fuse 13 A

L 3
L 1 L 2
Neutral

AC side
4 * 6 mm2
with 10 mm2 grounding

Over voltage protection

L 3
L 1 L 2
Neutral

AC side
4 * 10 mm2, with 10 mm2 grounding

Main switch

5 * 16 mm2

Δ U
Δ f   MERLIN power switch63 A with relais

kWh

Main fuse 50A

L 1
L 2
L 3
Neutral
Earth

Fig 4 Electrical configuration diagram
Source: Enecolo AG

Installation

The construction process went smoothly most of the time and according to schedule. As the installation involved about 30 apartments, the installers had to be sensitive to the daily needs of the residents and respect their wishes that no construction work be carried out on weekends. Construction work began with the removal of the tiles; the tiles were directly transported to the ground by a crane which also lifted the modules up to the roof, box by box (figure 5).

Next, new wood battens were mounted and lead sheeting was placed on the lower roof edge (figure 6).

After fixing new wood battens, stainless steel metal brackets were screwed onto the wooden battens. The Solrif modules were then inserted into the brackets (figure 7) and were placed on the refitted battens (figure 8).

Because of the universal profiles (no edges jut out) the Solrif modules can be placed very easily onto the roof structure. The metal brackets and the overlapping area of the Solrif modules secure the mechanical connections and fulfil the wind criteria against uplift. The site work complies with standard clay tile roofing procedures.

Figs 5–8 Installation
of the modules
Source: Enecolo AG

Fig 5

Fig 6

Fig 7

Fig 8

Successful approaches

Thanks to the easy and quick Solrif mounting system, the construction process on the roof was completed by four non-specialised PV workers in two weeks. This included the removal of the tiles, new battens, DC cabling, module mounting and flashing.

Problems during realisation

A few construction problems were encountered. The houses were built in the late 1970s, using a construction technique that is now obsolete. A concrete wall was built between the upper and lower roof by the two adjoining roofs 55 and 57, 61 and 63, (figure 2) as a fire safety structure. Because of the junction box, which did not fit well into the roof, most parts of the fire concrete wall had to be demolished. Additional work resulted, which was not foreseen and planned. Also, most parts of the roof were not levelled out. In some areas a double batten was used (thickness 48 mm) instead of a thickness of 24 mm. This helped to ensure that the Solrif modules were level. As a result, all construction details and work could be carried out according to the time schedule, so no delays or cost overruns resulted. The operation started on the date when the contract began with the Zurich power utility.

Performance characteristics

The estimated power production was calculated using the Meteonorm software. The tilt angle of the roof, its direction to south, the shading and many other data were taken into account. Based on 1997 values, the total power production was calculated to be approximately 43,270 kWh of green electricity per year.

Compared with the actual weather data, the Meteonorm base was multiplied and resulted in the predicted figures for the year 2000. Shown in figure 9 are the monthly predicted values (in total 44,143 kWh) versus actual performance data (in total 46,182 kWh). The performance was approximately 5 per cent better than predicted with savings of about 22 tonnes of equivalent greenhouse emissions (CO_2) in the year 2000. This was the result of no power failures during the year, less snow on the modules than expected and a perfect inverter performance with an annual efficiency of around 92 per cent. The performance ratio could not be calculated, because no pyranometer was installed on the roof. The owner dismissed an extensive monitoring program and demanded a monthly comparison of the predicted values versus the actual performance.

As the installation is a retrofit, the contribution to the building needs was not considered. The total solar energy production is fed to the grid and is being purchased for US$0.58/kWh (including subsidy) by the power utility of Zurich for 20 years. This rate was based on a cost analysis study and was negotiated between the client and the power utility.

Actual performance data Year 2000	Meter reading	Predicted 2000 (kWh)	Difference (%)	Actual data 2000 (kWh/m²)	Predicted 1997 (kWh)	Meteonorm Basis 97 (kWh/m²)
January	996	1,191	−16.4	30	1,032	26
February	1,974	1,786	10.5	45	1,747	44
March	4,212	3,335	26.3	84	3,295	83
April	5,500	4,764	15.5	120	4,525	114
May	6,386	6,153	3.8	155	5,875	148
June	6,804	7,265	−6.3	183	6,153	155
July	5,318	5,597	−5.0	141	6,788	171
August	6,622	5,796	14.3	146	5,716	144
September	4,560	4,367	4.4	110	3,970	100
October	1,970	1,826	7.9	46	2,302	58
November	951	1,112	−14.4	28	1,072	27
December	889	953	−6.7	24	794	20
Annual energy production	46,182	44,143	4.6	1,112	43,270	1,090
Annual energy yield (kWh/kWp)	871	832			816	

Fig 9 Predicted versus actual performance data

Project cost breakdown

	US$	
Solar modules	229,000	
Inverters	47,880	
Solrif frame	27,300	
Roof work	17,400	
Electro installation	42,800	
Engineering	28,350	
Miscellaneous	14,800	
Total installation cost ($)	**407,530**	**(7.68/Wp)**
Ongoing maintenance	1,000	
Savings for inverter replacement	2,700	
Monitoring	200	
Total annual cost ($)	**3,900**	

Of the project cost of US$407,530 (including value added tax), 75 per cent was paid by the owner and the remaining 25 per cent was subsidised by the Swiss federal office for energy.

The PV plant is operated within the EWZ Solarstock exchange and receives a feedback tariff of US$0.58/kWh (including subsidy). This tariff is contracted and ensured over 20 years operating time and will be adjusted to the national inflation rate. At an interest rate of 3.5 per cent and an annual energy production of 43,000 kWh, amortisation within 20 years is guaranteed.

Post-installation feedback

In the planning stage, many concerns were raised by the residents in relation to electro-magnetic resonance. Several open and seriously debated discussions took place between the residents and the building society. It was clearly shown that a PV installation is not comparable to a mobile phone transmitting station in relation to electro-magnetic pollution. Due to some remaining fears, it was decided to modify the AC installation (all cable grounded and earthed) and the inverters were placed on a copper metal sheet, which is also grounded. It is important that the project team carefully check and present the project and the design to the building owner and the residents so that questions and concerns can be addressed at an early stage.

Further problems were more from the technical side. Because the houses were built in the late 1970s, using a now-obsolete construction technique, additional unforseen work was required. Retrofit on a roof structure older than 30 years requires technical flexibility and a not-too-tight timetable.

Project team

No changes took place during the project. BP Solarex and the national distributor provided the necessary warranties for the functioning of the technology. A delivery contract was signed by all involved parties, fixing the price, specification of the products, time schedule, warranty conditions and the expected energy production.

The installation and construction went according to the time schedule and was controlled by the specialised PV engineering company Enecolo AG.

Company	Description	Tasks and responsibilities
ABZ	Building society	Client, investor and PV owner
Enecolo AG	PV engineering company	PV engineering, project design and controlling; project leader
Energieprojekte Kottmann	Installer	Installer, contractor and distributor of inverters
Ernst Schweizer AG	Manufacturer, supplier	Manufacturer of new integration system, supplier
BP Solarex	Manufacturer	Manufacturer of PV modules
Holinger Solar AG	National distributor	Distributor of PV modules

Fig 10 Project organisation
Source: Enecolo AG

SWITZERLAND: **STUDENT HOUSING**

IODATA

PROJECT:	2 PV façades of 7.2 kWp each
LOCATION/CITY:	Lausanne
COUNTRY:	Switzerland
TYPE OF PV BUILDING:	Façade-integrated
BUILDING TYPE:	Student accommodation
NEW/RETROFIT:	Retrofit

CLIMATIC CHARACTERISTICS

LATITUDE:	46° 30′N
LONGITUDE:	6° 36′E
ALTITUDE:	380 metres above sea level
CLIMATIC TYPE:	Continental (Temperature: winter average = 5 °C; summer average = 17 °C)
SUNSHINE HOURS:	Yearly average = 3.8 hours per day

General project background

The complex is composed of two parallel buildings of three floors each, connected to each other by a covered passageway. It was built in 1962 as the administrative centre for the 1964 Swiss National Exhibition. The buildings were then used for student accommodation and are now the property of the University's student housing foundation and the Swiss Federal Institute of Technology in Lausanne.

Although the buildings had been partially renovated on several occasions, a complete rehabilitation was required. The envelope was experiencing significant thermal losses and the services, subjected to intensive use, required easier maintenance and an upgrade to meet current safety requirements. In particular, the concrete façades had to be covered with insulation and metal cladding.

The retrofitting costs amounted to nearly CHF 7 million. The work was carried out over two, three-month periods during the summer holidays, in July, August and September of 1999 and 2000.

Fig 1 PV façade of the first building
Source: Direction des services industriels/Service de l'énergie (SILSE)

BiPV design process
Planning process, alternative designs

Due to its involvement in the European 'Heliotram' programme, the Energy Office of Lausanne was looking for large surfaces appropriate to host a photovoltaic installation to supply the trolley bus network. The Energy Office proposed the installation of a PV plant integrated into the south façades.

At that time, the Energy Office of Lausanne already had significant experience with roof-integrated PV. It had noticed that, despite initial promotion, many installations were quickly forgotten because they were not visible from the street. In addition, there was no cost reduction with previous installations, as PV modules were added to the building and did not replace structural components. The initial idea in this case was to mount sun shades in front of the south façades. However, this solution was abandoned in favour of an installation with vertical panels directly integrated into the façades. The solution was more visible, and showed an example of PV used as a structural component.

Decision process

The student housing foundation accepted the proposal of the Energy Office, attracted by the project's positive image, and no additional costs. Indeed, the Energy Office proposed to pay for the extra cost of the PV plant, provided that it received any revenue from the electricity production. It was agreed that the façade would be put to its disposal for 30 years and that the SILSE would be responsible for the energy efficiency of the HVAC plant.

Project organisation

The organisation of the work was not much affected by the PV installation project: the PV panels were mounted by the façade fitter who retrofitted all façades; a PV system provider delivered Solarex panels and they were connected by a PV installer to the main electrical boxes. From there, a utility company handled the connection with the trolley bus network.

To control the efficiency of the PV plant, the Energy Office installed a control system for online collection of operating data. This information is accessible remotely via modem allowing real-time control to achieve optimum performance of the PV plant. Once both façades were finished, trees that overshadowed the PV were cut down and new trees were planted between the two buildings.

BiPV design

The Energy Office proposed to use the 'Power-Wall-Laminate' PV façade system developed by Solarex. This concept has a pleasing appearance, with large modules of 240 Wp each. The blue Tedlar cells provide a uniform appearance of the PV surface and facilitate façade integration. The size of these standard laminates also fitted the building dimensions and could be easily integrated into the façade architecture. The fastening system is simple and was mounted with only minor modifications to the façade framework. The built-in ventilation spaces (ventilated façade principle) were kept for cooling the PV plant. The blue thermo-lacquered panels in the middle of each building were installed in order to improve the appearance and to give the impression of a taller building.

The European 'Heliotram' project was launched to show that PV could be connected to a DC network with better efficiency than with inverters and was a proactive solution for cities with electric public transportation. The idea was to avoid installing inverters, which generated losses and most of the failures. The electrical connections were calculated to deliver the specific voltage used for this network (680 V). Six strings of 10 PV modules were connected providing 2.4 kWp each. With this configuration, both PV façades had two extra modules that were connected to the grid via a small inverter.

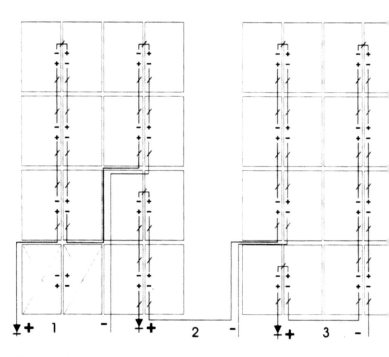

Fig 2 Electrical connection of one façade
Source: Direction des services industriels/Service de l'énergie (SILSE)

COMPONENT CHARACTERISTICS

PV system power	14.4 kWp
Type of building integration	Façade
Type of cell technology	Poly-crystalline silicon (Solarex)
Modular dimensions	187 x 111 cm
Weight	26 kg
Visual details	Blue Tedlar
Inverter	No inverter
Monitoring equipment	VNR Electronic SA

Fig 3 Module handling
Source: Direction des services industriels/Service de l'énergie (SILSE)

Fig 4 Lifting plant
Source: Direction des services industriels/Service de l'énergie (SILSE)

Fig 5 Module mounting on façade framework
Source: Direction des services industriels/Service de l'énergie (SILSE)

Performance characteristics

The Energy Office of Lausanne is monitoring the operation of the PV plant by means of a control device with remote data acquisition. The analysis of the first results highlighted that the production efficiency decreased under strong insolation. This phenomenon was studied in more detail and it was noticed that the PV panel temperature could reach high values during the day (figure 6).

An on-site visit showed that the openings planned for air-cooling had been closed by the PV installer. It is suspected that an over-conscientious worker had not followed the mounting instructions to achieve a completely watertight façade. This was immediately corrected by the façade manufacturer, and three months after the PV plant was commissioned, the problem had been identified and solved. Overall, the data analysis showed that the production efficiency was not much better without a DC/AC inverter. There were no electricity transformation losses but the operating mode is also not optimised by a maximum power point tracker.

Project cost breakdown

The total cost involved in the installation of the PV plants amounted to CHF160,000 which corresponds to CHF11,000/kWp. As the installation was part of the Heliotram program, the European Community took part in the funding, providing CHF4,000/kWp. The Swiss Federal Office of Energy's E2000 programme subsidised the photovoltaic installations at a rate of CHF3,000/kWp and the town of Lausanne contributed CHF3,500/kWp. The remaining CHF500/kWp was paid by the Energy Office of Lausanne. PV electricity production is distributed to the trolley network, of which the City of Lausanne is a main shareholder.

Production efficiency related to the PV panel temperature

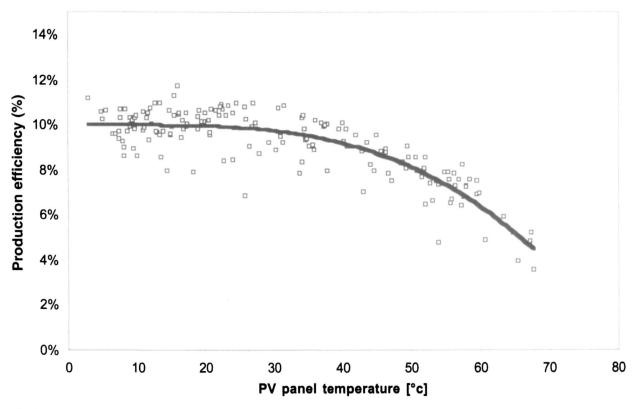

Fig 6 Production efficiency related to the PV panel temperature

Post-installation feedback

In this project, the City of Lausanne benefits from the electricity production for a very small investment. The project contributed to improving the acceptance of PV integration and allowed the connection with a DC network to be tested. The architectural aspect is very successful; the large PV panels are perfectly integrated in the façades. In addition, they replace building components such as the overcladding elements.

Two weak points were highlighted through the operation control measurements performed by the Energy Office:

- The electricity production efficiency decreased under strong insolation. A scrupulous worker had closed the openings planned for air-cooling;
- The performance is not improved with direct connection to the DC network. The lack of an inverter makes it possible to avoid transformation losses, but the operating mode is no longer optimised by a MPP tracker, which leads to a smaller DC efficiency.

Fig 7 PV façade detail
Source: Direction des services industriels/Service de l'énergie (SILSE)

UK: JUBILEE CAMPUS NOTTINGHAM UNIVERSITY

IODATA

PROJECT:	Jubilee Campus Nottingham University
LOCATION/CITY:	Nottingham
COUNTRY:	United Kingdom
TYPE OF PV BUILDING:	Roof-integrated
BUILDING TYPE:	Educational
NEW/RETROFIT:	New

CLIMATIC CHARACTERISTICS

LATITUDE:	52° 58'N
LONGITUDE:	1° 8'W
ALTITUDE:	30 metres above sea level
CLIMATIC TYPE:	Temperate (Temperature: winter average = 7 °C; summer average = 15 °C)
SUNSHINE HOURS:	Yearly average = 3.4 hours per day

Fig 1 View of faculty buildings from the south
Source: Ove Arup and Partners

Fig 2 Roof showing solar shading, cowls and PV arrays
Source: Ove Arup and Partners

General project background
The project

It is 50 years since Nottingham University gained its Royal Charter. During this time it has cultivated a reputation for promoting environmentally conscious design. True to the University's principles, the aim of this new campus is to be a model of sustainable development for the region. Its purpose is to reduce carbon dioxide emissions by 70 per cent, raise awareness of environmental issues among students and within the field of tertiary education, demonstrate the viability of sustainable industrial regeneration and achieve these objectives within existing funding structures and levels.

The design team won the commission for the £25.5 million project in a competition held in 1996. Funding assistance of £0.75 million for the low-energy building specification and solar-powered ventilation strategy was gained through the European Union's THERMIE programme as a demonstration project in 1998.

The site

The Jubilee Campus is constructed on a 7.5 hectare brownfield site (formerly the location of the Raleigh Bicycle factory) within a mile of the existing Beeston Campus. Flanked by an urban dual carriageway, low-rise social housing and warehousing, the development has introduced a landscaped oasis; a composition of faculty buildings and student halls of residence bordering a new lake.

The campus is planned to meet the needs of 2,500 students with 41,000 square metres of buildings including undergraduate and postgraduate accommodation, three faculty buildings, a central teaching building and a learning resource centre. The 13,000-square-metre lake forms a buffer between the suburban edge and the new buildings, maximising public amenity space and reintroducing wildlife. The site strategy focuses on optimising orientation and views of the landscape and giving pedestrians priority over cars. Buildings are sited to exploit the prevailing southwesterly winds and to optimise passive solar gains. The strip of woodland to the southwest has been reinforced to provide shade, shelter and cooling.

The building and its construction

The main faculty buildings each consist of three-storey wings. These are connected by full-height atria or open landscaped courts that act as social and environmental buffers. The structure has been kept simple, with floors of in-situ concrete on a 6 x 6-metre grid. Exposed soffits and columns provide thermal mass, as do green roofs, which are planted with tundra species. Where possible, materials from sustainable sources are chosen and those with high embodied energy are avoided. The faculty buildings are clad in western red cedar from a Canadian source, chosen for its progress towards achieving World Wildlife Fund/Forest Stewardship Council (WWF/FSC) certification. The client, initially sceptical about the use of timber for a university building, was persuaded by its use on other prestige projects, along with its excellent thermal performance.

The prefabricated timber cladding panels wrap around the buildings both externally and inside the atria. They form a 'breathing wall', with a middle layer of Warmcell insulation, which absorbs moisture and helps give the wall a U-value of 0.287. Panels lining the atria incorporate a layer of hessian-covered acoustic quilt behind the timber to absorb sound.

Ventilation

The ventilation strategy draws on previous research by Hopkins and Arup. While natural ventilation was the initial objective, it seemed unlikely that fresh air would be able to fully penetrate the intensively used faculty buildings without assistance. Instead they settled on a mixed-mode, low pressure mechanical system with heat recovery, which gave better energy performance than solely natural ventilation. Although it is anticipated that the buildings may be mechanically ventilated for most of the year, a control mode has been included for genuine natural ventilation, using windows alone to introduce fresh air and relying on the stack effect. The balance will be determined after the first year's monitoring of the building.

An efficient passive ventilation system requires careful consideration of the hidden details. The design hinges on efficient mechanical ventilation, which circulates 100 per cent fresh air all year. Corridors and stair towers are used as air plenums, reducing the energy needed to circulate air. The air flow passes through ducts which need to be as large as possible to reduce air pressure and so minimise the fan power required to move air through the buildings. This was achieved by designing 350 mm-deep floor voids and by spacing underfloor batts as far apart as possible. These ducts had to be unobstructed, so care was needed to avoid air leakage.

Lighting

Even on dull days, very little artificial lighting is needed in either the atria or the teaching spaces. Low-energy lighting is controlled by a stand-alone intelligent lighting system linked to passive infra-red movement detectors in each room, activated by people

Fig 3 Ventilation schematic
Source: Ove Arup and Partners

Fig 4 One of the atria showing the PV array incorporated into the glazed roof
Source: Ove Arup and Partners

presence and daylight. Peak electrical power input is 8 watts per square metre. The larger teaching spaces under the barrel-vaulted roofs are top-lit by Monodraught light tubes, which can be closed if a 'blackout' is required. The potential problem of glare in the teaching rooms was avoided by positioning horizontal timber louvres over the upper parts of the windows at risk. The louvres are painted white on their upper surface to act as light shelves. Automatic internal blinds shade the double-glazed atria spaces, and the southwest façades carry retractable blinds to reduce glare and solar gain.

BiPV design process

An early decision was taken to use photovoltaic power generation to drive the ventilation of the building when there is insufficient pressure differential to operate the purpose-designed rotating wind cowls and drive the air through natural convection.

Integrated building services

The structural elements of the building are used as thermal mass to provide comfortable conditions within the spaces as well as air paths for the ventilation system, to avoid conventional ductwork installations. False ceilings are avoided to reduce cost and material use. Using this approach the energy requirements for the fan power could be reduced from 3–4 watts per litre per second (W/l/s) to only 0.5 W/l/s, which is well below the Scandinavian standards and therefore provides the client with reduced operating cost. These targets were achieved by limiting the total pressure drop of the supply and extract system at normal operation to only 280 pascal.

Super-efficient ventilation system.

The ventilation system of the buildings is designed to minimise pressure losses by utilising ultra-low pressure elements within the air handling equipment, using the building structure as air paths and wind pressure to naturally ventilate the buildings. Large self-powered wind cowls exhaust air from the seminar spaces. Fresh air is taken into the air-handling units at a high level and heat is exchanged through large thermal wheels, providing only 50 pascal pressure drop and 83 per cent efficiency. Highly efficient indirect gas heaters warm up the air only as necessary to provide heating in winter; evaporative cooling provides cooling to the fresh air in summer and ensures comfortable internal conditions.

Zero-CO$_2$ ventilation system

To reduce the fan power, additional air dampers are included within the design of the air handling plants to bypass components if not required. Electrostatic filters are used to minimise pressure drop across the panels. Computer simulations indicated the energy requirements for the ventilation system to be approximately 50 MWh per year. The total energy demand of the ventilation plant during the year was then matched with the photovoltaic cells output mounted on the atria roofs, therefore providing a zero-CO$_2$ ventilation system.

One of the challenging aspects of the design and construction was the requirement to achieve an outstanding, innovative project in a very short time period and within a tight budget. To reduce the pressure on the budget and to enable the client to include innovative, 'green' equipment such as the wind cowls and the photovoltaic installation, an application was made to the European Commission. The application was successful and £750,000 was made available for the development of the zero-CO$_2$ ventilation system including the PV system.

BiPV design

The PV cells are an integral part of the atria roofs within the school of management, the department for computer science and the faculty of education. They provide shading to the spaces below and replace the glazing system with laminated glass panels with integrated square PV cells.

COMPONENT CHARACTERISTICS

PV system power	SMF atria 1: 12 kW
	SMF atria 2: 12 kW
	DCS: 18.5 kW
	FOE: 18.5 kW
Type of building integration	Atria roof-integrated
Type of cell technology	Mono-crystalline square cells
Modular dimensions	SMF atria 1: 56 modules
	SMF atria 2: 56 modules
	DCS: 72 modules
	FOE: 72 modules
	Number of cells per module: 88
Array dimensions	450 m^2 total
Visual details	Black, mounted between 6 mm heat-toughened glass (low iron)

Performance characteristics

The total energy output of the PV installation is 51 MWh per year with a peak output capacity of 53.3 kWp.

Project cost breakdown

The total cost of the main buildings was £25.5 million, an average of £900 per square metre (£1,225 for the learning resource centre and £880 for the faculty buildings). The design team has effectively shown that 'green design' need not mean high cost.

An EU THERMIE grant of £750,000 allowed for additional equipment to achieve the energy efficiency targets and also to fund monitoring in the first year of occupancy.

Post-installation feedback

On completion of the services installation, commissioning of the systems proved to be less time consuming because the systems were designed for low-pressure drop and simplicity. All control points were monitored by the building management system and linked back to the client's main campus offices.

A condition of the THERMIE grant was that the installed equipment be monitored and the achieved figures be reported back to the European Commission. These are currently monitored with more than 40 data-points on internal temperatures, water, electricity and gas consumption collected by the Faculty of the Built Environment. The information collected by the University is very useful and provides a good insight in the operation of the building. Monitoring is usually seen as an expensive luxury and therefore not applied on other projects, whereas here it was included as part of the THERMIE grant and therefore helps the client to understand and operate the building in the most efficient way.

Fig 5 Wind cowl tested at BRE laboratory to establish pressure coefficient
Source: Ove Arup and Partners

Fig 6 Atrium detail
Source: Ove Arup and Partners

UK: SOLAR OFFICE, DOXFORD INTERNATIONAL BUSINESS PARK

BIODATA

PROJECT:	Solar Office, Doxford International Business Park
LOCATION/CITY:	Sunderland
COUNTRY:	United Kingdom
TYPE OF PV BUILDING:	Integrated inclined façade
BUILDING TYPE:	Commercial
NEW/RETROFIT:	New

CLIMATIC CHARACTERISTICS

LATITUDE:	55° N
LONGITUDE:	1° 4′E
ALTITUDE:	30 metres above sea level
CLIMATIC TYPE:	Temperate (Temperature: winter average = 7 °C; summer average = 15 °C)
SUNSHINE HOURS:	Yearly average = 3.4 hours per day

General project background

The Solar Office is a new office building designed for Akeler Developments Ltd on the 32-hectare Doxford International Business Park, located near Sunderland in the northeast of England. It is occupied by a leading building society, Northern Rock. The brief for the building and its procurement follow the robust fast-track pattern that is now commonplace in speculative office development.

The design addresses all the environmental and energy conservation issues currently being addressed in buildings. The energy consumption target for the building when occupied by a tenant with conventional power requirements is 85 kWh/m²/y compared with a conventional air-conditioned office of more than 400 kWh/m²/y. Electricity generation is provided by a photovoltaic solar array, integrated into the building envelope. The 73 kWp array provides 55,100 kWh of electrical power per annum, which represents between one third and one quarter of the electricity expected to be used by the building over one year. In summer, when it generates more than is required, the surplus is exported to the national grid.

The Solar Office is the first speculatively constructed office building to incorporate BiPV and the resulting solar façade is the largest constructed in Europe to date. It is one of the few BiPV projects to adopt a holistic energy strategy.

Fig 1 The south façade of the Solar Office
Source: Dennis Gilbert Photographer

Project brief

Project background

The project came about through the developer asking the architect to look at the possibility of more radical low-energy schemes for the business park. This led the architect to suggest the possibility of introducing integrated renewable energy; in particular photovoltaics, a technology in which he had developed an interest and for which the practice had already carried out research and development work. The architect was a UK representative to Task 7 of the IEA PVPS Programme. The project developed concurrently with the Programme, enabling a two-way exchange between the experts (and their sub-tasks) and the design of the project.

Project organisation

Once outside funding had been found to cover the BiPV and associated research and development work, and the developer had accepted the proposal, he asked the architect to put a design team together. The core design team was made up of Studio E Architects, the building services engineers, the structural engineers and the master planners and landscape architects. None of these participants had previously been involved in a BiPV project. The developer meanwhile put in place the building contractor, project manager, agents for the trust fund and civil engineers. The research, development and testing element of the project involved various specialists including acoustic, wind effects, air tightness, commissioning and testing, daylight and dazzle, and computer graphics specialists. The PV installation package included the curtain wall installer, the curtain wall supplier, the cell manufacturer, the module fabricator, the inverter manufacturers, the specialist electrical and data-logging contractor and the display systems designers and suppliers. A university department was responsible for the monitoring.

The brief

The brief required the design of a speculative office building to meet the requirements of the commercial market. It also required the building to be designed to 'best practice' low-energy, environmentally sound principles. The client supported initiatives taken by the architect to include a worthwhile PV installation, but would only agree to include it in the brief if it was wholly funded from outside sources and it did not extend the design and construction programme.

Accordingly, the 4,600-square-metre, three-storey building was constructed to a 'shell and core' specification, and was fitted out to suit the specific requirements of the occupying tenant. The tenant was encouraged to operate the building in its low-energy 'passive solar' mode, but chose to augment this strategy by utilising the provision made for comfort cooling due to high project incidental cooling loads. It can, if necessary, be divided into up to six separate tenancies.

The whole building was designed and constructed over 15 months on a design-and-build basis. This means that the contractor was required to construct it within a fixed cost, to a fixed delivery date with the consultants novated to the contractor on completion of an approved outline design.

The overriding objective in terms of the environmental design was to find a synthesis between the low-energy measures and those needed for an effective photovoltaic installation.

The construction cost for the entire building was £4,225,000.

BiPV design process

Building layout

The building is V-shaped in plan with the extreme ends of the V splayed away from each other. A central core is located at the apex of the V. The building incorporates a 66-metre-long, south-facing inclined façade, at the centre of which is the main entrance. Behind the façade is a three-storey atrium and, between the façade and the splayed wings, is an internal passageway.

Setting back each wing in plan by 5° off south has very little effect on PV efficiency but does give the long elevation a faceted dynamic. The façade is around 950 square metres in area.

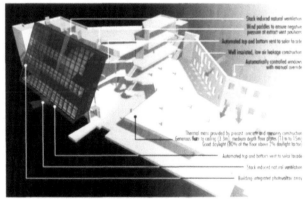

Fig 2 Cutaway perspective illustrating energy features

The building and its site

Key site issues related to layout, orientation and climate. It was found that the façade could be aligned to face due south and sloped at 60° to the ground without compromising internal planning. This configuration provides good solar radiation at this northerly latitude. The inclined and sealed façade overcame the potential problems of dazzle and of noise from passing traffic on the adjacent trunk road; office windows could be placed facing north, northeast and northwest, obviating the need for elaborate solar protection; placing the car park in front of the building ensured that the solar façade would not be overshadowed.

The site, being elevated and close to the sea, is very exposed and therefore subject to strong winds. Care was needed in the detailed design of the openings to ensure the building exploited the beneficial and not the detrimental effects of the wind.

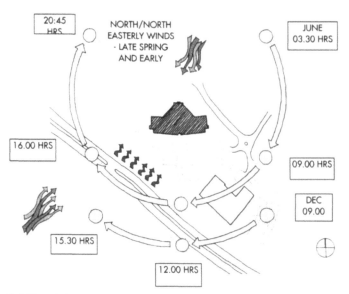

Fig 3 Site analysis

Potential conflicts

In some instances the optimisation of the photovoltaic power generation runs counter to low-energy design. Where possible, these apparent conflicts were reconciled to be mutually reinforcing, and where not possible, a balance was struck between respective requirements. Accordingly:

- Dangerous heat gains from the façade (only a portion of the energy in incoming light is converted to electricity, the rest is transmitted as heat) can be used in winter to assist in heating the building and, in summer, to ventilate the office space. Lack of thermal mass in the façade is countered to some extent by specifying a concrete roof slab in place of the normal trussed and pitched roofs used elsewhere in the park. The insulating properties of the solar façade are good in the context of glazing (U value: 1.2 W/m²/°C for the PV modules), but relatively poor compared to solid wall construction (U value: 0.4 W/m²/°C). Heat loss, however, was minimised by ensuring that leakage of air through the building envelope as a whole is exceptionally low.

- The façade incorporates over 400,000 opaque photovoltaic cells. The concentration of cell coverage was necessary to achieve the power output target. Bands of clear glazing have been introduced into the façade to allow views out and ensure good internal light levels. The balance between maximisation of PV power and maximisation of daylight (a requirement of at least 2 DF over 80 per cent of the office floors) was arrived at by modelling glazing permutations using a 1:40 scale model under an artificial sky. The risk of glare is minimised by the introduction of semi-transparent modules (modules that have transparent

front and rear surfaces and a lower cell count and are therefore able to let more daylight through) immediately above the clear glazed panels, and by provision for the introduction of locally controlled roller blinds capable of covering both the clear and semi-transparent modules.

Design for photovoltaics and for low-energy use, therefore advanced hand-in-hand, one augmenting the other.

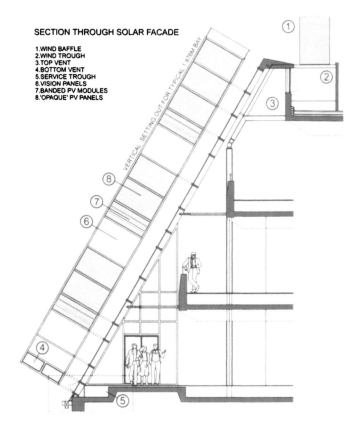

Fig 4 Detailed section through solar façade and single bay in elevation

Natural ventilation and cooling

The office depth was limited to 15 metres to allow for cross-ventilation. Openable windows with automated vents are provided in the north-facing façades. Consideration was given to introducing openable windows in the PV façade, but the difficulties of achieving weather-tightness on a 60° inclined façade and the cost and complexity of providing mechanised window-opening ruled it out.

The two options for natural driving forces are the wind and stack effect. (Stack effect is the rising of currents of air that are warmer, thus less dense and more buoyant than surrounding air). Both stack effect and wind-driven air movement are available here. Wind effects are typically several times more powerful than stack effects, especially for a relatively windy site such as this, with a mean wind speed of 5–6 metres per second.

The stack effect is promoted by the PV façade itself. As the temperature rises at the back of the façade, due to solar gain, a current of warm air rises to roof level, helping to draw air out of the adjacent office spaces. Mechanical vents have been installed at the bottom and top of the façade to help encourage this airflow and to keep the PV arrays cool.

Wind passing over a roof can create negative pressures (suction), helping to draw air across the floors and up out of the building. There is also a danger that wind can blow in through the vents and reverse this airflow. To ensure constant suction in windy conditions, the rooftop air outlets are located in a sheltered trough (to counter southwest and northeast winds) surmounted by transverse baffles to cope with winds blowing along the trough.

Fig 5 The PV array viewed from the main entrance
Source: Dennis Gilbert Photographer

Installation design

The basic strategy for the PV installation was determined in the early stages of the building design. The size of the PV array was derived from a balance of maximising power output, establishing a size that could comfortably be integrated into a three-storey, 4,600-square-metre office building, and determining the upper limit of any European Regional Development Fund grant available for the installation. Once the limit of the funding had been established and the prices for the installation had been returned, the extent of the array was adjusted to an area of 650 square metres with a nominal rating of 73.1 kWp.

The façades of the passive solar version of the building would have largely been constructed in masonry with about one-third of the façades comprising window openings. The windows, where

exposed to the sun, would have incorporated fixed external solar shading, motorised top lights, manually operated mid lights, provision for internal blinds and possibly a combined internal light shelf and glare screen.

The introduction of the solar façade required an entirely different strategy. It had to incorporate 600 square metres of 100 x 100 mm opaque solar cells. As described in the previous section, a balance had to be found between maximising solar irradiance, shielding the interior from unwanted solar gains, providing good internal daylight levels, providing views out, minimising glare, providing reasonable thermal insulation, and concealing the PV associated wiring and junction boxes – all within a tight budget. The outcome was a proprietary curtain wall/roof structure incorporating:

- horizontal bands of clear glazing;
- semi-transparent PV modules where the cells that make up the module are themselves banded and graded to allow diminishing intensities of daylight to enter the interior; and
- opaque PV modules where 80–90 per cent of the daylight is excluded by the tight packing of the cells.

Mono-crystalline or poly-crystalline modules were considered to be more suitable than amorphous silicon in this application due to their greater efficiency and their proven durability. The appearance of the cells was also a factor. The sparkle offered by the poly-crystalline cells determined their choice over the mono-crystalline, particularly taking into account the large area of the array.

Glass-on-glass modules were preferred to those with an opaque backing since the possibility of manipulating daylight when passing through them had the potential to enliven internal spaces and reduce glare and contrast.

It was important to the developer and designers that all cabling, junction boxes and switchgear were concealed from view. Accordingly, those tendering for the installation were asked to provide a curtain wall system that was capable of satisfying this requirement.

Nine different module designs, in terms of size, shape and cell density, were used (table 1). These were designed to meet the needs of physical integration and shading levels for different positions on the façade. All the modules are rectangular except G and H, which are trapezium-shaped to fit around the entrance area.

Module reference	Size (m²)	Rating (Wp)	Quantity in system
A	2.02	238	224
B	1.42	143	96
C	1.58	182	14
D	1.11	109	4
E	2.62	285	6
F	2.06	218	2
F	2.06	176	2
G	0.93	70	2
H	2.33	221	2

Table 1 Module parameters and quantities in the Solar Office PV array

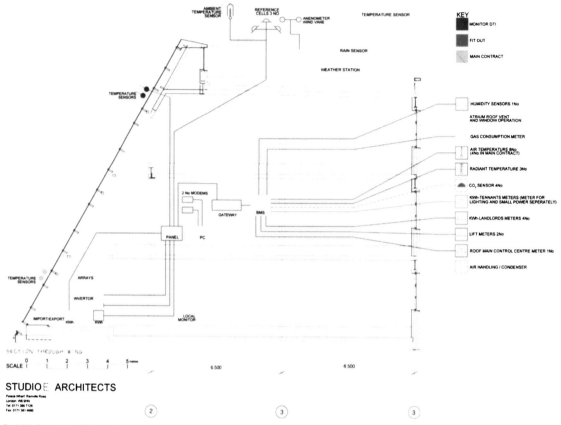

STUDIO E ARCHITECTS

Fig 6 Monitoring installation schematic
Source: Studio E Architects

The total array is split into four sub-arrays, two on each side of the entrance and with slightly different orientations. The two west sub-arrays are oriented 5° off south towards the west, while the two east sub-arrays are the same angle off south towards the east. This enhances the visual aspect of the façade, but in practice, results in a negligible difference in output between the different sides of the array.

There are two large sub-arrays, each consisting of 17 series strings and with an array rating of 35.6 kWp. Each of these feeds into its own 35 kWp inverter. The two small sub-arrays have a single series string and are rated at 0.94 kWp. They each feed into a 0.85 kWp inverter and are located around the building entrance. In all cases, the modules are strung down the façade to a junction box in the supply trench at the bottom.

The glazing build-up is:

- 5 mm heat-strengthened glass;
- 2 mm cast resin encapsulating solar cells;
- 4 mm heat strengthened glass (Parsol);
- 12 mm krypton-filled void;
- 6 mm laminated glass with low-E coating.

One option the designers considered was developing some form of cavity façade construction. This may have had the added benefit of improving PV efficiency, but there was no ready-made solution and not enough development time available. So instead they focused the CFD airflow modelling on refining the building section and improving dissipation of heat from the façade.

Fig 7 The solar façade showing the effect glass-glass modules can have on the interior spaces when the sun is out
Source: Dennis Gilbert Photographer

Grid integration

As noted earlier, the PV installation is grid-connected. The building's PV power output had to be quality-tested for constancy in voltage and frequency and acceptable variations in harmonic distortion before grid connection could be made.

Although the UK has a Non-Fossil Fuel Obligation requiring electricity companies to get involved in selling electricity generated without fossil fuels, there is no specific requirement for them to buy PV-generated electricity from grid-connected buildings. The project team found only one of the six regional electricity companies ready to do so – Northern Electric.

Even then, the standard contract started at 1 MW and the annual administration charge was £1,000 (later reduced to £500). Clearly, some new thinking, and particularly regulation, is needed to facilitate smaller, building-integrated renewable energy generators.

Planning approval

The Solar Office is located in an Enterprise Zone, a designation which, among other development incentives, waives the necessity of obtaining planning permission. Even so, the local planning authority was consulted. It had no objection to the design. On another site, the size and prominence of the array might have raised considerable controversy.

PV procurement

The client recognised that there was a risk attached to BiPV on this scale and was keen to minimise it. Several steps helped:

- Choosing a tried-and-tested PV system.
- Letting the PV installation as a single package of supply, installation, interfacing and commissioning.
- Given that the façade system provides sound cladding, in energy terms the PV is failsafe. If it fails it generates nothing – the electricity bill goes up 25–30 per cent, but the building stays weather-tight.

For the designers, the tendering process was more difficult than usual. For other construction packages there was considerable reuse of methods and assemblies from previous buildings on the park. None of these fitted the PV package, which had to be tendered as a performance specification. There is as yet little in the way of standards or precedents to draw on, so the specifiers began largely from scratch. For example, discussions arose over:

- Tolerance on alignment of cells within a module.
- The acceptability of bubbles in the resin that encapsulates the cells.
- The appearance of circuit tape at module edges.
- A few cells that were torn during lamination (though testing showed no loss of performance).

There was also an initial difference between the 12-year warranty required by the developer and the 10 years customarily provided by the supplier.

Maintenance

Maintenance of the façade is minimal: a wash down every six months externally and maybe once a year internally.

A cherry-picker was selected that is capable of reaching the inclined outer surface of the façade and also folding compactly enough to enter and service the atrium. For the office wings, the inside of the façade will be reached using a proprietary ladder-and-plank system.

Testing and commissioning

Commissioning the PV installation was carried out in two stages relating to the two facets of the façade. Following commissioning, the building services engineers carried out a series of tests to satisfy the builder that the system was operating satisfactorily. The weather conditions were particularly favourable over the commissioning period and the instantaneous peak performance of the system (73 kWp AC) was recorded.

The monitoring equipment, including sensors, cabling, data logger and control system, was installed by the façade contractors in parallel with the installation of the PV façade, or by the building contractors, as appropriate. Some of the data are routed through the Building Management System and thus interfacing with the data logging system was required.

The commissioning of the monitoring hardware took place at the same time as the commissioning of the PV system in March 1998 and included final decisions on the file format for recording of the data, setting up of the modem links, including password protection, and connection of the display terminal in the building.

The collection and assessment of performance data commenced on 5 March 1998, with the purpose of the data commissioning process being to:

- establish the procedure for downloading the performance data from the monitoring computer in the inverter room;
- develop handling routines for those data;
- inspect the initial data for availability and inconsistencies and advise the installers of any problems.

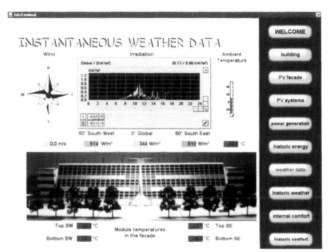

Fig 8 A page from the visual display illustrating instantaneous energy consumption
Source: Studio E Architects/Scheuhle

Performance characteristics
Estimated annual PV energy output

Incident radiation		
Solar radiation for the area (horizontal plane)		950 kWh/m^2/y
Installation factor (inclination, orientation, coating)	x	1.04
Density of solar radiation	=	**988 kWh/m^2/y**
Active area of solar cells	x	532 m^2
Total radiation on façade	=	**525,600 kWh/y**
Element efficiencies		
Efficiency of solar cells (at 25 °C)		x 0.14
Reduction due to operating temperature	x	0.90
Losses from cables	x	0.98
Average efficiency of PV inverter		x 0.85
Estimated annual energy output	=	**55,200 kWh/y**
System efficiency (estimated output/total radiation)		**10.5%**

Table 2 Estimated annual PV energy output

Project cost breakdown
Project funding and costs (in GBP)

The capital funding of the project was made up as follows:

Trust fund provided by City investors	2,763,785
European Regional Development Fund grant for capital expenditure	1,350,000
UK Department of Trade and Industry grant for design development, commissioning, testing and monitoring	111,215
Total	**4,225,000**

The cost of delivering the shell-and-core building was made up as follows:

Building construction	3,514,000
Landscape and infrastructure	365,000
Professional fees and other charges	346,000
Total	**4,225,000**

The cost of the fit-out was as follows:

Fit-out	600,000
Professional fees and other charges	54,000
Total	**654,000**

The rent charged for the Solar Office reflects rental levels for office development generally at Doxford International.

PV costs

The shell-and-core building cost was made up of costs relating to work undertaken by about 17 different trade contractors together with the main contractor's management costs.

The largest trade contractor's package in value was that of Dane Architectural Systems who were responsible for the installation of all the glazing including the windows, the stair tower glazing and the complete PV installation, but excluding the atrium roof lights. The value of the PV installation, including the curtain wall framework, the main entrance and lobby, the opening lights, the clear glazing, the cabling, the junction boxes, the inverters, the monitoring equipment and the graphic display monitor came to about £950,000. This is the equivalent of £200 per square metre of the gross floor area or £1,000 per square metre of the solar façade area. About half of this cost is attributable to providing the curtain wall framework, the entrance, the vision panels and the opening lights. This cannot be described solely as the balance of systems costs, since the façade provides for vents, vision panels and the main entrance, in addition to the solar array. The PV content, including the monitoring systems, therefore works out at about £470,000 or £100 per square metre of the gross floor area.

A benefit of BiPV that is invariably cited is the potential for offsetting a proportion of its costs against conventional components, such as glazing, cladding panels, roof finishes or sun shades, which would otherwise have been used in place of the modules. While a straight offset can be calculated in relatively simple applications, in a project such as the Solar Office, with its very large array, and the fact that it is not just integrated into the building envelope but into a comprehensive energy strategy, calculations of this kind are precluded.

Immediately adjacent to the Solar Office is an office building of similar size. It is designed to a very similar brief except for one aspect – the PV installation. In form and layout, it bears no resemblance to the Solar Office. If the Solar Office had been designed purely as a low-energy office building, it would not be focused towards the south in the way that it is – the floor plates would not have been stepped back and the atrium, although probably retained, would have been a very different shape and size.

A cost comparison between a comprehensively conceived PV building and a conventional one is not therefore between a building with PV and more or less the same building with the PV replaced by conventional glazing or cladding, but between a building that acknowledges the full implications of PV integration, such as the Solar Office, and a building similar to its neighbour, with (in this case) an enhanced passive solar specification. The Solar Office cost about £940 per square metre; the building next door, enhanced to include a concrete roof, a larger atrium, wind trough and baffles costs about £790 per square metre, a difference of £150 per square metre.

Clearly there are cost offsets to be made, the extent of which will vary from building to building. However, the real benefit of building-integrated photovoltaics is to do with 'commodity and delight'. Integration allows a holistic response to physical by-products of PV power generation such as the utilisation of its thermal characteristics as well as its electrical output. It facilitates economy of detailing, allowing solutions that resolve a number of demands such as power generation, glazing, solar control and cable management. It can also bring attributes that are exclusively its own, such as the quality and character of daylight within the building.

Estimated cost savings

Cost savings arising from the operation of the Solar Office compared to conventional office development were an important aspect of marketing the project. Cost savings compared with an air-conditioned office are in three areas:

- Reduction in energy use arising from energy-saving measures incorporated in the design of the building;
- Reduction in maintenance costs, mainly arising from the omission of air-conditioning plant; and
- Savings in grid-supplied electricity arising from the PV supply.

The building service engineers calculated that the Solar Office in its low-energy mode, with an occupant of average energy demands, might have an energy consumption demand of about 85 kWh/m²/year. They further calculated that savings arising from the building with this demand compared to a best practice air-conditioned office would offer the following savings:

	£
Reduced energy use from energy-saving measures	19,680
Reduced maintenance costs	32,000
Savings in grid-supplied electricity	3,320
Total	**55,000**

The price paid by Northern Electric for PV generated electricity was 2p/kWh.

Operating cost savings are based on the following:

- A treated building area of 4,000 square metres;
- A predicted cost of gas 2p/kWh; electricity 4p/kWh;
- A maintenance cost saving, over an air-conditioned building, of £8 per square metre.

Post-installation feedback
Electrical performance

Over the first two years the PV array has generated some 113,000 kWh/y (Table 3), which compares well to the design prediction. Most parameters – irradiation, ambient temperature, input and output power – were recorded as average values over 10 minutes. Data was downloaded from an on-site computer on a weekly basis and analysed by Newcastle Photovoltaics Applications Centre.

A total output of 94,790 kWh was recorded between March 1998 and May 2000. Some inverter problems experienced during the summer months masked the seasonal differences in performance that were expected, although there was a noticeable improvement in efficiency during the winter months. Efficiencies for the four PV subsystems have also been calculated. These consist of two large arrays connected to three-phase inverters, and two small arrays connected to single-phase inverters. The smaller subsystems exhibited consistently lower performance than the larger systems. Half the difference was assigned to the fact that the smaller subsystem array has a higher proportion of lower efficiency modules due to higher cell spacing. The remainder was assigned to a lower efficiency of DC/AC conversion. It is usual to de-rate inverters for climates where there is a significant percentage of operation under low light conditions. In this way the loss of output under occasional high insolation conditions is more than offset by the increased efficiency at low insolation conditions. In the Solar Office, the large inverters are rated at 98 per cent of the corresponding PV array capacity, while the small inverters are rated at 90 per cent of their corresponding PV capacity. Assuming similar low level light, the small inverters should give slightly higher subsystem efficiencies. However, the single-phase inverters have exhibited slightly different performances. This is partly due to different solar input thresholds and partly due to transparency differences between the modules. While the major difference between the two inverter sizes is due to low light-level performance, this had had very little effect on total performance as the subsystems with the small inverters represent a small proportion (2.6 per cent) of the array capacity.

System losses

Overall, the PV array efficiency at the Solar Office is calculated to be 11.3 per cent (compared to the design efficiency of 14 per cent), based on the watt peak ratings of the individual modules and taking into account their design levels of transparency. With system efficiency at 8 per cent (compared to the design efficiency of 10.5 per cent), about 30 per cent of the expected DC power generation must be due to system losses (cable losses were originally assumed to be 2 per cent of total output). Other losses are:

- inverter efficiency (originally assumed to be 85 per cent);
- insolation below inverter thresholds;
- module temperature (originally assumed to reduce output by 10 per cent);
- soiling (dependent on cleaning; 3 per cent assumed);
- angle of incidence (assumed 5 per cent);
- performance mismatch between modules and cables (possibly up to 12 per cent).

System reliability

The large inverters were the main cause of operational problems at the Solar Office, causing intermittent output. These were resolved and the system has since been operating without further problems. The only other reliability problems were caused by regular crashes of the monitoring computer, variously ascribed to corrupted software and clashes with the website software. Other than that, the system has operated with excellent reliability.

Comfort conditions

The building's occupation by Domainnames.com late in the monitoring programme means that a full user survey of the building and the relationship to the PV system has yet to be carried out. BRECSU has been granted funding to carry out an analysis of the building under their Best Practice Programme, but this is conditional on the building running in a passive solar or mixed-mode regime.

Due to the tenant's high incidental cooling loads, a viable air volume system with mechanical cooling has been installed. It is hoped that this can be modified to run in a mixed-mode format.

Month	Monitoring Fraction (%)	Measured AC Output (kWh)	Estimated Loss due to Inverter (kWh)	Estimated Loss due to Monitoring (kWh)	Corrected Output
Apr 98	97	4,607	0	142	4,749
May 98	85	5,411	0	955	6,366
Jun 98	100	4,240	927	0	5,167
Jul 98	100	4,558	485	0	5,043
Aug 98	100	3,277	1,710	0	4,987
Sep 98	100	2,740	497	0	3,237
Oct 98	100	3,267	710	0	3,977
Nov 98	77	2,330	0	695	3,025
Dec 98	0	0	0	1,377	1,377
Jan 99	74	1,987	0	698	2,685
Feb 99	100	3,687	0	0	3,687
Mar 99	100	2,787	807	0	3,594
Apr 99	100	4,170	1,147	0	5,317
May 99	100	4,165	1,606	0	5,771
Jun 99	100	4,927	0	0	4,927
Jul 99	100	6,469	0	0	6,469
Aug 99	100	5,114	0	0	5,114
Sep 99	100	5,630	0	0	5,630
Oct 99	100	3,743	0	0	3,743
Nov 99	0	0	0	2,711	2,711
Dec 99	0	0	0	4,008	4,008
Jan 00	51	1,138	0	1,093	2,231
Feb 00	100	3,793	0	0	3,793
Mar 00	100	4,712	0	0	4,712
All	83	82,752	7,889	11,679	102,320

Table 3 Corrected outputs for the Solar Office's PV system

Conclusion

The project clearly demonstrates that BiPV can be part of a purely commercial development. It shows that the PV industry has developed to a point that it can provide the product within a timescale and to a performance standard that is acceptable to a developer constructing speculatively.

What the industry is unable to do yet is deliver the product at a cost that can be accommodated in the parameters of commercial office development. Had separate funding of the photovoltaic installation not been available, it would not have been included in the project.

A fundamental objective behind the design was to combine low-energy measures with PV power. There are few PV projects where this approach has been pursued with such rigour. It became clear at a very early stage that pursuit of energy efficiency and renewable energy generation from PV would throw up conflicts in design resolution. The very size of the installation required to generate a worthwhile amount of electricity created considerable challenges in respect to its successful integration. The project shows that passive solar and renewable energy provision can be reconciled and synthesised to create a striking building.

Fig 9 South façade of the Solar Office, showing the array, main entrance and wind baffles
Source: Dennis Gilbert Photographer

USA: 4 TIMES SQUARE

BIODATA

PROJECT:	14 kWp PV system, integrated in skyscraper curtain wall façade
LOCATION/CITY:	Manhattan, New York
COUNTRY:	USA
TYPE OF PV BUILDING:	Façade-integrated using custom-sized BiPV glass laminate
BUILDING TYPE:	Commercial – office building
NEW/RETROFIT:	New

CLIMATIC CHARACTERISTICS

LATITUDE:	40° 47'N
LONGITUDE:	73° 58'W
ALTITUDE:	57 metres above sea level
CLIMATIC TYPE:	Humid Continental (Temperature: January average = 0 °C; July average = 24 °C)
SUNSHINE HOURS:	Yearly average = 4.6 hours per day

Fig 1 4 Times Square
Source: Janna Johansson-Pereira

Project brief

This 48-storey skyscraper at the corner of Broadway and 42nd Street was the first major office building to be constructed in New York City in the 1990s. To raise the environmental standards of high-quality urban buildings, the Durst Organization, which owns this skyscraper, is implementing a wide range of strategies that result in healthy, energy-efficient buildings. Kiss + Cathcart Architects designed the PV system for the building in collaboration with Fox & Fowle, the base building architects. By doubling as a building material, the PV curtain wall is one of the most economical photovoltaic building components installed to date in an urban area. Energy Photovoltaics, Inc., of Princeton, New Jersey, developed custom-made PV modules to fit the building's rigorous aesthetic, structural, and electrical criteria.

he traditional view is that photovoltaic systems are economical only for remote cabins and telecommunications, but this is changing. As the first major commercial application of building-integrated photovoltaics (BiPV) in the United States, 4 Times Square points the way to a future in which clean, silent, solar electricity will be generated on a large scale where it is needed most – at the point of greatest use. The next major market for PV in the developed world should be where electricity costs are high and high-quality buildings are being built – in other words, in urban centres like New York City. In addition to the BiPV system, 4 Times Square incorporates a number of other sustainable building technologies in its design. Oversized windows enhance daylighting and reduce the need for artificial light; the HVAC system uses CFC- and HCFC-free gas-fired absorption chillers instead of conventional, fossil fuel-powered electrical systems; fuel cells produce electricity by burning natural gas, and their by-product, hot water, is recycled for hot water and perimeter heating. Environmental considerations also shaped the construction process – a waste management plan salvaged and recycled materials during demolition, and the use of paper was minimised as the structural engineer and the steel contractor checked tens of thousands of drawings digitally.

BiPV design process

Kiss + Cathcart obtained special approval from the Bureau of Electrical Control of the New York City Building Department.

The client wanted a technology that was economical, not immediately, but in the near future. The design balanced aesthetic concerns with a low cost per square metre, which was comparable to the cost of the materials already specified.

Because BiPV was incorporated into the design at the end of the construction documents phase, the installation was made to harmonise with the already established design concept. The PV modules were built to the same size specifications as standard glass. This project demonstrates the flexibility of BiPV technology in accommodating existing configurations. Shading issues were a special concern, because of the building's height, and because it is located in one of the densest urban environments in the world. The designers studied the local massing and obtained projections of future construction at the site. They determined that the optimum area for the PV skin would extend from the 37th to the 43rd floor on the south and east façades. At that height, because of setback requirements, even the tallest new construction would cast only fleeting shadows on that part of the building.

PV modules were chosen which met the client's standards for cost efficiency, and performed well at the high temperatures they would encounter in this installation. They also were chosen to be shadow-resistant to minimise the effects of shading from other parts of the façade.

COMPONENT CHARACTERISTICS

PV system power	14 kWp
Projected system output	13,800 kWh per year
PV efficiency	6%
Type of building integration	Façade-integrated custom-sized BiPV glass laminate
Type of cell technology	Amorphous silicon
Modular dimensions	Gross surface area: 287 m²
Array dimensions	2.60 m²
Weight	65.9 kg/m²
Visual details	The PV modules are visible as dark bands of glazing on the exterior of the building
Inverters	Four: 2 x 6 kW (Omnion); 2 x 4 kW (Trace)
Monitoring equipment	Power output monitored through kW/h readouts and recorders

Installation

Installation was a very smooth and straightforward process, despite the fact that Times Square is one of the most highly regulated and physically difficult building environments in the country. The BiPV panels were pre-placed into the panelised curtain wall, and the pre-assembled façade was delivered and installed on-site just as any other conventional wall. It took two weeks off-site pre-assembly and two weeks on-site installation to complete the BiPV construction. Electricians then connected the wires into the PV electrical system.

The PV modules are attached to the building structure in exactly the same way that standard glass is attached. The glass units are attached with structural silicone adhesive around the back edge to an aluminium frame. An additional silicone bead is inserted between the edges of adjacent panels as a water seal.

The south and east façades of the 37th through the 43rd floor were designated as the sites for a photovoltaic 'skin'. The PV modules, which are custom-sized BiPV glass laminate, replace conventional spandrel glass for the south and east façades. There are four different sizes of modules, and they correspond to the spandrel sizes established earlier in the design process.

There is a separate electrical system for each façade. Each system consists of two inverters, one 6 kW inverter and one 4 kW inverter. The larger inverters serve the two large PV modules, which have electrical characteristics that are slightly different from those of the two smaller ones. Using two inverters per side enables the system to perform more efficiently. The inverters are located in a single electrical cabinet at the core of the building. The AC output of the inverters is transformed from 120 V to 480 V before being fed into the mains electricity.

Performance characteristics

The PV area is less than one per cent of the available building spandrel area, and the PV contribution to the building's energy demand is less than one per cent. The BiPV system will pay back the energy used to produce it in eight months. The PV system is performing slightly better than anticipated.

Lessons learnt from the project

Because the BiPV system was only considered after the design of the core building had been completed, crucial opportunities to optimise the system's cost and performance were missed. The cost of the PV panels could have been lowered through the use of more optimal dimensions, and the design of the electrical system could have been simplified, both in the PV wiring and in the building interconnection.

This was the first custom-sized thin-film BiPV project in the United States, and it is the highest-visibility BiPV application in the country. The success of getting project approval in the stringent regulatory environment of Times Square has shown that BiPV can be used in the toughest environments. Conservatism and fear have been the biggest barriers to the acceptance of BiPV by the building community. This high-visibility project, financed through the private sector, has given the building community more confidence in the broad applicability of BiPV systems.

The major lesson learned is an organisational one. Kiss + Cathcart's experience suggests that it is still too soon to treat BiPV systems as a standard part of the building process. Specialists are needed to oversee design, construction, and commissioning of BiPV systems.

Project cost breakdown

The project received support from the New York Sustainable Energy Research and Development Authority (NYSERDA).

Savings in terms of replacement of building components with PV: US$105 per square metre

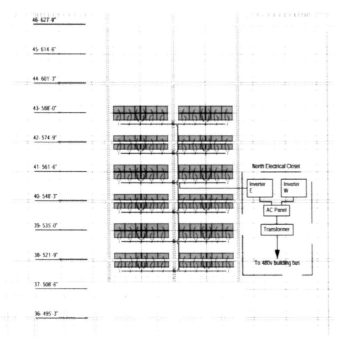

Fig 2 Electrical configuration schematic
Source: Kiss + Cathcart, Architects

Fig 3 View of curtain wall illustrates that BiPV panels (dark panels) can be mounted in exactly the same way as conventional glazing (lighter panels)
Source: Janna Johansson-Pereira

CHAPTER FOUR

NON-BUILDING PV STRUCTURES

Introduction

Photovoltaic (PV) technologies are particularly suited to being innovatively integrated within urban environments and conventional street furniture. Urban architecture, buildings, streetscapes, parklands and water features resonate with the social, economic and cultural traits of a city, and are often designed as components of that city's image. The use of solar power as a functional feature and creative expression in public spaces, and as a means of engaging city communities in the technology, is becoming increasingly common as city authorities strive to deliver sustainable solutions. The use of PV in urban spaces is being successfully achieved in numerous cities around the world. In many cases, this is being achieved through the positive integration of PV within its urban context, while enhancing the architectural quality of the photovoltaic material and respecting its technological efficiency. This section provides a detailed discussion and examples of PV non-building structures (PV-NBS) in the built environment including design process and performance issues. These are presented in later sections of this chapter and include:

- *Urban street equipment:* parking meters, information signs, ticket vending machines, meteorological stations, tourist information points/maps;
- *Barriers:* fences, road noise barriers, gates, handrails;
- *Shelters and kiosks:* petrol stations, bus stops, phone booths, parking, umbrellas, information stands, pavilions, telephone boxes, toilets, news-stands;
- *Single 'aerial' structures:* street lights, street signs, commercial signs, road sign-posts; and
- *Multi 'aerial' structures:* screen road signs, screen publicity structures.

Design considerations

Each PV-NBS categorised above faces particular design considerations that effectively have to resolve the following key issues:

Dimensions: the scale and shape of the NBS directly impacts on PV power performance and is often dictated by the urban context in which it is placed and the purpose it plays as a non-building feature.

Structural requirements: this aspect depends also on dimensions but is more related to the mechanical performance of the structure in terms of material type, wind loading and static load characteristics.

Functional requirements: the technical solutions and construction systems are determined by the function that the NBS should provide (such as water-tightness, wind protection, shading devices, safety and equipment control).

Wear and tear: the way in which people use these structures defines certain specific characteristics of different typologies, such as durability, accessibility, material strength, control and security.

Peak power, storage and surface requirements: the quality and quantity (number and type) of functions provided by the structure will determine the electrical needs and surface requirements, which in turn determine the final dimensions of the structure.

Storage capacity: operation criticality is an important aspect, and as such, the need for batteries to store the produced energy and related aspects such as ventilation, overheating, safety and maintenance, must be considered.

Implementation issues and challenges

When PV modules are to be installed in city areas, invariably at street level, the likelihood of conditions such as shading and low irradiation is heightened. The non-building structure (NBS) might already exist and the orientation may not offer an optimum position. Similarly, physical limits to available surface area and aesthetic constraints could reduce the available power production to a level that is insufficient for the electrical needs of the actual installation. Invariably, this can be avoided through sensible planning, although careful consideration of micro-climate influences in urban environments should be taken. The built form and activities within it disturb the natural local climatic equilibrium, creating pockets of high ambient temperatures and humidity, haze and smog, and wind perturbations between buildings and streetscapes. Unpredictable street wind gusts can cause problems for some installations, especially to the wind load resistance of large module surfaces relative to the supporting structure. High ambient temperature, humidity and reduced direct irradiation impact upon PV module degradation tolerance and performance, depending on PV system type. Further, soiling from bird droppings or the build up of traffic pollutant residues can have detrimental effects on PV performance. Good design and maintenance strategies can help mitigate these threats.

- *Technical performance*: maintenance, repair and replacement. Occasional module failures should be possible to deal with by repairing or changing the affected modules without destroying or dismounting the rest of the NBS elements.
- *Aesthetic appearance*: installation in special areas. When PV has to be installed in some locations such as historical centres, parks and gardens, non-urban areas, or wild or natural landscapes, the integration in the environment should be carefully considered.
- *Visual impact*: some of the PV-NBS do not have an attractive design at all, but are only elements of technical equipment, often resulting in the rejection of this type of installation.
- *Vandalism*: this is a serious problem in cities and costs ratepayers and service providers considerably. Damaged street furniture, graffiti, wrecked urban equipment and broken cars and shop windows can be an all too common sight after, for instance, weekend nights, concerts, or sporting events. This problem cannot be solved by the PV community alone, and should be handled in the best way with appropriate design strategies. The problem with theft occurs mainly when PV modules are mounted in public areas. People who steal modules do not see the connection between modules and their owner. This makes stealing easier, even for the less hardened of criminals. The high price of modules and existence of a second-hand market for this kind of product has great influence.
- *Cost*: this is not a specific problem of NBS but needs to be addressed as, although using more sophisticated designs and materials could solve some of the previously mentioned problems, high-cost PV components can threaten project affordability.

PV-NBS design strategies

PV-NBS with well thought through design strategies can help mitigate unwanted threats from thieves and vandals, ensure public safety compliance and maintain desirable operating performance. Some thoughts on this are provided below:

Protection

Hiding: integration of the module into the NBS shell so that it will not be seen.

Smart designs: using odd/very special module sizes, colours, voltages, which render the module useless or difficult to use in everyday applications.

Safe mounting systems: using rivets or screws that require special tools to operate.

Protection systems: making it difficult to reach the module by using fences, posts, and other obstacles.

Technical improvements

Strong/flexible materials: using extra-strong/flexible materials such as different plastic covers instead of glass.

Strong mounting systems: using extra strong joints and connections between the modules and the structural components to resist wind loads.

Electronic remote operation: using electronic remote-control and localised systems that can be incorporated into the modules and are necessary for operating the system. These could also avoid theft because it would not be possible to use the module without the remote-control device.

Module etching: etching the module with a specific code or sign, which could be identified and recognised on later inspection.

Energy efficiency: using energy-efficient consumption devices such as sensors or light-emitting diode (LED) lamps.

Cleaning systems: using machines for efficient cleaning of modules (cleaning systems in Sweden, for illuminant posts along roads in wintertime, employ high-pressure air or water devices).

Reflectors: applying reflecting irradiation elements or surfaces in order to increase the available solar gain of the PV modules when local conditions are not optimal; and

'Elastic' structures: using systems that can absorb wind loads through the elasticity of their structural elements.

Fig 1 PV installation on a bus stop sign and on a phone box
Source: (left) Energibanken, Sweden, (right) ECoCoDE, Sweden

Design flexibility and adaptability

Multi-oriented systems: making flexible mounting systems to allow the modules to be moved in the optimum direction independently of the orientation of the NBS.

Dismounting systems: ensuring no permanent joints and connections have been used, making maintenance, repair and replacement of modules easier.

Independent systems: ensuring a separation between PV mounting components and NBS in order to allow for mounting and dismounting to operate independently of each other.

Expandable systems: creating a design that easily allows the addition of new PV modules to the NBS. Addition of new PV modules as a function of the power needs should be possible.

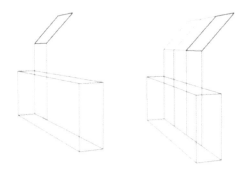

Fig 2 Design for expandable systems
Source: Mark Snow, University of NSW, Australia

Aesthetic and practical improvements

Integration: integrating modules into the NBS shape. The battery, the support element and the PV modules are integrated into the NBS design.

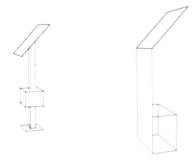

Fig 3 Integration of module into NBS
Source: Mark Snow, University of NSW, Australia

PV aesthetic characteristics: using the aesthetic performance of the PV cells and modules such as texture, colour, laminate materials, and module framing alternatives.

Double function: combining several simple functions in the same NBS is a simple way of reducing installation costs. To some extent, this is a given for many NBS applications, such as shading structures or road barriers.

Standardisation: using a design that facilitates the use of standard modules and components in order to reduce the cost.

Posts: using a post for the module to reduce shading concerns or to make it difficult to reach the module.

Instructions/manuals: providing mounting and dismounting instructions incorporated in a visible place in the NBS to guide maintenance personnel and prevent inappropriate operation.

Information: using environmental, economic and other information in an effort to convince people not to steal modules, and to use them correctly.

Urban street equipment
Parking meters

Costs and time associated with extending conventional electricity supply, digging up roads, disrupting traffic movement and installing underground power lines can often be less sensible than a solar alternative. The integrated solar NBS solution is easily and quickly installed without the disruptions often associated with new electrical services. Aesthetic benefits may accrue by eliminating the need for overhead power lines and system reliability also provides what might be essential emergency power supply at times of grid blackouts.

The PV-equipped version of the standard pay-and-display ticket machine has provided aesthetic and technical integration of the solar cells. The solar module on the Parkline 2001 meter (figure 4) features poly-crystalline cells with a surface area of 190 square centimetres and a peak power of 23 Wp. The PV parking meter is equipped with a 12V/60 Ah battery. The battery box is stored behind a strong door panel. The elements for maintenance and replacement are easily accessible, but only to authorised staff with a special key.

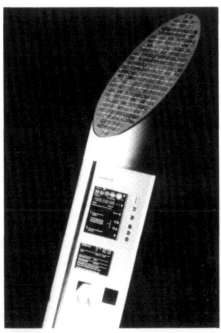

Fig 4 Parkline 2001 meter; country of manufacture: Germany
Peak power: 23 Wp
Storage capacity: 300–400 tickets per day
Source: Dambach-Werke Gmbh, Germany

DG Line PV parking meter

These are stand-alone PV vending machines with the solar generator integrated in the structure. This design helps deter theft and vandalism because the PV generator is well-engineered and secured to the machine.

Fig 5 DG Line parking ticket vending machine; country of manufacture: France
Peak power: 5–10 Wp
Storage capacity: 3 weeks at 100 tickets per day
Source: ECoCoDE, Sweden; Schlumberger

Mesap-Universal 2000 PV parking meter

This photovoltaic-powered vending machine offers a variety of PV mounting solutions that can be located so that they are completely invisible from the ground and difficult to access simply.

Fig 6 Mesap-Universal 2000 – vending machine for parking vouchers with different PV mounting solutions; country of manufacture: France
Peak power: 9–23 Wp
Storage capacity: 3 weeks at 100 tickets per day
Source: Schlumberger; Fraunhofer ISE Freiburg, Germany

TOM 94 PV parking meter

The standard ticket vending machine is equipped with a PV cell especially resistant against impact. The consumption is very low and the battery provides about two months of autonomy.

Fig 7 TOM 94 PV parking meter;
country of manufacture: Switzerland

PV-illuminated bus timetable

Energy-efficient light-emitting diodes (LEDs) illuminate the bus timetable and are activated by a movement sensor. Energy consumption is 3–70 mW and thus the PV cell required is small and can be simply glued on top of the sign.

Fig 8 Bus information system by Ebo Scherpenzeel beheer BV;
country of manufacture: the Netherlands
Source: ECoCoDE, Sweden

PV-powered timetable lighting

This public information device is equipped with a red blinking button active from dusk until daybreak, with each press of the button illuminating the timetable for a 20-second interval. The electricity generated during the day is stored in a nickel cadmium (NiCd) rechargeable battery. The electrical power consumption of the illumination unit has been reduced to about 0.6 W, by including a planar light guide. Extremely bright, red LEDs are used as the light source. Multi-coloured route maps are illuminated with different coloured LEDs.

Simple PV building signage (figure 10) can also be used to illuminate a business address at night.

Fig 10 PV-illuminated business sign; country of manufacture: Germany
Source: Fraunhofer ISE, Germany

Fig 9 PV-powered timetable lighting; country of manufacture: Germany
Source: Fraunhofer ISE, Germany

PV street information

The standard information screen is equipped with a PV cell specially placed to protect the post boxes against rain. The aesthetical integration has also been considered in the 'infoconcept' structure, combined with different standard designs.

Fig 11 Infoconcept PV-powered street information board
Source: ECoCoDE, Switzerland

Solar electric sunflowers

Nestled atop a hillside in northern California, 36 solar electric sunflowers represent an elegant combination of art and technology. The client requested an unconventional and artistic installation. They got just that. The solar electric sunflowers look and act like nature's very own sunflowers. Using a two-axis tracking system, the sunflowers wake up to follow the sun's path throughout the day, enabling the system to produce enough energy for 8–10 homes.

Fig 12 PV 'sunflower' integrated into the landscape, detail and general views
Source: Solar Design Associates, USA

arrier systems

PV wheel block

This PV wheel block model has been carefully designed for highly visible areas such as parks, pedestrian streets, public and historical buildings. It is powered by 0.25 Wp PV cadmium telluride (CdTe) cells integrated into the top of each pole and with a range of 1–10 Wp. In order to reduce the energy needed, the lighting system is operated with LED.

Fig 13 PV-powered wheel block, detail and general view; country of manufacture: Japan
Source: New Energy and Industrial Technology Development Organisation (NEDO), Japan

PV pedestrian safety handrail

This handrail is powered by 10 Wp PV amorphous silicon (a-Si) modules integrated in the shape of the steel profile. With an interesting aesthetic appearance, the product is designed for use in pedestrian streets and bridges, and to protect people from road traffic hazards. Again, by using lights operated with LED the energy requirements are significantly reduced.

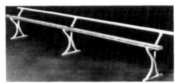

Fig 14 PV-illuminated handrail; country of manufacture: Japan
Source: NEDO, Japan

PV road noise barriers

The world's first PV noise barrier over 100 kWp was built in 1989 near Chur; the second was built in 1995 at Giebenach (near Basle) in Switzerland. A similar installation was built in 1992 along a railway line in the southern part of Switzerland, but without a noise protective function. Other smaller installations were piloted in Austria and Germany during 1992.

As a result of an international ideas competition in 1995 and the introduction of integrated PV sound barrier concepts, a further six installations of 10 kWp each were built commencing in 1997 through to 1999 in Germany and Switzerland, utilising different technologies to show typical advantages in different situations. The Netherlands boasts one of the longest and largest PV-NBS installations at 220 kWp, stretching 1.6 kilometres along a motorway near Amsterdam, and generating around 176,000 kWh of electricity per year (table 1 and figure 17).

Location	Country	Year	Capacity (kW$_p$)	Euros* per kW$_p$
Chur (A13)	Switzerland	1989	103	16,500
Gordola (rail)	Switzerland	1992	103	13,500
Seewalchen (A1)	Austria	1992	40	13,200
Rellingen (A23)	Germany	1992	30	n/a
Giebenach (A2)	Switzerland	1995	104	8,800
Saarbrücken (A6)	Germany	1995	60	8,000
Utrecht (A27)	Netherlands	1995	55	n/a
Ammersee (A96)	Germany	1997	30	n/a
Zürich (A1)	Switzerland	1997	30	n/a
Ouderkerk (A9)	Netherlands	1998	220	n/a
Lille	France	1999	70	n/a
Gleisdorf (A2)	Austria	2001	101	n/a
Safenwil (A1)	Switzerland	2001	80	n/a
Freising (A92)	Germany	2003	499	n/a

*Assume 1 US dollar = 1 Euro

Table 1
Source: SOLARCH, University of NSW, Australia

103 kWp Chur, Switzerland case study

The 103 kWp installation from Chur in the Swiss Alps (figures 15 and 16) is the oldest operable PV-NB installation. The site produces on average 108,000 kWh net energy per annum after deducting 1,863 kWh (1.7 per cent) for powering the monitoring system and inverter. For the 103 kWp system this provides an annual yield of 1,035 kWh/kWp. The site is oriented 25 degrees east of south at an optimal tilt angle of 45 degrees to the latitude of the region. It covers 800 linear metres of road, top mounted flush on a 2-metre vertical structure, and adds 968 square metres to the noise barrier area. It comprises 2,208 Kyocera poly-crystalline PV modules and one 100 kilovolt amps (kVa) Siemens inverter with a power conditioning efficiency of 94 per cent. After eight months planning and an eight-week installation timeframe, the site commenced generation in December 1989. In its first 10 years and three months, the site covered 38,157 hours of generation time resulting in an overall output production of 1,105,712 kWh.

Fig 15 A13 noise barrier, Switzerland
Source: Denie Lenardc, www.pvresources.com

Fig 16 A13 noise barrier, Switzerland, detail view
Source: Denie Lenardc, www.pvresources.com

Retrofit designs are currently the most common PV-NBS approach and provide additional area to an existing noise barrier structure. In instances where existing noise preventative measures no longer contain noise pollution levels through increased traffic volumes, a well-designed PV retrofit solution could assist in achieving compliance levels and generate electrical power simultaneously.

The top-mounted flush PV-NBS design shown in figure 17 uses the existing noise barrier as a support structure. The height and orientation of mounting minimises any incidence of headlight or sun glare and reduces the threat of theft or damage of the PV modules.

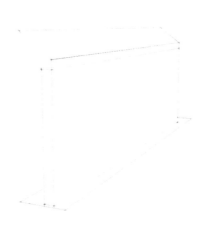

Fig 17 Top-mounted (flush) design: diagram and example, A27, Netherlands
Diagram: Mark Snow, University of NSW, Australia
Source: Riesjard Schropp

top-mounted overhang design (figure 16) offers greater PV surface area per linear metre of barrier wall. While not conclusive, the overhang could provide a beneficial noise reflection component away from urban zones. The diagrammatic PV-NBS design representations do not assume the PV modules face the road. There may be advantages in orienting the PV away from the roadside but that will depend on road orientation and surrounding natural and man-made objects that could augment unwanted shading influences.

Shingle retrofit designs utilise a larger surface area for solar power generation as shown in figure 19. Innovative tilt and orientation is required to reduce shading effects. Such a structure would demand careful consideration of public safety, and strategies such as siting the installation close to 24-hour toll gates or surveillance areas to minimise the threat of theft and vandalism.

Fig 18 Top mounted (overhang)
Source: Mark Snow, University of NSW, Australia

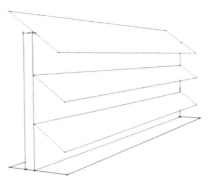

Fig 19 Shingle design and A6, Germany
Diagram: Mark Snow, University of NSW, Australia
Photography: Thomas Nordmann, Switzerland

Fig 20 PV noise barrier with semi-transparent bifacial modules; details and general view
Source: ASE GmbH and Kohlhauer GmbH

Fig 21 PV noise barrier concept
Source: Martin Van Der Laan (M. ART)

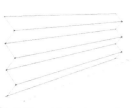

Fig 22 'Zig-zag' design along a railway line in Switzerland
Diagram: Mark Snow, University of NSW, Australia
Photography: Thomas Nordmann, Switzerland

Bifacial PV noise barrier

This innovative system allows the application of bifacial PV modules along motorways with north–south orientations. One side generates power in the morning and the other in the afternoon as the sun tracks a path across the sky. A 10 kWp system has been installed in Wallisellen, Switzerland (figure 20), where the PV bifacial modules replace the existing concrete wall. This application shows the high level of integration possible as the PV module functions as the noise protection structure with the incorporation of dummy elements, where necessary, for a better visual appearance. The wall, with a total length of 120 metres, has a power density of 80 W/m. For a standard 3-metre wall it could be raised to about 200 W/m, even allowing for a certain level of transparency. Given the vertical nature of the modules, careful headlight glare studies may be required to avoid unwanted reflections.

Two other integrated design approaches are presented in the 'zig-zag' design of figures 22 and 23 and the 'cassette' design of figure 24. Both offer a more effective method of reflecting noise but encourage structural shading. There are trade-offs between material and PV content and an effective noise abatement structure. The cassette design, while demanding higher material usage, is particularly effective in abating noise as the cassettes act as a sound absorption unit with the upper surfaces ideal for mounting the PV modules.

Fig 23 Different designs of 'zig-zag' PV noise barrier
Peak power: 10 kWp
Source: ARGE Crimmitschau (DLW Metecno)

Fig 24 'Cassette' design, Germany
Diagram: Mark Snow, University of NSW, Australia
Photography: Thomas Nordmann, Switzerland

Fig 25 Bahnhof 2000 station PV roof
Source: Atlantis, Switzerland

'Zig-zag' PV noise barrier

The concept of the stacked alternating planes allows for the combination of PV, transparent surfaces and noise absorbing surfaces. By choosing the distances between the cells of the PV modules, different degrees of transparency are allowed. The transparency could also be obtained by alternating PV and glass planes. The modules are tilted approximately 75° from the horizontal plane. Further prefabrication of the whole system will lead to a reduction of the mounting time and, therefore, the cost.

In general, the noise absorption level of PV-NBS depends on the share of the glass surface. With the zig-zag and cassette constructions, very high noise absorption levels are possible. For pure glass surfaces (such as the bifacial module) a high level of reflection would be anticipated. Careful consideration of surrounding housing would have to be taken in this instance. As for noise damping (behind the wall), this is similar to a conventional noise barrier, as long as the weight of the structure exceeds a certain minimum.

Shelters and kiosks
Railway station PV roof

This photovoltaic-powered station roof and showcase (figure 25) is a nice combination of a newly designed 'standard' roof and well-integrated PV modules placed in a public location. The solar modules provide both the required energy and a watertight roof covering. This is especially interesting because of the emphasis on architectural demands that have been considered in the roof integration design.

Fig 26 Railway station PV canopy at Morges, Switzerland
Source: EPFL – LESO, Switzerland

PV railway station canopy

The structure of this railway platform canopy consists of a horizontal steel tube with a roof, glazed on both sides of the tube. This is intended to let the daylight through to the central part of the platform but also to mark the border between the waiting and the boarding area. About half of the glazing area contains PV cells. Special attention has been paid to ensure compliance with Swiss standard requirements regarding wind and snow loads. Investigations with regard to electromagnetic fields revealed no direct threat to passengers below. Experience of dealing with bird soiling and pollution from train engine brake dust suggests that preventative measures to stop birds landing should be incorporated into the design. Either regular maintenance inspections and cleaning, or positioning of the panels to encourage natural cleaning during rainfall periods can alleviate performance losses from unwanted soiling.

Fig 27 PV bus shelter with semi-transparent curved modules composed by poly-crystalline cells (Röhm)
Source: Rhm/Solacryl GmbH, Germany

PV bus shelter

The Plexiglas semitransparent solar modules provide the bus shelter lighting. By combining solar cells with highly transparent Plexiglas, daylight can be used for natural lighting in addition to artificial light sources. This kind of module is sandwiched between two sheets of Plexiglas in a permanently elastic structure. The encapsulation method allows the PV modules to be designed at will. The resulting product is lightweight, weather resistant and durable. These modules could, therefore, replace any type of existing sheets in entrances or canopies without changing the geometry of the structure in question.

Fig 28 Alternative PV bus shelter designs
Source: NEDO, Japan

PV umbrella

The structure can be inclined from 0–20°, to respond better to the local solar conditions, and give optimum shading. The umbrella consists of a steel structure capable of supporting a maximum of five photovoltaic panels of 1 square metre each, accommodating around 600 Wp of nominal power. The design of the large square base, besides counterbalancing the structure, also offers the placement of several accessories, such as the seats, that hide the batteries, a table, planters, and a small light fixture. The project allows the use of coloured fabric or polycarbonate at the four corners of the cover, so that the object may blend into the environment in which it is used; tourist villages, or as displays for company logos are two examples.

Fig 29 Rendering of the PV umbrella
Source: Officine di Architettura, Cinzia Abbate, Italy

ENEL PV canopy

The shelter was designed for the Italian energy utility, ENEL, and selected for its wide range of possible uses in all those areas where excavations for electrical cabling could be particularly difficult, such as parks, archaeological sites, recreational areas, and sea resorts. The structures can be inclined from 0–20°, by tilting specially designed capitals on the two supporting columns.

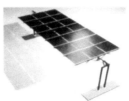

Fig 30 ENEL tilting PV canopy
Source: Officine di Architettura, Cinzia Abbate, Italy

Fig 31 PV shelter canopy design
Source: Officine di Architettura, Cinzia Abbate, Italy

ENEL PV kiosk

This PV kiosk is designed to accommodate a small office space and an exhibition area for temporary public events. The kiosk is modular; each module is conceived as a bay of three PV modules. Depending on the space available for the temporary event, and the electrical requirement of the kiosk, it is possible to assemble using as many bays (from a minimum of three) as desired. The small version of three bays incorporates nine PV modules (120 cm width x 120 cm length) and is sufficient for indoor lighting, the energy of a computer, and other exhibit displays. The wiring of the PV system is visible through the wooden composite pilaster of the kiosk. The two small seats located on the side of the entry hide the batteries. Both the UV-resistant Plexiglas roof and the fabric used for the exterior walls can be silkscreened with publicity or logos of the events.

PV car shelters and petrol stations

Car shelters either for parking or when filling up at petrol stations are excellent structures for integrating PV. With the growth of electric battery hybrid petrol vehicles and fuel cell technologies, there is a clear logic in charging or refuelling vehicles using direct PV electricity, or in the production of hydrogen to supply fuel cells. The US Department of Environment car fleet has adopted a PV electrical car battery system where the vehicle is recharged while sheltering from direct sunlight. Car shelter examples are shown at left.

Fig 32 Design of a standard flexible, lightweight structure for a carport roof
Source: NEDO, Japan

Fig 33 PV car park canopy at Sacramento Airport
Source: NEDO, Japan

Fig 34 19 kWp plug-in-the-sun mono-crystalline roof mounted system Olney, Maryland
Source: BP Solar

Fig 35 10 kWp curved thin-film integrated design, Alcalá de Henares, Spain
Source: BP Solar

BP Solar has initiated a US$50 million, 3.5 MW two-year program to install a maximum of 20 kWp on 200 of its 17,900 BP-branded retail PV petrol stations in 11 countries. This follows a successful trial of 19 sites in Europe, Australia, Malaysia and the USA. A 19-kWp system at Olney, Maryland (figure 34) supplies as much as 40 per cent of the petrol station's daytime electricity needs. BP Solar has redesigned its new petrol station canopies as curved surfaces to incorporate PV panels, either at the time of construction or at a later date (figure 35).

Other transport-related PV-NBS that can be used as a design option includes its use on garages (figure 37) or as a solar charging device for electronic vehicles (figure 38).

Fig 37 PV structures that could easily double as garage shelters for cars
Source: Demosite, Switzerland

Fig 36 70 kW canopy at Noichi Zoologial Park, Kochi, Japan
Source: NEDO, Japan

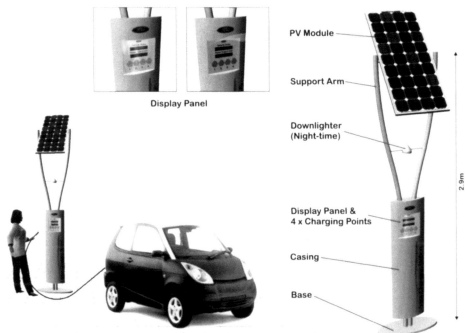

Display Panel

PV Module

Support Arm

Downlighter
(Night-time)

Display Panel &
4 x Charging Points

Casing

Base

2.9m

Fig 38 The 'solar charge pole' is a charging device for TH!NK electric vehicles (now owned by Ford). The PV on the pole itself powers the digital display and the downlighter at night.
Source: Mike Rhodes, PV Systems and ESD consulting, UK

Fig 39 Sainsbury's Superstore petrol outlet, UK
Source: Solar Century, UK

Fig 40 Public transport
information board
Source: Fraunhofer ISE, Freiburg

Fig 41 Solight PV streetlight by
Engcotec: Model FRO (NEDO)
Source: NEDO, Japan

Fig 42 Pole-mounted
PV below the
streetlight
Source: NEDO, Japan

Fig 43 Pole-mounted
PV above the
streetlight
Source: NEDO, Japan

Fig 44 Sydney Olympic PV pylons
Source: Carl Michael Johannesson, Sweden

Fig 45 Pylons at night
Source: Rob Largent, University of NSW, Australia

Single aerial structures
PV street sign

A photovoltaic-powered public transport information board is shown in figure 40. It is safe from theft of modules, since the PV component is an integral part of the board. The system is very efficient in energy use, since light for the board is based on LED and will only operate after pressing the button (and switches off after 20 seconds).

PV street lighting

The 'Solight' model (figure 41) is designed to be used in highly visible areas such as parks, pedestrian streets, public and historical buildings. The PV streetlight is made of high-grade steel, with special attention paid to the aesthetic design of the support structure and to the integration of the PV modules. It is equipped with energy-saving lamps of 27 W and a maintenance-free battery, controlled by an IMS (Intelligent Management System). The battery and the IMS system are incorporated inside the battery box under the lamp case. The two 50 Wp PV modules are fixed in with special anti-theft screws.

Multi-aerial structures

Multi-aerial structures are increasing in popularity as distinct features of urban landscape furniture.

Olympic boulevard PV lighting towers case study

The Sydney Olympic boulevard pylons are primarily urban sculptures designed to bring a sense of movement and festivity to the central pedestrian spine of the Olympic precinct. The towers serve as indicators of scale and direction within the very large boulevard as well as displaying signage, generating electricity, providing power outlets for activities in the boulevard, providing shade, providing night illumination and conveying graphic information through digital screens. The visibility of the PV is also designed to showcase solar energy production and screens at boulevard level display the amount of energy being generated to passersby.

The 19 photovoltaic light pylons line the 1.5-kilometre Olympic boulevard alongside the 110,000-seat Olympic stadium, and the country's largest indoor 'Superdome' arena where 20,000 spectators were able to watch, amongst other sports, basketball and gymnastics. Each lighting pylon marks the name of a past host Olympic city, is 30 metres in height and features a large concrete leg structure to counterbalance a steel tower.

All modules face away from the stadium towards true north (southern hemisphere) and provide the necessary electricity to power floodlights directed upwards onto a mirror which spreads light diffusely below. The PV-backing base layer is a blue fluorescent colour which illuminates the PVs and produces a stunning effect at night. Expected generating output per year is 160,000 kWh which matches the lighting demands of the public boulevard area.

he arm of the tower supports a 20-metre horizontal steel truss anopy of PV modules, collecting solar energy for night-time ghting use and providing shade during daylight hours. The PV ructures use 1,520 BP Solar laser grooved laminates in total. ach tower consists of five rows of 16 solar panels with a ominal capacity of 6.8 kWp per tower, generating approximately 3 kWh per day and equivalent to the power needs of two ouses. DC electricity is fed to six inverters connected to the ain electricity power grid by Sydney electricity utility company, nergyAustralia.

nother example of a grid-connected PV system is the Dutch lectricity company PNEM's use of 'solar sails' for promotional urposes. Problems with theft, vandalism, wind and aesthetics ere solved in an innovative way through the use of water as natural barrier. This made the structure inaccessible, and by esigning an open structure, the wind load was decreased. pecial attention was paid to the aesthetics, with bright coloured anels placed between the PV modules (figure 46).

ther innovative designs

he final figures in this chapter present innovative designs either s concepts, or in the case of the PV cube in Santa Ana (figure 54), s a completed project. The scope of using PV-NBS is unconfined s long as adequate solar access can be sourced to generate sable electricity either at point-of-use or fed directly into the lectricity grid.

he PV sunflower feature (figure 47) is prominently positioned in he grounds of an Austrian school, maximising its exposure to tudents, parents and passing public.

he PV planter (figure 48) is intended as a life-support system for rare tree. The PV provides power for water pumps and soil nonitoring sensors.

Fig 46 'Solar sail' grid-connected PV system
Source: ECN, the Netherlands

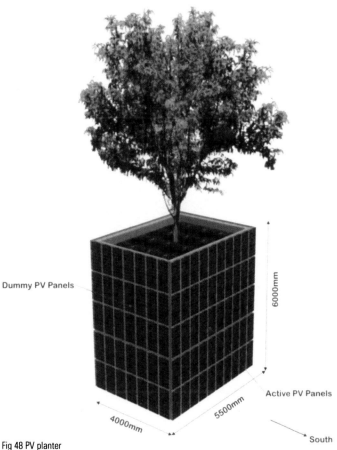

Dummy PV Panels

Active PV Panels

6000mm

5500mm

4000mm

South

Fig 48 PV planter
Source: Mike Rhodes, PV Systems EETS, Wales

ig 47 PV sunflower in school grounds, Gleisdorf, Austria
Source: Heinrich Wilk, Austria

Fig 50 Eye-catching PV advertising billboard with rotating sections
Source: Heinrich Wilk, Austria.

Fig 51 PV solar path that charges during the day and illuminates at night, Demosite, Japan
Source: Henk Kaan, ECN

Fig 49 PV sunflower
Source: NEDO, Japan

Fig 52 Man-made 7 kWp PV tree in Gleisdorf, Austria, contrasting its historical building context and providing both a main urban feature and shading for the street market below
Source: Heinrich Wilk, Austria

n operation since 1993, Steckborn boosts the first BiPV church in he world. Supported by the Swiss Federal Office of Energy, this 9 kWp installation comprises M65 Siemens modules (figure 53).

Fig 53 19 kWp PV church tower
clock, Steckborn, Switzerland
Source: Thomas Nordmann, Switzerland

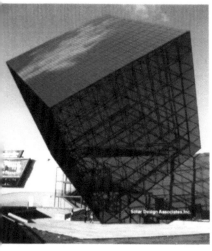

Fig 54 PV cube,
Discovery Science Center,
Santa Ana, California
Source: Solar Design Associates, USA

Conclusion

PV-NBS solutions are creating a new image and inspiration for city environments. The variety of examples provided in this chapter emphasises the extent to which solar power can be utilised to meet important infrastructure requirements in urban areas, with economic and aesthetic benefits. Importantly, it is also bringing the technology into the public realm and increasing understanding and appreciation of the role that integrated PV products can play. The more NBS applications enter the urban marketplace, the more likely that solar access needs and support structure functions will be better interpreted and accepted by urban planning authorities and other stakeholders. This can only help to alleviate institutional barriers that might confront the approval of PV-NBS.

Fig 55 The solar globe is a PV-powered advertising hoarding. The PV provides power to rotate the advert 'strip' and to light it from the top and bottom lips of the strip at night. The globe also incorporates a weather monitoring station.
Source: Mike Rhodes, PV systems EETS, Wales

CHAPTER FIVE

BIPV POTENTIAL AND DESIGN TOOLS

Introduction

This chapter commences by discussing the relevance of building surface orientations and their potential for harvesting solar power. A BiPV potential study at a city scale from IEA participating countries is presented including prospective opportunities. A review and analysis of BiPV design tools and consideration of related issues complement this, including key issues associated with pre-system design and model simulation approaches. Evaluation of this kind is an important process in understanding both the various design options available and making a case for BiPV applications. It also helps designers, planners, decision makers and other related practitioners establish prospective benefits and constraints on a case-by-case level.

How much sunshine can a building harvest?

BiPV faces a variety of solar architectural factors and requirements in order to achieve high-quality building integration and solar yield. Three important questions arise:

- What is the best solar yield?
- What is a good solar yield?
- What is the optimum solar yield?

The main types of building integration and associated tilt angles can be distinguished and related to the time/period of production and consumption of solar electricity. These include annual, seasonal and daily/hourly solar energy yield.

Type of integration	Angles to be discussed
1 Flat roof	0° and other for trade-off discussion
2 Sloped roof	eg 30°, 45°, 60°
3 Façade	90°
(shadovoltaic)	(0° – 90° for PV sun protection)

Fig 1 Integration typologies and associated tilt angles
Source: NET Nowak Energy & Technology Ltd., Switzerland

In general, one of the most critical factors to be considered is the orientation of the building envelope elements towards the sun. The solar yield for flat and sloped roofs as well as for façades is highlighted under the following topics:

- Annual solar yield for the building envelope: how does the orientation of the building envelope influence the solar yield?
- Seasonal solar yield for the building envelope: how does the orientation of the building envelope influence the solar yield through the seasons?
- Daily/hourly solar yield for the building envelope: how does the orientation of the building envelope influence the solar yield throughout the day?
- Trade-off on flat roofs: how does the type of integration influence the solar yield? What is the best/optimal use of flat roofs?
- BiPV potential in cities: how much solar electricity can a city produce?

In this chapter, the annual solar yield is discerned for a selection of cities across the world (figure 2). One location, Zurich in Switzerland, is researched in order to better understand the value and variation of seasonal and daily/hourly solar yield, the trade-off associated with flat roof integration and finally, the potential is assessed.

Annual solar yield for the building envelope

Solar radiation is unequally distributed on surfaces of different orientation. It is obvious that surfaces oriented to the equator and tilted by the degree corresponding to the local latitude usually promote optimal annual solar yield. For locations such as Los Angeles, at 33° 5'N or Sydney at 33° 5'S, sloped roof areas tilted by some 30° yield the maximum annual solar input possible. But this only partly reflects the BiPV opportunities. What is less commonly understood, but essential to keep in mind, is that a wide range of surface orientations obtain a very usable annual solar yield at only a few per cent less than the maximum. Further, these surfaces offer particularly interesting features with respect to the seasonal and daily/hourly solar yield.

Some basic principles and trends for BiPV and solar yield, with reference to the annual input, can be drawn:

- The best solar yield is usually obtained on surfaces oriented to the equator and tilted by the degree corresponding to the local latitude (although in some cases a lower tilt is optimal, for example Zurich, with a latitude of 47° 4' exhibits a preferred tilt of 30° from the horizontal plane).

- Good solar yield is also available for a very wide range of surface orientation, typically inclined from 0° up to 60° or even more, and/or deviating from the equator direction by at least 60°.

Figure 2 reflects the solar yield for flat and sloped roofs as well as for façades. The relative solar yield (per cent of maximum local annual solar input) facilitates the comparison of the different building surfaces and the various locations. The following characteristics are provided in figure 2:

- Distribution of the annual incident solar energy on a tilted surface as a function of slope (tilt) and azimuth.
- Relative solar yield for flat roofs.
- Best tilt for sloped roofs (best azimuth is usually 0° = equator-facing direction).
- Best azimuth for façades.
- Good solar yield for sloped roofs (tilt: 30°) and façades.

	Solar diagram	Description	Flat roof	Sloped roof	Façade
Stockholm	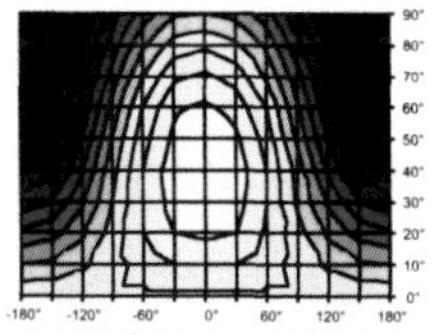	Latitude: 59.1°N Longitude: 17.6°E Altitude: 5 m Best solar yield: 1145 kWh/m²	Solar yield: 84%	Best tilt: 40° Good yield area on x-axis: 0° to 85° Good yield area for 30° tilt: −90° to +90°	Best azimuth: 0° Best yield: 75% Good yield area: −80° to +80°
Zurich	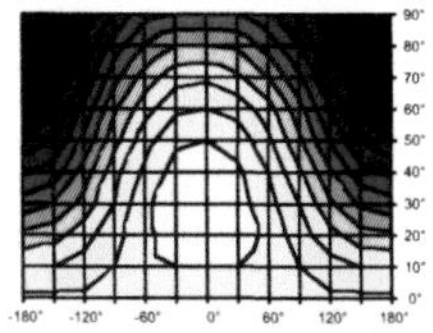	Latitude: 47.4°N Longitude: 8.6°W Altitude: 556 m Best solar yield: 1167 kWh/m²	Solar yield: 91%	Best tilt: 30° Good yield area on x-axis: 0° to 75° Good yield area for 30° tilt: −110° to +110°	Best azimuth: 0° Best yield: 65% Good yield area: −100° to +100°
Tokyo	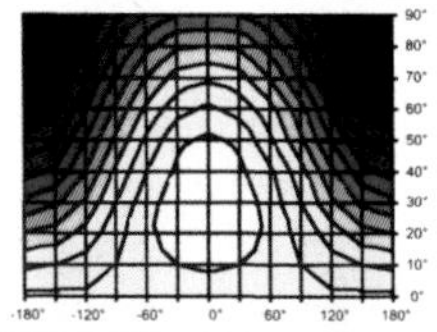	Latitude: 35.3°N Longitude: 139.5°E Altitude: 5 m Best solar yield: 1350 kWh/m²	Solar yield: 91%	Best tilt: 26° Good yield area on x-axis: 0° to 75° Good yield area for 30° tilt: −105° to +105°	Best azimuth: 0° Best yield: 65% Good yield area: −90° to +90°
Los Angeles	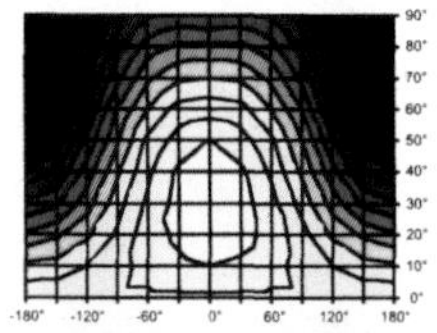	Latitude: 33.5°N Longitude: 118.1°W Altitude: 10 m Best solar yield: 2103 kWh/m²	Solar yield: 89%	Best tilt: 28° Good yield area on x-axis: 0° to 70° Good yield area for 30° tilt: −100° to +100°	Best azimuth: ±30° Best yield: 61% Good yield area: −95° to +95°
Sydney	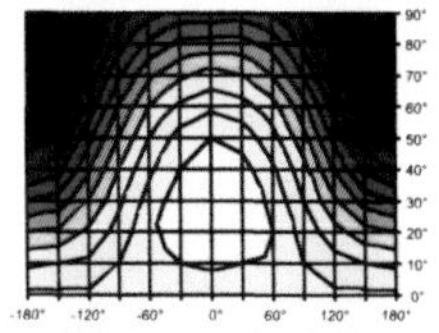	Latitude: 33.5°S Longitude: 151.2°E Altitude: 5 m Best solar yield: 1744 kWh/m²	Solar yield: 91%	Best tilt: 30° Good yield area on x-axis: 0° to 70° Good yield area for 30° tilt: −105° to +105°	Best azimuth: ±30° Best yield: 63% Good yield area: −95° to +95°
Mexico City	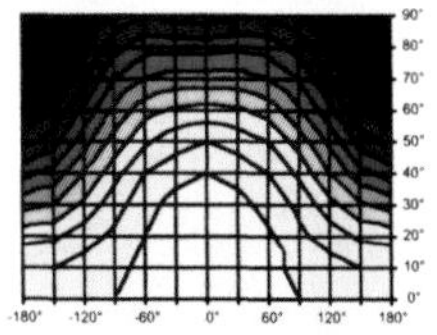	Latitude: 19.2°N Longitude: 99.1°W Altitude: 2277 m Best solar yield: 1903 kWh/m²	Solar yield: 95%	Best tilt: 18° Good yield area on x-axis: 0° to 60° Good yield area for 30° tilt: −120° to +120°	Best azimuth: ±60° Best yield: 55% Good yield area: −115° to +115°
Singapore	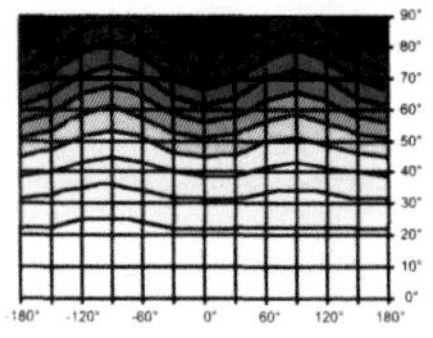	Latitude: 1.1°N Longitude:104.1°E Altitude: 5 m Best solar yield: 1626 kWh/m²	Solar yield: 100%	Best tilt: 0° Good yield area on x-axis: 0° to 45° Good yield area for 30° tilt: −180° to +180° (all around)	Best azimuth: ±90° Best yield: 54% Good yield area: −20° to −160° and +20° to +160°

Solar yield: | 90–100% | 80–90% | 70–80% | 60–70% |

Azimuth: 0° for equator direction, + values for west orientation, − values for east orientation, 180° for pole direction

Fig 2 Annual solar yield for different locations on the globe
Source: NET Nowak Energy & Technology Ltd, Switzerland; Meteonorm for absolute maximum values, PVSYST 3.1 for relative values

185 •

Some basic principles and trends for BiPV and solar yield referring to the latitudes can be drawn. Exceptions to these trends are due to specific local meteorological conditions such as a high share of diffuse light, a characteristic of overcast Zurich days.

The higher the latitude of the location, away from the equator line:

- The lower the relative solar yield of flat roof areas.
- The higher the tilt of best yield.
- The wider the tilt spread for south- or north-facing roof areas (depending on hemisphere) able to access good solar yield.
- The tighter the azimuth spread for roof areas (tilt 30°) obtaining good solar yield.
- The higher the relative solar yield of façades.
- The tighter the azimuth spread for façade areas obtaining good solar yield.

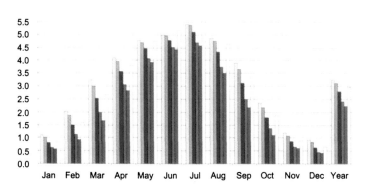

(a) □ South ■ SW / SE ■ West / East ■ NW / NE ■ North Horizon

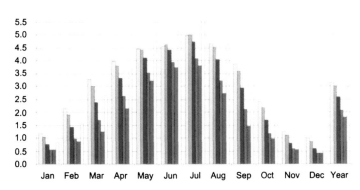

(b) □ South ■ SW / SE ■ West / East ■ NW / NE ■ North Horizon

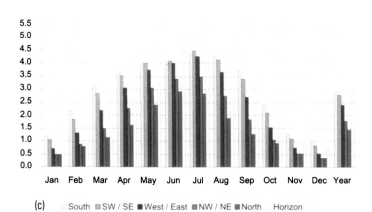

(c) □ South ■ SW / SE ■ West / East ■ NW / NE ■ North Horizon

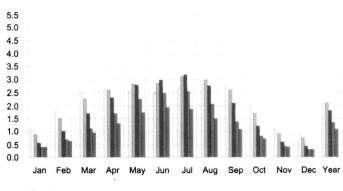

(d) □ South ■ SW / SE ■ West / East ■ NW / NE ■ North Horizon

Fig 3 Average daily global irradiation in kWh per square metre for roof areas tilted by a) 30°, b) 45°, c) 60° and d) for façades during the year in the urban area of Zurich, Switzerland, 47° 2'N 8° 3'E, 413 m, albedo = 0.2
Source: PVSYST 3.1 simulation

Seasonal solar yield for the building envelope

Seasons are a product of the tilting of the earth's axis as it orbits around the sun. The sun's height consequently varies according to the season and is more significant at higher latitudes. Figure 4 demonstrates a northern hemisphere example for determining the angle of incidence of a tilted PV module relative to the position of the sun at a given location and time of year. How the solar yield follows the seasonal changes is shown in the example for Zurich (figure 5). Average daily global irradiation values are given for each month of the year and for surfaces facing south, southwest and southeast, west and east, northwest and northeast and north.

Some basic principles and trends for BiPV and seasonal solar yield can be drawn. The higher the tilt of the surface:

- The higher the solar input in winter.
- The lower the solar input in summer.
- The tighter the azimuth spread for areas obtaining good solar yield.

The seasonal solar yield, both absolute and relative, is strongly correlated with surface orientation. Optimal or good annual solar yields can differ from optimal or good seasonal solar yields. Roof surfaces of higher tilts and façades oriented slightly away from north in the southern hemisphere and south in the northern hemisphere, achieve better solar yields in winter, such as in areas of moderate latitudes.

A clear-sky model can be applied for a more precise energy harvest calculation. Using the clear-sky model for global irradiation, the potential daily solar input and intensity can be calculated and compared between days throughout the year.

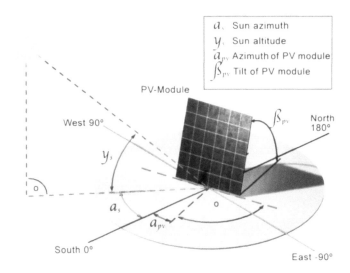

Fig 4 Angle of solar incidence on a tilted PV module
Source: Martin Van Der Laan (M. Art)

kWh/ m²/day	21 Jun	21 May/ 21 Jul	21 Apr/ 21 Aug	21 Mar/ 21 Sep	21 Feb/ 21 Oct	21 Jan/ 21 Nov	21 Dec	Average Year
0°	8.47	8.07	6.84	5.07	3.28	1.93	1.46	5.04
30°	8.21	8.14	7.81	6.91	5.48	3.95	3.28	6.37
45°	7.50	7.54	7.60	7.22	6.10	4.63	3.91	6.50
60°	6.36	6.54	6.93	7.09	6.34	5.01	4.29	6.24
90°	3.36	3.66	4.58	5.56	5.65	4.81	4.24	4.70

Fig 5 Values of daily solar yield (clear sky global irradiation) of south-facing areas in Zurich, Switzerland per day
Source: NET Nowak Energy & Technology Ltd, Switzerland, using PVSYST 3.1 Simulation tool

Horizontal area – flat roof

The sum of the solar irradiation per day and the length of sunshine follow the sun's height over the year, and in summer, produces almost 16 hours of sunshine and 8.5 kWh per square meter of irradiation. Minimum is 8 hours of sunshine and only 1.5 kWh per square metre of irradiation. Values for spring/autumn are exactly in-between: 12 hours of sunshine resulting in 5 kWh per square metre of irradiation.

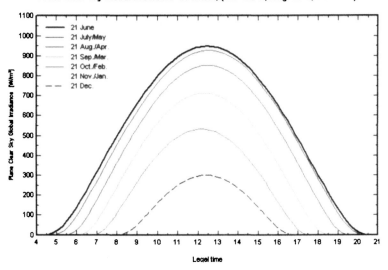

Inclined area – sloped roof tilt at 30°

The sum of the solar irradiation per day and the length of sunshine have high values during a longer period in summer. Due to the tilt of 30°, the values are multiplied by 2 in winter. The maximum in length is around 14 hours of (direct) sunshine and around 8 kWh per square metre of irradiation during a fairly long period in summer. Minimum is also 8 hours of sunshine but the sum of irradiation is over 3 kWh per square metre.

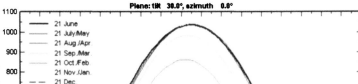
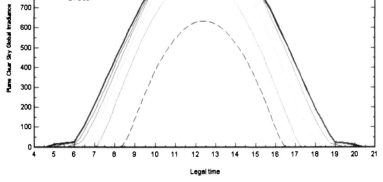

Inclined area – sloped roof tilt at 45°

The sum of the solar irradiation per day and the length of sunshine have fairly high and stable values during half of the year. The maximum in length is around 12 hours of (direct) sunshine and around 7.5 kWh per square metre of irradiation during a fairly long period in summer. The values in winter are 8 hours of sunshine and the sum of irradiation yields around or over 4 kWh per square metre.

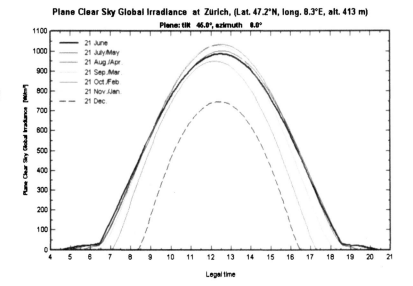

Inclined area – sloped roof tilt at 60°

Maximum values for the solar irradiation per day are not in summer but in spring and autumn. The areas then yield around 7 kWh per square metre. The length of sunshine is fairly stable during more than half of the year. The sun shines for about 11 hours per day. Compared with differently oriented areas, the areas have best values in winter with up to 5 kWh of irradiation. Again, the length of sunshine is a minimum of 8 hours.

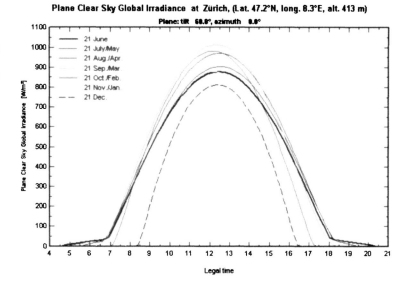

Plane Clear Sky Global Irradiance at Zürich, (Lat. 47.2°N, long. 8.3°E, alt. 413 m)
Plane: tilt 60.0°, azimuth 0.0°

Vertical area – façade

The sum of the solar irradiation per day is almost a reversal of the values found with horizontal areas. Maximum values of 5.5 kWh per square metre are in (early) spring and in (late) autumn. In summer, the values are only 3.5 kWh per square metre. The length of (direct) sunshine is around 10 hours. The values in winter are around 4.5 kWh per square metre of irradiation and, of course, 8 hours of sunshine.

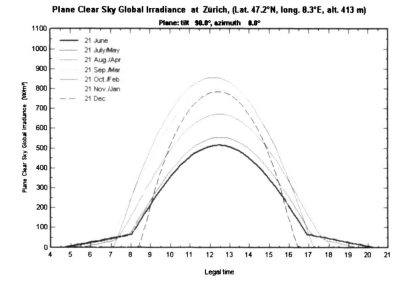

Plane Clear Sky Global Irradiance at Zürich, (Lat. 47.2°N, long. 8.3°E, alt. 413 m)
Plane: tilt 90.0°, azimuth 0.0°

Fig 6 Solar yield (clear sky global irradiation per day)
of south-facing areas in Zurich, Switzerland
Source: NET Nowak Energy & Technology Ltd, Switzerland,
using PVSYST 3.1 Simulation tool

Daily/hourly solar yield for the building envelope

The sun's height also depends on the rotation of the earth, hence differently oriented surfaces obtain solar irradiation at different intensity levels at different times through the day. Values of global solar irradiation on a clear 21st June (see box) in watts per square metre are given for façades and for surfaces facing south, southwest and southeast, west and east, northwest and northeast and north. Some basic principles and trends for BiPV and solar yield can be drawn from seasonal variations. The closer the azimuth of a slightly inclined roof surface to the west or east:

- the earlier or later in the day that high(er) solar yield can be obtained;
- the lower the peak output;
- the closer the azimuth of the façade surface to the west or east.

The daily/hourly solar yield – both absolute and relative – is strongly correlated with the orientation of the surface. Best/good daily/hourly solar yields can partly differ from best/good annual and/or seasonal solar yields. If solar electricity production is supposed to match specific daily/hourly loads, for example, before or after noon, then roof surfaces of higher tilts and façades facing east or west may have better solar yields. This may be particularly true (almost) all year round (locations close to the equator) or for seasonal periods (locations in moderate latitudes). Some specific applications are possible with bifacial elements, that is, two photovoltaic active areas, one surface facing east, the other facing west.

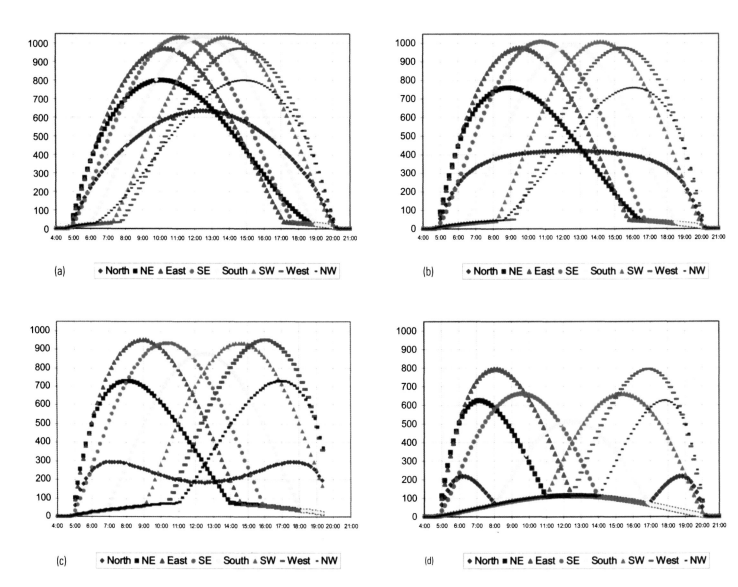

Fig 7 Global solar irradiation on a clear 21st June (local time) in watts per square metre for roof areas tilted by a) 30°, b) 45°, c) 60° and d) façades in Zurich, Switzerland
Sources: NET Nowak Energy & Technology Ltd., Switzerland, using PVSYST 3.1 for raw data.

Trade-off on flat roofs

The trade-off on flat roofs is illustrated for Zurich. There are two principles concerning best use of flat roof areas:

- Best yield per square metre of active area is obtained at a tilt angle of about 30–45°, the sum of optimal annual solar electricity production. However, the PV modules cause shading and only a part of the area can be used, hence implying a lower ratio 'module area/roof area'.

- Most roof area can be used by horizontal integration of the PV modules. The roof area available can be used 100 per cent as the PV modules have no shading effects, hence implying a ratio 'module area/roof area' of one. However, the solar yield is a little less than 90 per cent of the maximum possible.

For optimisation of the solar yield and the ratio 'module area/roof area', a further element has to be considered: the shading angle limit. The greater the angle of the PV module/array and the closer the PV arrays are, the higher the shading angle and the more the solar yield is affected by shading. That is why the (limit for the) shading angle has to be considered. A very low shading angle hardly affects the solar yield, that is, a shading angle limited at 10° reduces the solar yield by far less than

kWh/ m²/day	North	Northeast/ northwest	East/west	Southeast/ southwest	South
0°	8,466	8,466	8,466	8,466	8,466
30°	6,749	6,985	7,768	8,231	8,207
45°	5,291	5,786	7,145	7,621	7,500
60°	3,537	4,756	6,436	6,717	6,357
90°	1,899	3,409	4,740	4,422	3,366

Fig 8 Sum of the global solar irradiation on a clear 21st June in watts per square metre for areas in Zurich, Switzerland

Source: NET Nowak Energy & Technology Ltd., Switzerland

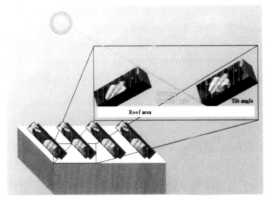

Fig 9 Sketch showing the relevant elements for trade-off discussion about flat-roof use

Source: NET Nowak Energy & Technology Ltd., Switzerland

Local and solar time

In most locations, there will normally be a difference between solar and local time. Solar time is determined by the position of the sun. At noon it is directly overhead, with sunrise and sunset occurring at symmetrical times either side of noon. Local (or clock) time is determined by the local time zone and is taken at a reference longitude. For each degree of difference in longitude between the actual and reference, there is a 4-minute time difference.

Thus, to convert solar time to local time, a simple formula is used:

Tlocal = Tsolar + ((Longitude – Longituderef) * 4)

The table below details the solstice and equinox dates for both hemispheres:

DATE	Southern Hemisphere	Northern Hemisphere	Description
Summer Solstice	22 December	22 June	Sun at its highest noon altitude
Autumn Equinox	21 March	21 September	Sun rises due east, sets due west
Winter Solstice	21 June	21 December	Sun at its lowest noon altitude
Spring Equinox	21 September	21 March	Sun rises due east, sets due west

one per cent. The reduction is three per cent for each step increase of shading angle up to 20°, 25° and finally 30° where global yield loss equals ten per cent. Higher shading angles cause much higher losses and should be avoided (figure 10).

A strict shading angle limit implies a lower ratio 'module area/roof area', that is, 5° for both the shading angle and the tilt of the active area result in a ratio of 0.5. The higher the tilt of the modules mounted in arrays, the lower the ratio is (horizontal axis). By accepting a higher shading angle, the ratio can be improved (vertical axis). To limit the solar yield losses, the shading angle should not exceed 30° (figure 11).

In order to optimise roof area utilisation and yield, the yield per roof area is applied. A simple multiplication of the annual irradiation in kWh per square metre of active area and the ratio 'module area/roof area' results in the annual irradiation in kWh per square metre of roof area. Of course, these figures do not take into account all financial and architectural aspects. They nevertheless reflect the two principles. Low tilt angles yield high irradiation values – best options appear to be horizontal integration but dirt and water tend to stay longer on horizontal areas and can provoke corrosion. On surfaces tilted by 5°, rainwater flows away and takes the dirt with it. Allowing losses due to shading, a fairly good value is obtained, that is, the active area tilted by 5° with a shading angle of 20° has a very high ratio of 0.81 and 1077 kWh of irradiation per square metre of active area. The solar yield is around 90 per cent of the maximum annual solar yield for the active area; the loss of some ten per cent is half due to sub-optimal orientation (tilt) and half due to the shading angle. Furthermore, the angle of 5° is less problematic for aesthetic reasons. Of course, optimal integration cannot be generalised but has to be considered specifically with every object (figure 12).

Annual irradiation in kWh/m² of active area for Zurich, Switzerland, 47° 2'N, 8° 3'E, 413 m

Limit shading angle	Tilt angle	0°	5°	10°	20°	30°	45°	60°
	0°	1,088	1,110	1,143	1,155	1,167	1,166	1,143
	5°	1,087	1,109	1,141	1,152	1,163	1,163	1,141
	10°	1,083	1,105	1,137	1,148	1,159	1,159	1,137
	20°	1,056	1,077	1,109	1,119	1,130	1,130	1,109
	25°	1,028	1,049	1,079	1,090	1,100	1,100	1,079
	30°	989	1,009	1,038	1,048	1,058	1,058	1,038
	45°	832	849	874	882	890	890	874

Fig 10 Annual irradiation in kWh per square metre of active area for Zurich, Switzerland
Source: PVSYST 3.1 Simulation

Ratio 'module area/roof area' for Zurich, Switzerland, 47° 2'N, 8° 3'E, 413 m

Limit shading angle	Tilt angle	0°	5°	10°	20°	30°	45°	60°
	0°	1	0	0	0	0	0	0
	5°	1	0.50	0.34	0.21	0.15	0.11	0.10
	10°	1	0.67	0.51	0.35	0.27	0.21	0.18
	20°	1	0.81	0.68	0.53	0.45	0.38	0.35
	25°	1	0.84	0.74	0.60	0.52	0.45	0.43
	30°	1	0.87	0.78	0.65	0.58	0.52	0.50
	45°	1	0.92	0.86	0.78	0.73	0.71	0.73

Fig 11 Ratio 'module area/roof area' for Zurich, Switzerland
Source: PVSYST 3.1 for raw data

Annual irradiation in kWh/m² roof area for Zurich, Switzerland, 47° 2'N, 8° 3'E, 413 m

Limit shading angle	Tilt angle	0°	5°	10°	20°	30°	45°	60°
	0°	1,088	0	0	0	0	0	0
	5°	1,087	554	388	242	174	128	114
	10°	1,083	740	580	402	313	243	205
	20°	1,056	872	754	593	508	429	388
	25°	1,028	881	799	654	572	495	464
	30°	989	878	810	681	614	550	519
	45°	832	781	751	688	650	632	638

Fig 12 Annual irradiation in kWh per square metre roof area for Zurich, Switzerland
Source: PVSYST 3.1 for raw data

Building category	Number of buildings	BiPV (roof area) potential with solar yield > 0.9 in square metres	%	BiPV (roof area) potential with solar yield > 0.8 in square metres	%
Residential buildings (1 unit)	9,877	90,758	3	331,651	7
Residential buildings (> 1 unit)	16,631	952,464	36	1,747,631	38
Residential/commercial buildings	8,016	374,694	14	758,050	16
Administrative buildings	1,013	197,067	7	236,845	5
Commercial buildings	3,119	416,707	16	595,995	13
Tourism buildings	486	37,904	1	58,996	1
Industrial buildings	1,952	272,730	10	376,673	8
Agricultural buildings	453	26,615	1	33,046	1
Buildings for cultural use	115	19,083	1	29,564	1
Buildings for sports	671	82,112	3	82,112	2
School buildings	722	90,780	3	140,395	3
Hospital buildings	230	10,979	0	99,090	2
Church buildings	234	24,591	1	49,441	1
Other buildings	3,488	71,609	3	78,585	2
All buildings	47,007	2,668,093	100	4,618,072	100

Fig 13 BiPV (roof area) potential in the city of Zurich for 14 building categories
Source: NET Nowak Energy & Technology Ltd., Switzerland

BiPV potential in cities

By looking at the annual solar yield, it can be seen that a wide range of surface orientations, and thus building surfaces, obtain a solar energy harvest of 90 per cent of the maximum possible. A considerable share of the building envelope yielding 80 per cent of the maximum annual solar input achieves very favourable seasonal and/or daily/hourly values. These areas can therefore make a valuable contribution to the optimisation of solar electricity production and supply.

The assessment of the BiPV potential in several countries showed that per 100 square metres of ground floor area, an average of 40 square metres of roof area and another 15 square metres of façade area are available for BiPV use. About half of the unsuitable roof area is typically due to construction, historical and shading elements, and half due to insufficient solar yield. Façade areas contribute about one-seventh of the solar electricity production potential. The ratio 'solar electricity production potential/electricity consumption' is around 30 per cent for several countries. For cities, this ratio happens to be smaller despite the considerable building stock present in these areas as electricity consumption is often proportionally higher per square metre of ground floor area.

Nevertheless, the BiPV potential is considerable as the example of a case study in the city of Zurich illustrates (figure 13). Here, the BiPV roof area potential for the existing building stock is presented, with suitable areas being defined as having at least 80 per cent and 90 per cent of the maximum annual solar yield. About 20 per cent of the gross roof area (13.7 square kilometres) is suitable and has a fairly high solar yield; another 14 per cent has a fairly good yield. About 31 per cent of the roof area is unsuitable for construction elements and another 22 per cent for shading; the remaining roof area has to be eliminated purely on the grounds of insufficient solar yield.

New constructions and, above all, renovations, represent an interesting dynamic BiPV potential. Taking average values, there are about 30,000 square metres of suitable roof area on new constructions and another 40,000 square metres within the category of renovations. This area is considerable and would permit the installation of several MWp of solar power each year. The solar electricity production potential with today's 'common' technology is 270 GWh/y for the 90 per cent yield roof area and another 170 GWh/y for the 80 per cent yield roof area, totalling 440 GWh/y. These values roughly correspond to one tenth and one sixth, respectively, of the actual electricity consumption of Zurich.

Solar architectural rules of thumb for BiPV potential	Solar architectural rules of thumb for BiPV potential on *Roofs*	Roofs and façades	Solar architectural rules of thumb for BiPV potential on *Façades*
Ground floor area	1 m²	Base of BiPV potential in relative terms	1 m²
Gross area	1.2 m²	Ratio 'gross area/ground floor area'	1.5 m²
	60%	Suitable building envelope parts taking into account construction, historical and shading elements, including vandalism factor	20%
Architecturally suitable area	0.72 m²	Ratio 'architecturally suitable area/ground floor area'	0.3 m²
	55%	Suitable building envelope parts taking into account sufficient solar yield	50%
Solar architecturally suitable area	0.4 m²	Ratio 'solar architecturally suitable area/ground floor area' (utilisation factor)	0.15 m²

Fig 14 Rules of thumb for PV potential calculations
Source: NET Nowak Energy & Technology Ltd., Switzerland

International BiPV potential findings

Based on the case studies and further data sent by the partners of the participating IEA countries, some rules of thumb can be derived, as shown in figure 14.

The methodology can be used in order to collect essential global figures and generate corroborated BiPV area potential data.

Architecturally suitable solar areas on buildings

As mentioned, the architecturally unsuitable part of most buildings is 40 per cent for roofs and 80 per cent for façades. About half of the remaining architecturally suitable building area does not obtain any good solar yield, with 45 per cent for roof areas and 50 per cent for façade areas. Finally, the ratio 'architecturally suitable solar area/ground floor area' (called the utilisation factor) can be calculated. A ground floor area of 100 square metres results correspondingly in 40 square metres of solar-architecturally suitable roof area (utilisation factor of 0.4) and in 15 square metres of solar-architecturally suitable façade area (utilisation factor of 0.15).

It can be stated that the *relative* values reflected in the utilisation factors vary less and are more coherent on an international level between countries and world regions. The *absolute* figures for the BiPV potential in square metres vary much more, even when the ground floor area per capita is considered. That is why the rules of thumb in relative terms are valuable and operational global key figures, whereas in absolute terms the ground floor areas can be aggregated for different regions on the globe with similar building stock characteristics.

Rules of thumb for architecturally suitable solar building envelope area in absolute terms for central western Europe

The ground floor area can be aggregated for an area, in this case central western Europe. A statistically typical building for a person living in central western Europe has about 45 square metres of ground floor area. Half is used for residential purposes, 7 square metres for the primary sector, 6 square metres each for the secondary and tertiary sectors and the rest for other purposes.

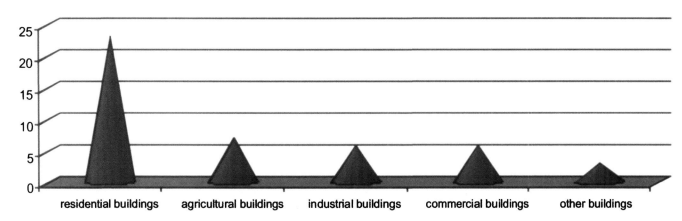

g 15 Use of the ground floor area of a statistically typical building in central western Europe (45 square metres)
ource: NET Nowak Energy & Technology Ltd., Switzerland

pplying the corresponding overall utilisation factor of 0.4 for roofs and 0.15 for façades (for the building stock), the solar-rchitecturally suitable building roof and façade areas per capita re calculated for central western Europe (figure 16).

BiPV potential for central Western Europe for roof and façade areas in square metres per capita (approximate figures)

Roof areas		Façade areas
9.0	Residential buildings	3.5
3.0	Agricultural buildings	0.5
2.5	Industrial buildings	1.0
2.5	Commercial buildings	1.0
1.5	Other buildings	0.5
18.0	All buildings	6.5

ig 16 BiPV potential for central western Europe for roof and façade areas in square metres per capita
ource: NET Nowak Energy & Technology Ltd., Switzerland

here are 18 square metres of roof area per capita potentially usable for PV with a good solar yield. Typically, the roof area characterised by very high solar yield (including only surfaces with more than 90 per cent of the maximum yield) is half of the roof area with good solar yield. Additionally, there are 6.5 square metres of façade area per capita fulfilling solar architectural equirements, and hence potentially usable for PV. About a third of the BiPV area potential is attributed to roof areas and a quarter o façade areas.

The statistical building is – compared to the ground floor area of 45 square metres per capita for central western Europe – much larger in the US and Australia, where the ground floor area is about double. This is mainly due to much higher values for available residential building areas. Consequently, the share of ground floor area for residential purposes is larger and makes up around two-thirds. Japan has just 20 square metres of ground floor area.

Determination of the BiPV potential for selected IEA countries

By linking the average figures of solar-architecturally suitable area per capita to country-specific features, (mainly population size and annual solar irradiation) the solar electricity potential can be calculated (figure 17).

More precisely, the formula ingredients are:

- Building type: residential, agricultural (primary sector), industry (secondary sector), commercial (tertiary sector), other and total (all building stock).
- Available area per capita: average figures/standards. The values are given in square metres.
- Utilisation factor (suitability in relative terms) of 0.4 for roofs and 0.15 for façades.
- Population size: number of people living in the country in millions.
- Solar yield: weighted average relative yield of good areas per geographical unit (countries).
- Solar irradiation: country-specific weighted value for the maximum annual solar input in kWh/y/m².
- Global conversion efficiency: ratio of 'electricity output/solar irradiation' (simplified ratio: generally 10 per cent).
- Production of solar electricity: product of the factors described above in TWh/y.

Applying this calculation scheme leads to the following figures for the solar electric BiPV potential (figure 19).

Fig 17 Calculation scheme for the production of BiPV solar electricity
Source: NET Nowak Energy & Technology Ltd., Switzerland

BiPV area potential (in km²)		Residential buildings	Agricultural buildings	Industrial buildings	Commercial buildings	Other buildings	All buildings
Australia	Roof	373.50	22.50	6.00	16.5	3.75	422.25
	Façade	140.06	2.81	2.25	8.25	1.41	158.34
Austria	Roof	85.65	17.13	15.19	17.45	4.20	139.62
	Façade	32.12	2.14	5.70	8.73	1.58	52.36
Canada	Roof	727.20	36.36	60.60	133.32	6.06	963.54
	Façade	272.70	4.55	22.73	66.66	2.72	361.33
Denmark	Roof	50.88	14.84	10.60	10.60	1.06	87.98
	Façade	19.08	1.86	3.98	5.30	0.40	32.99
Finland	Roof	78.28	21.01	19.16	8.45	0.41	127.31
	Façade	19.08	1.86	3.98	5.30	0.40	32.99
Germany	Roof	721.78	164.04	229.66	164.04	16.40	1,295.92
	Façade	270.67	20.51	86.12	82.02	6.15	485.97
Italy	Roof	410.26	113.96	136.75	91.17	11.40	763.53
	Façade	153.85	14.25	51.28	45.58	4.27	286.32
Japan	Roof	753.88	40.48	75.89	91.07	5.06	966.38
	Façade	282.71	5.06	28.46	45.54	1.90	362.39
Netherlands	Roof	127.48	42.70	52.75	35.80	0.63	259.36
	Façade	47.81	5.34	19.78	17.90	0.24	97.26
Spain	Roof	251.97	78.74	55.12	55.12	7.87	448.82
	Façade	94.49	9.84	10.67	27.56	2.95	168.31
Sweden	Roof	134.52	36.11	32.92	14.51	0.71	218.77
	Façade	50.45	4.51	12.35	7.26	0.27	82.04
Switzerland	Roof	67.12	21.90	21.05	12.80	15.36	138.22
	Façade	25.17	2.74	7.89	6.40	5.76	51.83
United Kingdom	Roof	601.88	71.09	61.61	168.24	11.85	914.67
	Façade	225.70	8.89	23.10	84.12	4.44	343.00
United States	Roof	6,791.83	322.91	602.76	2,260.36	118.40	10,096.26
	Façade	2,546.94	40.36	226.04	1,130.18	44.40	3,786.10

Fig 18 BiPV area potential for roofs and areas of some selected IEA countries, sorted according to residential, agricultural, industrial, commercial and other buildings (in square kilometres) and fulfilling good solar yield criteria (80 per cent of the maximum local annual solar input, separately defined for sloped roofs and façades and individually for each location)
Source: NET Nowak Energy & Technology Ltd., Switzerland

Solar electricity BiPV production potential	Potential production of solar electricity (TWh/y) on *roofs*	Potential production of solar electricity (TWh/y) on *façades*	Potential production of solar electricity (TWh/y) on *building envelope*	Actual electricity consumption (in TWh)	Ratio 'solar electricity production potential/ electricity consumption'
Australia	68.176	15.881	84.057	182.24	46.1%
Austria	15.197	3.528	18.725	53.93	34.7%
Canada	118.708	33.054	151.762	495.31	30.6%
Denmark	8.710	2.155	10.865	34.43	31.6%
Finland	11.763	3.063	14.827	76.51	19.4%
Germany	128.296	31.745	160.040	531.64	30.1%
Italy	103.077	23.827	126.904	282.01	45.0%
Japan	117.416	29.456	146.872	1,012.94	14.5%
Netherlands	25.677	6.210	31.887	99.06	32.2%
Spain	70.689	15.784	86.473	180.17	48.0%
Sweden	21.177	5.515	26.692	137.12	19.5%
Switzerland	15.044	3.367	18.410	53.17	34.6%
United Kingdom	83.235	22.160	105.395	343.58	30.7%
United States	1,662.349	418.312	2,080.661	3,602.63	57.8%

Fig 19 Solar electricity BiPV potential fulfilling a good solar yield (80 per cent of the maximum local annual solar input, separately defined for sloped roofs and façades, and individually for each location/geographical unit)

Source: IEA for electricity consumption for 1998

The role of BiPV design tools in whole building design

Design rules for photovoltaics in buildings are not yet well established. Indeed, given the diversity of possibilities for PV integration and the number of building typologies available for PV, such rules may not even be easy to formulate. At the same time, PV is a valuable building and energy component that deserves special attention in the building design process. Design and analysis tools are important aids to the architect and enable the PV designer to develop the best design solution for PV and to maximise the resulting benefits. PV is a new technology for buildings and may still involve design caveats for traditional building designers. In addition to the general solar surface aspects, PV in buildings has technology-specific features that the designer should be aware of, such as shading and temperature sensitivity. Good PV design tools account for these features and are very useful in avoiding unnecessary design mistakes.

Design tools can also have an important educational function. When design tools are used consistently in the PV design process, they help the designers and architects to obtain a more profound design insight and thus improve their professional competence with PV. For advanced PV designers, tools represent an important aid to develop and test new BiPV concepts before their practical realisation. In such instances, design tools can deliver considerable savings in the development costs of BiPV. They also have a use in the analysis of commissioned projects, often helping to interpret performance outcomes or ambiguities found between planned and measured PV performance.

Design tools are useful throughout the whole building process from the very first sketching of the BiPV to prediction and performance analysis. Based on practical experience, the important tasks in which design tools are particularly useful are:

- pre-design of BiPV systems;
- system optimisation;
- technical planning of the BiPV;
- analysis of system performance.

In many cases the pre-design phase may play the most central role in the BiPV design process as crucial decisions are usually made at the very beginning of the project. For example, architectural integration approaches, selection of appropriate solar technologies and whole BiPV project evaluation are often made before the actual building planning or detailed planning. The early design may include a simple performance analysis, searching for insight on possibilities and even a first crude optimisation of the BiPV system. The level of accuracy required from the tool is clearly less than for technical planning of the PV system.

Finally, when moving into the technical planning phase of PV system design, the questions to be addressed become more detailed. For example, the following tasks need to be accomplished:

- electrical and architectural integration design;
- PV array design;
- custom designs for PV integration;
- accounting for real conditions at site (such as shading and temperature);
- estimation of PV production dynamics, requiring more sophisticated and accurate assessment tools.

It is also important to distinguish between freestanding PV, and PV that is integrated into the building. Obtaining the full benefits from PV in buildings often requires going beyond maximising only the electricity production, which may be the case for free-standing PV systems; it can involve stronger integration of the PV into the building envelope, or handling PV as a building component or even as a building material. These needs place new requirements on the BiPV design tools compared to PV tools in general.

When designing sustainable, ecological or low-energy buildings, the architect and designer need to consider the building as a whole to optimise the overall energy concept. There are several technology options and alternatives available in addition to PV that may require consideration in parallel. From the BiPV point of view, it is important to use building energy design tools that not only consider traditional energy concepts but also are able to properly account for the contributions of BiPV systems. Design situations often have to evaluate the importance of PV as part of the whole building design or building energy system design process. Examples are:

- Where there is a trade-off between different technology options and especially other solar technologies.
- Where there is interaction between PV and other energy flows in buildings.
- Where there is the use of multifunctional PV modules for daylighting, electricity, and heat production.

What tools are available for what purpose?

The design tasks encountered in a BiPV project range from early design to detailed design. The tools available to the designer and architect range from the simple to the sophisticated. The level of sophistication required from the BiPV design tool depends very much on what purpose the tool is used for. One may find simple design guidelines or rules of thumb, computer-aided design (CAD) and simulation tools. A sophisticated tool could, in principle, handle several of the different tasks in the PV design process.

The first design task faced often relates to determining the size of the PV array and estimating the electricity produced. All PC-based PV design tools are capable of doing this, but there are even easier ways to establish a rough estimate. A first very crude estimate of the yearly electricity output of the PV system can be obtained using the following formula:

$$Q_{PV} = \eta \times I_{tot,rad} \times A_{PV}$$

where:

Q_{PV} = the yearly electricity output of the PV system [kWh];

η = the average efficiency of the PV system;

$I_{tot,rad}$ = the yearly total solar radiation on the PV surface [kWh/m²];

A_{PV} = the surface area of the PV system [m²].

The availability of solar radiation on the PV module depends on the location and array orientation. In high latitudes, the total insolation on the collector plane can range between 1,000–1,400 kWh per square metre (latitude 60–45°); 1,400–1,700 kWh per square metre (latitude < 45°); and up to 2,000 kWh per square metre at the equator. Accurate solar radiation values may be found from handbooks and meteorological services and datasets. The PV surface area comes from the BiPV design and from the architect's decision.

Example. In central Europe, the average solar radiation on the collector plane is around 1,300 kWh per square metre. The expected yield per square metre of a crystalline silicon PV system would be 12–15 per cent x 1,300 kWh per square metre, or 130–150 kWh per square metre of PV area. If the PV area is 30 square metres, then the total PV production is 10–12 per cent x 1,300 kWh per square metre x 30 square metres = 3,900–4,680 kWh per year.

The next stage of complexity of design tools would include more detailed factors on the PV technology and building typology. To obtain a more reliable estimate of the PV output, the designer may need to account for the orientation and inclination of the PV modules, temperature effects on PV production, losses in cabling and protective devices, efficiency of the grid-connection (inverter), and so on. Many of these phenomena vary during the day and season and their dynamic nature cannot be handled with simple tools. Dynamic design tools are often based on mathematical PV system and subsystem models, enabling step-by-step simulations of the PV performance for a given set of input parameters. The input parameters may easily be varied in such models to calculate another case and thus the designer may swiftly investigate a range of possible PV system designs as well as the sensitivity of the main PV design to variations or uncertainties in the input data. PC tools for performing a yearly simulation of the PV system are readily available, but most of these have weak points in describing the BiPV-specific features and are best suited to free-standing PV systems.

A flow chart of a step-by-step simulation of BiPV systems is shown in figure 20. The simulation starts with the determination of the global solar radiation on the PV array for which an hourly weather data set is needed. At this stage, the shading effects of nearby obstructions should also be accounted for. Simple pre-design tools seldom account for shading and more sophisticated tools are required to deal with obstructions. The simulation then proceeds through the different subsystems and components, ending up with the net PV power. When calculating the PV electricity output, determining the PV module temperature may become important in hot climates as the output decreases with temperature. Typically, the power output decreases by 0.5 per cent per degree (°C) increase in module temperature. In some cases, the amount of PV electricity that cannot be directly used on site in the building needs to be estimated as the surplus will be fed into the grid. This is of particular interest when the surplus needs to be minimised, for example, if the local utility does not provide any financial compensation for the surplus. Thus at each step, specific issues need to be considered properly to reach good ccuracy in the estimation of the PV power. Not shown here are he secondary effects of PV on the building energy performance: or example if the PV modules are used as shading elements, the uilding cooling load may decrease. Rooftop modules may rovide additional insulation that positively affects both the ooling and heating demand.

n addition to dynamic simulation of the PV system performance, esign tools for BiPV systems should consider the architectural nd structural use of PV elements and the associated phenomena hat affect the PV performance. Thus, instead of a free-standing V module, one may foresee the following typologies relevant for he BiPV design:

- BiPV surfaces in cool climates.
- BiPV surfaces in warm climates.
- BiPV as shade elements.
- Multifunctional BiPV (such as PV/thermal systems).

ach of these BiPV designs may need a special treatment in the esign tool and should not be generalised with one model only. his is mainly because the PV module temperature will differ in ach case. For example, a shading element corresponds to a free-tanding module with ambient air cooling both sides of the nodule, whereas a hot BiPV façade element is tightly integrated nto the building envelope, is less ventilated and consequently will how a much higher module temperature. In addition, these BiPV onfigurations often have a large surface area, such as the roof or açade, which may be subject to shading from adjacent objects. hading of the PV modules is of special concern as the output nay drop significantly even with small shadows.

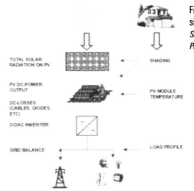

Fig 20 Flow chart of a step-by-step simulation of BiPV systems
Source: Helsinki University of Technology, Peter Lund

Design tools that can handle the BiPV-specific features identified above are still rare. The PVSYST design tool developed at the École Polytechnique de Lausanne and the University of Geneva in Switzerland, and tested within the IEA Task 7, represents an interesting approach for advanced design. PVSYST is capable of dealing with a sophisticated shading geometry and BiPV, and allows for accurate engineering planning of the PV system.

BiPV design tools are particularly relevant in the pre-design phase of a building project and when analysing BiPV amongst other energy technologies and systems. The pre-design analysis or optimisation is followed by a detailed design described earlier. In the whole building design tool approach, the interplay and interaction between different components, for example, BiPV and the building thermal performance, is taken into account. Thus, the PV is not only considered as an electricity production unit, but one that may also interact with the heating, cooling, and daylight energy flows of the building as is often the case in real BiPV systems. The designer may seek the best combination of different technologies using a variety or combination of different criteria (figure 21).

Technologies and Systems	Decision Criteria and Factors
Energy Conservation	**Economy**
insulation/glazing	construction cost (investment)
heat recovery	operating and maintenance cost
heat pumps	lifecycle cost
advanced control systems	annual energy cost
air infiltration	**Environmental impacts**
natural/mechanical ventilation	CO_2 emissions
Passive solar heating	SO_2 emissions
direct gain systems	lifecycle emission (direct and indirect)
solar walls	Material and resources use
sunspaces/atria	**Energy**
daylighting systems	heating
natural cooling	cooling
Active solar heating	daylight
air collectors	electricity
water collectors	annual energy use
heat storage	monthly energy use
Photovoltaic systems	daily energy use
freestanding PV modules	hourly energy use
building-integrated PV	indirect energy use
multifunctional PV	**Comfort/health impacts**
PV shading elements	

Fig 21 Example of technologies and decision criteria relevant to BiPV pre-design tools in the whole building context
Source: Helsinki University of Technology, Peter Lund

MODEL NAME	PLACE OF ORIGIN	TOOL CHARACTERISTICS	YEAR
ALLSOL	SOLPROS, Helsinki University of Technology, Finland	BiPV design & whole building tool	1998
ASHLING 7.0	National Microelectronics Centre, Cork, Ireland	Advanced PV & BiPV design tool	1997
BIPVSIM	University of NSW, Sydney, Australia	Advanced PV & BiPV design tool	1997
Energy 10	National Renewable Energy Lab (NREL) Sustainable Buildings Industry Council, Lawrence Berkeley National Laboratory, and the Berkeley Solar Group., USA	Pre-design & whole building tool	1997
GOMBIS 7.0	TU- Berlin Institut für Energietechnik, Germany	Simple PV sizing	1996
INSEL 6.0	University of Oldenburg, Germany	PV technical design	1991
ITE-BOSS	ZSW Baden-Wurttemberg, Stuttgart, Germany	Scientific simulation tool for PV	1994
PHANTASM PVHYDRO PV-SDHW	University of Wisconsin, Madison, USA All three tools use the model platform TRNSYS 14.0	Scientific simulation tool for PV	1998 1995 1996
PHOTO	Helsinki University of Technology, Finland	PV technical design	1989
PV	Oberösterreichische Kraftwerke AG, Linz, Austria	Simple PV sizing	1994
PV Calc 1.05	Oberösterreichische Kraftwerke AG, Linz, Austria	Simple PV sizing	1994
PVCAD 1.1.1	Institut für Solar Energieversorgungstechnik (ISET), Germany	Advanced PV & BiPV design tool	1999
PV-DesignPro 4.0	Sandia Lab and Maui Solar, USA	Advanced PV & BiPV design tool	1999
PV DIM 1.3	Genec, France	Pre-design and sizing of PV	1994
PV F-CHART	University of Wisconsin, USA	Simple PV calculation	1985
PV FORM 4.0	Sandia National Lab, Albuquerque, New Mexico, USA	Pre-design and sizing of PV	1997
PV Node	ZSW Baden-Württemberg, Stuttgart, Germany	Pre-design and sizing of PV	1992
PVS 2000	Fraunhofer Institute FhG-ISE, Freiburg, Germany	Pre-design and sizing of PV	1993
PVSHAD	Oldenburg University, Germany	Pre-design and sizing of PV	1995
PV*SOL 2.12	Valentin Energy Software, Berlin, Germany	Advanced PV & BiPV design tool	1999
PVSYST 3.2	GAP University of Geneva, Switzerland	Advanced PV & BiPV design tool	1997
PV TAS	Oldenburg University, Germany	Pre-design and sizing of PV	1994
PV WATTS	NREL, USA	Pre-design and sizing of PV	1999
RETScreen 2000	Natural Resources Canada's CANMET Energy Diversification Research Laboratory (CEDRL)	Pre-design and sizing of PV	1999
SHADE	ZSW Baden-Württemberg, Stuttgart, Germany	Pre-design and sizing of PV	1992
SMILE 1.1.17	TU-Berlin, Institute for Energy Technology, Germany	Pre-design and sizing of PV	1993
Solar Pro 3.49	LaPlace Systems, Japan	Advanced PV & BiPV design tool	2000
SOLARSIZER	CREST, USA	Pre-design and sizing of PV	1996
SOMBRERO 2.01	University GH Seigen, Seigen, Germany	Shading analysis	1997
SOMES 3.2	Utrecht University, The Netherlands	Pre-design and sizing of PV	1997
SUNDI 1.2	TU-Berlin, Institute for Energy Technology, Germany	Shading analysis	1995
WATSUN PV	WATSUN Sim Lab, Waterloo University, Ontario, Canada	PV pre-design and technical design	1997

Fig 22 BiPV design tools and characteristics
Source SOLARCH, University of NSW, Mark Snow

Examples of PV design tools

The growth in research, development and deployment of PV systems has stimulated numerous computer-based tools for predicting operational characteristics and sizing new system technologies. While there are a large number of simulation tools (Figure 22) from technical and electrical engineering backgrounds, predominantly from Germany and USA, very few consider the BiPV situations directly, or the effects of surrounding shade influences. Reviews of existing PV tools have identified the following limitations:

- Emphasis on stand-alone/individual building applications.
- Limited shading predictions.
- Poor interface with other building design tools for both input and output of data.
- Limited considerations of economic and cost implications for various PV system options.
- Limited use in considering BiPV as part of an overall energy strategy within an urban context.

PVSYST 3.2, PV CAD 1.1.1, PV*SOL, BIPVsim, Solar Pro and PV Design Pro-G offer direct BiPV simulation functions and shade analysis. The majority of tools are practical for only stand-alone fields of application and do not offer simple input procedures through Windows-based platforms. In addition, there are also a number of assumptions about local climate data, sky irradiation modelling and PV component specifications, which limit the functionality of the tool to basic preliminary design enquiries. Many of the tools use solar geometry algorithms from Liu and Jordan (1960), and Erbs (1980) to calculate diffuse radiation when only global radiation figures striking a horizontal surface are available, or when only direct radiation has been monitored.

BiPV tools should not assume that diffuse irradiation is uni-directional (isotropic) from all parts of the sky. Isotropic sky-vault modelling leads to an underestimation of total radiation reaching tilted surfaces and is a common limitation identified in the PV design tool review. A clear sky does not appear to be the same blueness from the horizon to the upper zenith point as the sun's position and the variation of atmospheric thickness affects the concentration of diffuse irradiation. BiPV models will often use non-isotropic sky conditions from calculations by Richard Perez, known as the Perez model, which account for this sky characteristic. This can prove important, especially if PV systems such as amorphous silicon cells more readily harness diffuse irradiation conditions.

Examples of a number of BiPV specific tools are displayed in figures 23–26 showing their graphic user interfaces and salient features.

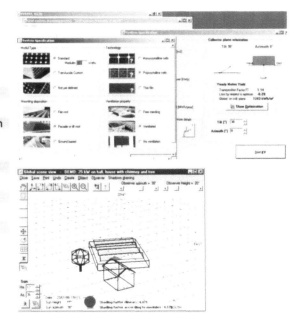

PVSYST 3.2 University of Geneva, Switzerland

Tool focus	Stand-alone, DC grid connected and BiPV Grid
Weather data	Meteonorm, US TMY2, ASCII file input conversion
Sky simulation	Non-isotropic including user defined albedo values
Shading	3-D elementary model
Website	www.pvsyst.com
Demo version	yes
Languages	English and French
Price	CHF 700 single – CHF100 for lab license of 10

Fig 23

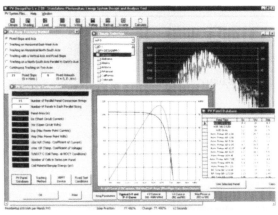

PV-DesignPro 4.0 Maui Solar Energy Software Corporation, USA

Tool focus	Stand-alone, PV water pumping and BiPV Grid
Weather data	2,000 global locations
Sky simulation	Non-isotropic
Shading	3-D model
Website	www.mauisolarsoftware.com
Demo version	no
Languages	English and Spanish
Price	US$140

Fig 24

PV*SOL 2.12 Valentin Energiesoftware, Berlin, Germany

Tool focus	Stand-alone, PV water pumping and BiPV Grid
Weather data	Meteonorm, 2,000 global locations
Sky simulation	Non-isotropic
Shading	3-D model
Website	www.valentin.de
Demo version	yes
Languages	German and English
Price	US$315 (normal version); US$430 (professional version)

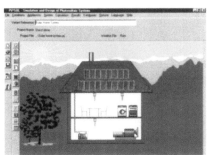

Fig 25

Solar Pro 3.49 LaPlace Systems, Kyoto, Japan

Tool focus	Stand-alone, PV pumping systems and BiPV Grid
Weather data	2,000 global locations
Sky simulation	Non-isotropic
Shading	3-D graphics tool
Website	www.lapsys.co.jp/pro/pro_english.html
Demo version	Yes
Languages	Japanese and English
Price	US$7,000

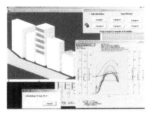

Fig 26

Fig 27 Illustration of the final design of the Ekoviikki PV project based on the pre-design. The PV modules (25 kWp) are integrated into the balconies.
Source: YIT Ltd., Finland

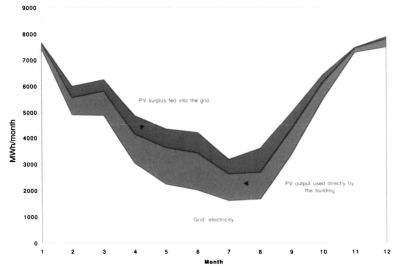

Fig 28 Monthly electricity balance of a 25 kWp PV system in the Ekoviikki building
Source: Helsinki University of Technology

Example of the PV pre-design in the whole building context

The Ekoviikki PV building development is an instructive example of using PV design tools in the whole building context. Ekoviikki is a major ecological building site near Helsinki (60° N), Finland, where 64,000 square metres of residential buildings have been designed according to strict sustainable and ecological criteria. The Ekoviikki project employs a novel integration scheme in which the PV modules have been integrated into the balconies, that is, the PV modules have the same function as the balcony sheets. Each residential building has some 1,500 square metres of living area (figure 27).

Prior to the decision to utilise PV in Ekoviikki, a comprehensive pre-design was necessary. The pre-design of the PV building in Ekoviikki included a range of questions that the architect and construction company put forward prior to accepting PV into the building. The issues relevant in the Ekoviikki case included:

- Estimation of the PV output for different PV array sizes, orientations and inclinations, and finding an optimum solution for the building site.
- Estimation of economics of PV and comparison with other sustainable energy options.
- Matching of PV output with building load as the utility buy-back rate was unclear; with the building-optimised design, about 75 per cent of the PV production is directly used by the building and 25 per cent is fed into the grid, mainly in the summer months (figure 28). This was accomplished by integrating the PV modules vertically into the balconies. A vertical surface shifts the maximum solar yield away from the northern summer with low electricity consumption to spring and autumn with higher consumption level.
- Comparison of the lifecycle costs, emissions and energy use of different sustainable energy options (figure 29), including the PV lifecycle (LC), includes investment/construction and 40 years of operation. The solar heat (domestic hot water) and wind power options have been optimised for the building. The reference energy system is natural gas.

If a positive outcome could be found for PV in each of the above questions, then building companies might be more willing to consider installing PV systems. The pre-design tasks that were performed here included energy, economic and environmental analysis of the PV system and other sustainable energy options and accounting for lifecycle aspects.

Based on the outcome of the pre-design analysis done with the ALLSOL whole building tool that includes different renewable energy options and PV, the local building company chose a BiPV system for the Ekoviikki building. Accounting for the lifecycle aspects of the technical systems instead of investments only proved to be one of the key factors favouring PV because the technical lifetime of PV modules without any moving parts is in general longer than that of mechanical energy systems.

The pre-design also allowed optimisation of the BiPV system. The final pre-design included integration of the PV modules into the balconies with several technical advantages: better matching of PV production with the electricity demand, lower module temperatures increasing the PV output, and avoidance of obstructions and shading of the PV modules. In addition, the cost savings were achieved as the PV modules replaced the ordinary balcony glazing.

Conclusion

This chapter identifies some of the key aspects in PV design assessment and lists a number of useful computer simulation tools for this purpose. Design tools are proving to be an essential component for evaluation of the prospective BiPV potential of a given site location. Influences of shade effects, different PV system characteristics and the relationship of PV generation to the energy needs of buildings require early investigation in the pre-design phase. Often, seeking expert advice to complete PV potential evaluations during early project stages can pay dividends in the longer term.

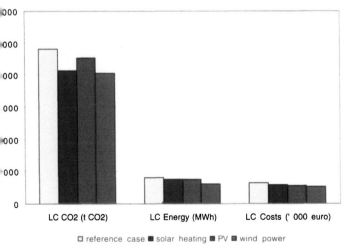

Fig 29 Lifecycle analysis of different energy options considered for the Ekoviikki project.
Three viewpoints on lifecycle: costs, CO_2 emissions and energy
Source: Helsinki University of Technology

CHAPTER SIX

ELECTRICAL CONCEPTS, RELIABILITY AND STANDARDS

Introduction

Photovoltaics is the direct conversion of light energy to electrical energy. Hence the name: *photo* relating to light, and *voltaics* indicating the electrical aspect. The conversion is a direct process. Unlike other solar-based technologies, there is no intermediate thermal stage in a photovoltaic power supply. It is an electrical phenomenon, an important consideration when integrating photovoltaic power systems into buildings. This chapter discusses electrical concepts, reliability issues and relevant standards applicable to building-integrated photovoltaic (BiPV) applications.

Electrical basics

Electrical power is measured in units of watts (W), or more commonly in typical applications, in terms of kilowatts (kW). Power is determined by two factors, the voltage of the source and the amount of current supplied. Simply stated:

Power (watts) = Voltage (volts) x Current (amperes)

In a hydraulic analogy where one is determining the power of a waterfall, the voltage can be thought of as the height of the waterfall and the electrical current as the water flow. Significantly, the same amount of power can be supplied by a high current and a low voltage or conversely, a low current and a high voltage.

Another relevant electrical concept is that of distribution losses, that is, power losses in electrical wires due to resistance, which is a function of the current flowing through the wire. Again stated mathematically:

Power Losses (watts) = Current (amperes)² x Resistance (ohms)

In other words, losses are very sensitive to the current. That is why electric utilities select very high voltages and a relatively low current when transmitting power over long distances. In distributing power from a building-integrated photovoltaic system to the internal building loads, there are limits to the voltage level allowed for safety reasons. This has implications for system losses and is discussed further below.

Electrical PV basics and the solar spectrum

The basic building block of a photovoltaic power supply, that converts light energy into electricity, is the photovoltaic cell. A typical photovoltaic cell is a thin wafer of silicon, either a single crystal or multi-crystal silicon cell measuring around 10 x 10 centimetres. The cell is a diode, allowing current to flow in one direction but not the other. The voltage generated across a silicon cell, front to back surface, is approximately half a volt. The voltage is almost constant, varying only slightly with temperature. Cooler temperatures actually increase voltage and therefore power, dismissing the widespread belief that photovoltaics, a solar-based technology, is not appropriate for cooler climates. The current generated by the cell varies directly with the amount of light. Under bright sunlight, a typical cell produces approximately 3 amperes of direct current. It shuts off at night if no natural or artificial light is present.

The sun emits a continuous spectrum of electromagnetic radiation. This solar spectrum (figure 1) is comprised of photons with different levels of energy or wavelengths. The spectrum of light reaching the earth may vary considerably, both in intensity and colour content (and in directional properties – direct/diffuse). This is usually expressed as an 'Air Mass' (AM) number, which is an indication of the path-length the light has travelled through the earth's atmosphere. AM0 is the spectral distribution of sunlight outside the earth's atmosphere and is known as the Solar Constant. If the sun is exactly overhead and the sky is clear, the spectrum will be (close to) AM1, indicating the light has crossed one atmospheric thickness. For standard reference and measurement purposes, AM1.5 is used, since it is close to the typical energy distribution of light reaching the earth's surface.

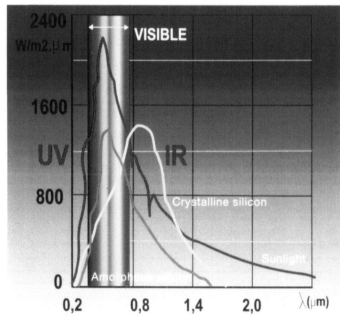

UV = ultra-violet
IR = infra-red
_ = wavelength
μm = nanometer (10⁶ of a metre)
W/m² = watts per square metre

Fig 1 Solar spectrum relative to PV silicon type
Source: Demosite, Switzerland

Photons associated with shorter wavelengths (and therefore higher wave frequency) of light, near the violet end of the spectrum, have greater energy than longer wavelengths (or lower frequencies) near the red end of the visible spectrum. Invisible ultra-violet light provides PV cells with an even higher level of photon energy. Conversely, invisible infra-red photons, which we feel as heat, provide a lower energy level than in the visible light band. Sunlight is made up of direct irradiation (emanating directly from the sun itself) and diffuse irradiation (reflected and scattered light from the sky). On clear-sky days, the direct component will dominate and on overcast days the majority of solar irradiation will come from diffuse irradiation. Useable irradiation by a PV cell will comprise direct and diffuse sources but also ground reflectance, otherwise referred to as 'albedo'. As represented in figure 1, amorphous silicon's spectral range makes better use of shorter wavelength photons associated with diffuse or overcast weather conditions than does crystalline silicon. Crystalline silicon, however, responds more readily to photon power towards the infra-red spectrum. Some thin film photovoltaic modules use layers of different materials, to capture as wide a range of available spectral energy as possible.

Wiring of cells to form modules and arrays

The low voltage and current of an individual cell limit its practical applications. Cells are therefore wired together to produce a more useful source of electricity. When cells are wired in series, positive terminal to negative terminal, much like the batteries in a flashlight, the voltage increases. When cells are wired in parallel, that is all positive terminals are joined together and all negative terminals are joined together, the current increases. Typically, the cells may be wired in a series of 36 cells (Chapter 2 figure 12) to produce a voltage of 17–18 volts, the recommended charging voltage for a lead acid storage battery. The string of cells is encapsulated under a plate of glass and the sealed unit is referred to as a photovoltaic module. Modules are produced with outputs ranging from a few watts to several hundred watts. A module with a single string of 36 cells, as described above, would have a power rating of about 50 watts (17 volts x 3 amperes).

As with cells, photovoltaic modules can be connected in any combination to generate the desired voltage and current. When modules are interconnected electrically, the resulting structure is referred to as a photovoltaic array. This is the component of interest to the architect when integrating a photovoltaic system into a building. A photovoltaic array may be rated in tens of watts up to hundreds of kilowatts, depending on its surface area and the efficiency of the cells and modules used.

Shadowing can significantly reduce the output of a PV module. In a module, solar cells are connected in series and the electrical current has to flow through all cells. If one cell is covered or shadowed, it produces a smaller current than the others and limits the output of the entire module, depending on the exact interconnection design. This is similar to the analogy of liquid flow in a hose, as shown in figure 2. Voltage relates to water pressure and current is the amount of water passing through. If the hose nozzle is closed, the voltage (pressure) remains high and the current is zero. Shadow restricts the flow of current through cells and modules.

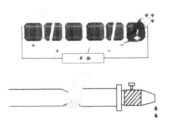

Fig 2 Electrical circuit analogy
Source: Demosite, Switzerland

Standard test conditions and rating of modules

PV modules are rated under a defined set of conditions known as Standard Test Conditions (STC). These conditions include the temperature of the PV cells (25 °C or 77 °F), the intensity of radiation (1 kW/m²), and the spectral distribution of the light (AM1.5). This is aimed at simulating noon on a clear sunny day with the sun about 60° above the horizon, the PV module directly facing the sun, and an air temperature of 0 °C (32 °F).

Figure 3 shows a typical IV curve for a silicon PV module operating at STC. This condition is sometimes called 'one sun' or 'peak sun' and assumes no shading. Key points on the curve include:

short-circuit current (I_{sc}) which is the point on the curve where the voltage and the power output of the solar cell is zero;

open-circuit voltage (V_{oc}) where the current and the power output of the solar cell is zero;

maximum power point (MPP) the point on the IV curve of a PV module, string or array, where the product of current and voltage and thus, power is maximum. This point is also referred to as the 'knee' of the IV curve.

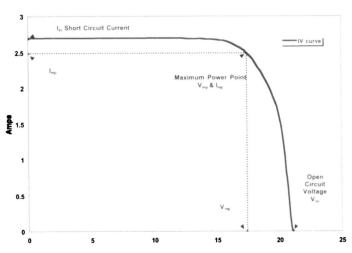

Figure 3 A typical IV curve
Source: SOLARCH, University of NSW, Australia

The conversion efficiency (Ce) of a PV module (PV) is the ratio between the optimal electric power (P_o) delivered by the PV module and total solar insolation (S) received at a given cell temperature ($_T$). This can be calculated by dividing the optimal electrical power (P_o) in watts (W) by the surface area of the active PV module area (PV_A) in square metres (m²) multiplied by total solar insolation (S) in watts per square metre (W/m²). Thus:

$$Ce = \frac{P_{oT}}{PV_A.S}$$

Mismatch occurs in PV arrays when cells with different IV characteristics are interconnected. Different IV curves may result from either the properties of the cells, or different conditions such as light intensity or temperature. Mismatch can result in poor performance of the module, since the performance of the 'good' cells is limited by that of the 'bad' cells or even one bad cell. The open circuit voltage is the sum of the open circuit voltages of the individual cells.

PV yield can be used to compare the performance of PV installations of different sizes to each other. This is calculated by taking the energy produced by the system (kWh) divided with the installed peak power (kWp). For example, if a system with an installed peak power of 3 kWp produces 300 kWh over a month, the calculated yield for the month is 100 kWh/kWp. This does not take into account the difference in irradiation at different locations. If system performance is required for different systems and irradiation conditions, the calculation of Performance Ratio (PR) provides a performance for the system regardless of the installed peak power and irradiation. The calculation divides the energy produced by the system with the installed peak power and the irradiation as follows:

$$PR = \frac{\text{produced energy (kWh)}}{\text{Installed peak power (kWp) x irradiation (kWh/m}^2\text{)}}$$

Thus, an installed peak power of 3 kWp produces 300 kWh over a month. If the irradiation for the month was 125 kWh/m² this gives a performance ratio of 0.8 or 80 per cent.

Electrical design concepts
Maximising solar exposure

Since the power produced by a photovoltaic array is directly proportional to light intensity, it would seem obvious that the energy produced by a photovoltaic system can be maximised by increasing the array's exposure to the sun over time. However, this may not always be possible. In the real world, factors such as array orientation, shadows from nearby objects and soiling of the surface of individual modules will reduce the amount of sunlight to which the photovoltaic array is exposed.

The ideal photovoltaic array would follow the sun as it makes its daily path across the sky. The array surface would always be normal to the sun's rays. There are in fact, small pedestal-type array support structures that can track the sun by one mechanism or another. Typical applications for trackers are for water-pumping photovoltaic systems. The benefits are the greatest in the summer when the sun makes its widest path across the sky. Under these conditions, a tracker can improve daily energy production by up to 40 per cent compared to a fixed array.

Fig 4 Example of a two-axis tracker
Source: University of NSW, Australia

There are very few building-integrated photovoltaic installations that could incorporate a two-axis tracker. Some canopy or louvre designs, which can track in a vertical direction either on a daily or a seasonal basis, have been tried. However, most building-integrated photovoltaic arrays are fixed and in many instances, the orientation of the array is dictated by the building design or orientation. Chapter 5's discussion on PV potential reiterates that beneficial PV power generation does not necessarily require optimally oriented surfaces.

Shading impacts

The current produced by a photovoltaic source, such as a building-integrated array, is proportional to the light level to which it is exposed. However, when only part of the array is exposed to light, the current is not proportional to the area exposed. As explained earlier, shading of even a small portion of a photovoltaic surface can have a dramatic effect on its electrical power output, because the photovoltaic modules are wired in series to increase the system voltage. This means that the current in individual strings travels through each module. In this 'chain' of modules, the current is only as high as the current produced by the weakest module. A single shaded module will limit the current produced by the unshaded modules.

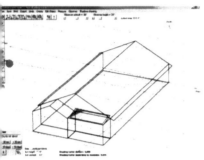

Fig 5 Using PVSYST to simulate in yellow, modules affected by shading
Source: University of NSW, Australia

The effect will depend on the contrast of the shadow, that is, the ratio of direct and diffuse sunlight, but shading just 10 per cent of a photovoltaic array can reduce the total output by more than 50 per cent, as presented diagrammatically in figure 6. Any design of a building-integrated photovoltaic array should make every attempt to avoid shadows.

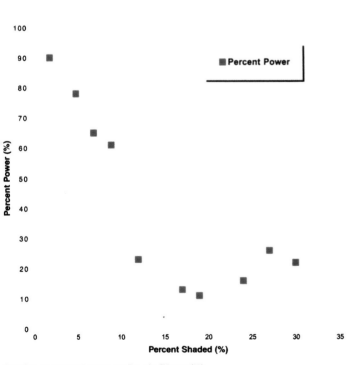

Fig 6 Drop in power relative to shading of a PV array (%)
Source: Per Drewes, Canada

Snow and ice load

In northern climates, snow or ice cover can be an issue. Photovoltaic power supplies are known to operate in some rather harsh environments. Arrays providing power for mountain-top radio repeaters can be covered in thick ice for weeks. Building-integrated photovoltaic arrays in friendlier environments may also experience times when they are covered with snow. In most cases, sunlight will penetrate the layer of snow and will warm the dark photovoltaic surface. A film of water then forms, allowing the snow to slide off the sloped array. An important array design consideration is therefore to minimise any bottom protrusions of the support structure that will prevent this natural sliding of the snow cover.

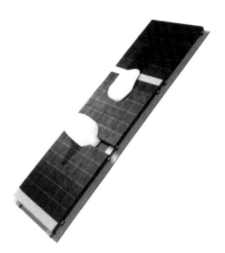

Fig 7 The spaces between PV panels should not encourage the accumulation of snow or ice
Source: Martin Van Der Laan (M. ART)

Temperature

While light exposure is the single most important factor determining the power generated by the photovoltaic array, system designers should also consider the temperatures that the array experiences. High temperatures will reduce the operating voltage of crystalline silicon modules and therefore the photovoltaic performance. Note the effect of panel temperature on production efficiency for the Swiss Student Housing, (see Case Study, Chapter 3). Extreme temperatures can also damage photovoltaic modules. Since the photovoltaic arrays are exposed to direct sunlight, there is a natural tendency to create heat. A well-designed photovoltaic array installation should include provisions for venting the array back surface to remove this heat. Photovoltaic cell temperatures should ideally be kept below 70 °C. As a European rule of thumb, the following losses (%) may occur with façade-mounted systems:

- No air gap – 10% loss;
- 5 cm air gap – 5% loss;
- 15 cm air gap – minimal loss from temperature effects.

Building-integrated PV electrical components

In a typical utility-interconnected PV system, the array is connected to an inverter or power conditioning unit which converts the DC power into AC power. This AC power is then fed into the building's internal distribution system or into the public grid (figure 8).

Inverter or power conditioner

The inverter, often called a power conditioner, is the component that changes the power output from the photovoltaic array into conventional electrical power as supplied by the local utility. It is the link to the outside world and performs three functions:

1 The inverter controls the operation of the photovoltaic array. As the sun rises in the morning, it connects the photovoltaic array to the utility system. As light and temperature change throughout the day, the inverter adjusts the array current and voltage levels to maximise the efficiency of the photovoltaic array, tracking the maximum power point. Finally, as the sun sets in the evening, it disconnects the array from the utility system. This may be described as the power tracking function of the inverter.

2 The second function of the inverter is to change the direct current (DC) from the photovoltaic array to alternating current (AC) with a frequency and voltage matching the supply from the local utility. This is the inverter function.

3 The inverter also functions as a safety component. Most building-integrated photovoltaic systems will be utility interconnected. They are categorised as distributed resources (DR) or embedded generator and at times will be able to feed power back into the utility distribution systems. This means that the inverter must meet certain safety and power quality requirements. A DR must not feed power back to any utility distribution system experiencing a power outage. During periods of normal operation, power fed to the utility must meet standards for voltage, frequency and harmonic content. Safety and power quality issues are the main concern of utilities. The inverter will monitor various parameters such as voltage and frequency and will shut down if they deviate from a set range.

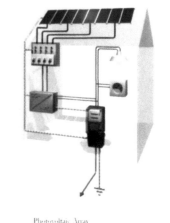
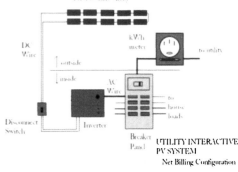

Fig 8 The electrical structure of a residential PV system
Source: Martin Van Der Laan (M. ART)

The inverter may be considered to be an electronic component. It will require periodic inspection and maintenance. Often the inverter incorporates a display panel indicating power production or fault conditions. It should therefore be installed in an accessible location and, unless designed for outdoor exposure, should be located in a dry and temperate environment. Inverters come in a variety of casings; some of them are rather sober, while some are designed for public display. In some cases the appearance may influence the product selection and hence the system design.

Fig 9 BP Solar 2400 W inverters
Source: ECN, the Netherlands

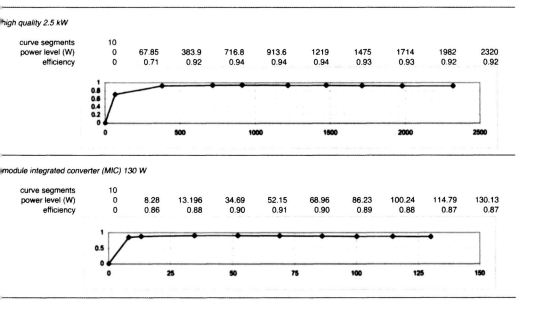

high quality 2.5 kW

curve segments	10									
power level (W)	0	67.85	383.9	716.8	913.6	1219	1475	1714	1982	2320
efficiency	0	0.71	0.92	0.94	0.94	0.94	0.93	0.93	0.92	0.92

module integrated converter (MIC) 130 W

curve segments	10									
power level (W)	0	8.28	13.196	34.69	52.15	68.96	86.23	100.24	114.79	130.13
efficiency	0	0.86	0.88	0.90	0.91	0.90	0.89	0.88	0.87	0.87

flat inverter response (used when assuming no inverter)

Inverter efficiency	1
power level (W)	0
efficiency	1

Fig 10 Inverter efficiency curve
Source: University of NSW, Australia

Most inverters are installed next to the building's electrical service panel. Some inverters are known to generate some background noise. The sound can be irritating when the high frequency switching coincides with certain psychologically annoying frequencies. Noise may therefore be a factor in selecting the location of the inverter.

The range of input current and voltage to the inverter must match the range of output currents and voltages from the array. Also the number of modules in an array has to be a multiple of the number of modules in a string. Inverter performance is influenced by the DC power output from the PV modules. Figure 10 shows how the inverter efficiency reduces with loss in DC power.

Electrical configuration

Photovoltaic systems have a highly modular structure and can be electrically configured in a variety of ways. The building blocks of a solar generator, the PV panels and their rated peak power of typically between 50 W and 300 W can be configured to maximise output efficiency. Figure 11 presents a number of electrical configuration concepts.

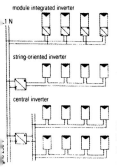

Fig 11 Different array concepts: central inverter, string oriented inverter and AC modules, connected to the 230 V low-voltage grid
Source: Heinrich Wilk, Energie AG, Austria

Central inverter design

The traditional array structure consists of modules connected in series to form a string. A number of strings are connected in parallel to form an array. One central inverter transforms the DC electricity to AC current that can be fed into the grid. Large building-integrated PV projects like the 2,000 kWp PV Munich Trade Fair Centre or 200 kWp PV power station, Auf dem Kruge in Bremen have been implemented this way. The large inverters can be installed in cabinets in a central system room where repair work and maintenance can be done in a convenient way. If several big inverters are cascaded into a master–slave configuration the efficiency could reach very high values, up to 95 per cent.

String oriented inverter concept

Another approach is to connect each single string of modules to one inverter. Partial shading and different thermal characteristics will not trigger losses across the whole PV array with this design. Examples can be found in the Germany Mont-Cenis Academy case study. This concept often leads to rather high voltage levels on the DC-side that can cause safety problems with the solar modules. It is good practice to mount the inverter close to the modules to reduce the length and cost of DC cabling. According to the German manufacturer SMA, watertight units can be installed outside buildings on rooftops or façades. Future maintenance records will show the reliability of these systems.

Multi-string inverter design

The main advantage of the string inverter is the optimal power tracking done for a single string of PV modules. The rated power of the inverter is limited to approximately 2 kW because solar modules cannot withstand very high voltages. To gain the cost advantages of larger units and maintain the optimal string inverter, several DC/DC converters with maximum power point tracking (MPPT) combine the power of the different strings to a DC bus. The multi-string inverter is very useful when PV strings of different rated power or different orientations are combined, or to mitigate partial shading.

AC modules

With AC modules, also known as module-integrated converters (MICs), very small inverters are integrated in the junction box of solar modules (figure12). The typical system size is 100–300 W. This design is advantageous when partial shading cannot be avoided or small systems are used. Mass production could lead to a substantial unit cost reduction. Reliability and maintenance are more significant issues because the devices are exposed to a wide range of temperature and humidity conditions when mounted on a roof top or a façade. Ambient temperatures between –20 °C and +90 °C are quite common in these applications. This means extra stress on electronic components, influencing the useful life. Keeping inverter boxes watertight is of high priority. Repair work can be rather an effort because of the distributed installation of many small units in a large array field. It will be necessary to develop systems that immediately detect the failure of any of the small inverters.

Table 1 summarises salient points from each inverter concept. In general, a central inverter costs a little less than string inverters of the same capacity, but has higher installation costs due to the other components involved such as the wiring and junction box. String inverters, however need a better monitoring concept compared to a central inverter. Currently, the cost balance is slightly in favour of the string inverter concept.

System wiring

The photovoltaic modules in the array are joined together in parallel strings. As described earlier, the number of modules in the individual strings determines the system voltage and the number of parallel strings determines the current. Electrical basics also state that the current determines resistance losses in a wire. Large array currents will therefore require a heavy wire to minimise resistance losses while transmitting the power from the photovoltaic array to the inverter. The array is often roof-mounted while the inverter may be located in a basement electrical room. Long separations result in high wire losses. Heavy wire will reduce these losses but is expensive and often difficult to install. High operating voltages lower the current and therefore the resistance losses, but in most building-integrated applications, electrical codes limit the allowable level to 600 volts. It is recommended for the wiring on the DC side to use double-insulated UV stable cables, either 2.5 mm^2 or if needed 4 mm^2. The cables must also tolerate temperatures up to 60 °C.

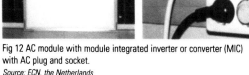

Fig 12 AC module with module integrated inverter or converter (MIC) with AC plug and socket.
Source: ECN, the Netherlands

Fig 13 Module wiring and a white junction box
Source: ECN, the Netherlands

Inverter concept	Central	String	Module
Cost	Approximately US$0.6/wp	Approximately US$0.7/Wp	Approximately US$1/Wp
Dimension/weight	Large and heavy	A4 size and approximately 10 kg	A5 size and approximately 1 kg
Installation	Junction box, more wiring, more effort	Very easy and fast	Little more than string inverters
Shading influence	High	Not much	Not noticeable
Monitoring	No big issue	Some effort necessary	Extensive
Reliability	High	Moderate	Moderate – Low

Table 1 Comparison of inverter concepts

the strings making up a PV array are connected to a junction box, which is preferably located close to the PV array and not exposed to the natural elements. A junction box should have similar features to a cable, such as UV resistance, and temperature. As each string is controlled by a standard fuse (5 x 20 mm) the junction box should be accessible if needed. A junction box has dimensions of around 30 x 20 x 10 cm (w x h x d) for a residential application (3 kWp) and for a bigger installation (greater than 10 kWp) around 60 x 80 x 20 cm (w x h x d). It is also a desirable design feature to carefully store wire runs that are easily identifiable, as in figure 14, or to cleverly create a decorative wiring feature as long as safety is not compromised.

PV overvoltage devices and DC isolating devices (an array switch) are needed to electrically separate the PV array from the inverter if high voltages occur, for example, due to nearby lightning strikes, or when needed for maintenance. It is important to note that these devices are operating under DC conditions and the operating current will vary at any time. The devices must be able to support the PV array short current (if surge-protected, then current does not pass through this device) and the open DC voltage.

Standard modules often come with screw terminals in their junction box. However, special plug/receptacle connectors offer an advantage for roof-mounted modules. They can be connected by electrical laypersons, for example roofers, and thus help to streamline the installation. Modules for mullion transom façade systems are often custom-tailored and their terminal wires often leave the module at the edge. If these wires need to be field-connected, crimping is the most reliable method of connection. The crimp connection should be protected by a layer of shrinkable tube fitted with inside adhesive.

Grid interconnection

The total photovoltaic system is connected to the utility system at what is referred to as the point of common coupling or interconnection. In a residential system, this is normally the building breaker panel. The photovoltaic system will be connected as a load to the panel. Most jurisdictions require the photovoltaic system to be connected via a dedicated line. That is, there are no other loads on that particular breaker. Depending on the design of the interconnection, it may also be necessary for the breaker to be rated for bi-directional operation. The breaker and the array switch serve to isolate the inverter for servicing.

Monitoring and metering

It is important to have some simple method of monitoring whether or not the system is working. The inverter is an electronic component and can have failures. If a central inverter is not properly monitored and breaks down, solar energy is not being converted and the PV owner loses out. Today, most inverters offer access to the inverter data via dial-in modem. With a suitable PC program, the user can automatically receive an error message in case of a failure or check the performance of the inverters. Alternatively, if the inverter is located in an easily visible location, a warning light should show malfunction.

Net metering

An increasing number of utilities offer 'net metering' for small residential customers. Under this arrangement, a single billing meter is allowed to run forwards and backwards. Whenever power produced by the photovoltaic system exceeds the customer's load, the excess feeds back to the utility. The photovoltaic system can thereby reduce the customer's electricity bill. However, the utility may not pay for any surplus showing at the end of the billing period. This arrangement reduces metering costs and in effect, allows the customer to sell the energy to the utility at retail rates. It also simplifies the utility's billing procedures for very small generators. Net billing is not really a technical issue, other than in the choice of a suitable meter. It is a commercial arrangement and in many instances, is becoming a policy mandated by governments.

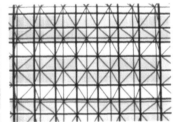

Fig 14 Designs to house cabling (wiring) efficiently and decoratively
Source: ECN, the Netherlands

Fig 15 Overvoltage protection device
Source: ECN, the Netherlands

Fig 16 Breaker and interconnection
Source: ECN, the Netherlands

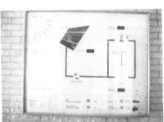

Fig 17 Net meter and PV output displays
Source: ECN, the Netherlands

Protection

In grid-connected systems, several protection devices are installed to safeguard the installation against abnormal circumstances. Within the crystalline silicon module, bypass diodes are usually set parallel to every group of 12–18 cells to avoid so-called hot spots. The spots can occur as a result of shading or dirt when a part of the cells does not generate power but dissipates power from the other cells and consequently heats up. The bypass diodes divert the module current and prevent hot spots.

It is customary to install blocking diodes and surge protectors to every string, although their usefulness and desirability are subject to debate. The diodes prevent reversal of the string current, as may be caused by earth faults somewhere in the string. In case of nearby lightning, surge protectors will limit the voltage to reasonable values. Some utilities will require an additional isolation switch that is lockable and accessible to utility personnel only. The exact requirements vary from country to country and even among utilities of the same country.

Of more importance are the safety aspects of a PV installation. It is crucial to remember that while a generator can easily be switched off, a PV array cannot be switched off. As long as the sun shines on it, the physical process of producing electricity continues. This aspect must be included in protection guidelines for installers and operators. Many guidelines are still in draft stage and will take time to be approved. Most draft guidelines represent state-of-the-art knowledge and should be applied at all installations. All electrical issues are under the standard body of the IEC (International Electrotechnical Commission) and can be ordered from www.iec.ch.

Protection against electrical shock has been considered under *Electrical installations of buildings Part 7-712* guidance and is based on standard IEC 60346-4-41. This offers excellent advice on how to protect against electrical shock. Key points include:

- If the DC voltage (open circuit at 25 °C and 1000 W/m² irradiation) does not exceed 120 V then no special measures are required.
- If the DC voltage (open circuit at 25 °C and 1000 W/m² irradiation) does exceed 120 V then one has to use solar modules with protection class II and consequently use of class II equipment for the DC side (such as cables, switches, junction box).
- Protection against indirect contact of the PV generator may be omitted where the PV generator is located within an area accessible only to skilled or instructed persons. In such cases, easily visible warning signs shall be placed in the vicinity of the PV generator. All junction boxes (PV generator and PV array boxes) shall bear a warning label indicating that active parts inside the boxes may still be live after isolation from the PV inverter.

The preferred locations for the installation of PV arrays are unused building surfaces, such as roofs or façades. This means they are often exposed to direct strikes of lightning or the induction of overvoltage after nearby strikes. The lightning protection concept should have neither a negative influence on the demands made on the availability of the systems, nor on the protection of the system and the operators or other persons. The installation depends on the existing lightning protection structure (LPS) on the building. Figures 18 and 19 show a recommendation that is practised in most European applications.

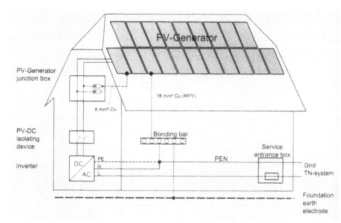

Fig 18 LPS without external protection
Source: Enecolo AG, Switzerland

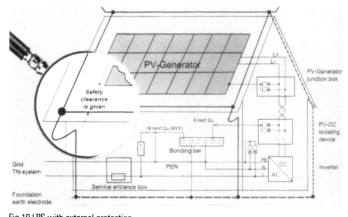

Fig 19 LPS with external protection
Source: Enecolo AG, Switzerland

An external LPS protects against direct lightning strikes, but a safety clearance between the PV installation and the external LPS has to be taken into account. A direct connection from the PV array to the external LPS is also recommended and reduces the risk of inviting a lightning strike into the building.

DC side electrical configurations

Some appliances, such as televisions, radios, and computers often function internally using DC supply by transforming the 240 volt AC grid voltage to, typically, a low-level 12 or 24 DC voltage. Utilising PV energy in this way involves two steps of energy conversion from DC to AC and back from AC to DC again, with inherent energy losses. The power demand of some appliances is high and therefore cannot be supplied by low-voltage DC distribution system (due to losses in DC cables). Also, DC switches are more expensive than AC switches and DC appliances are not widely available. Only at the 12 V level is there a broad range of devices available based on the leisure and camping market. These consumer articles are not produced to the highest quality and energy efficiency standards.

There are however, a number of PV applications using DC. An emerging market for PV is that of Uninterruptible Power Supply (UPS) systems. UPS systems traditionally comprise inverters, switches, control circuitry and a means of energy storage (such as batteries) for maintaining continuity of electrical supply to a load in the case of a grid power outage. Many large buildings have UPS systems to ensure availability of power for essential functions such as sewerage and communications during blackouts.

Standard UPS system with PV

In standard UPS systems, the battery is kept fully charged, prepared for a grid power outage. Charging is done by the grid-connected rectifier. Photovoltaic panels may contribute to charge the battery. The advantage of this concept is that some electricity consumption from the grid is replaced by PV energy. During longer grid outages, operation of the whole system is possible at a reduced power level in emergency mode.

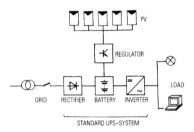

Fig 20 Standard UPS system supported by photovoltaic electricity
Source: Heinrich Wilk, Energie AG, Austria

Advanced UPS system with PV

Figure 21 shows the flow diagram of an advanced UPS system. In standard UPS units all electricity passes through rectifier, battery and inverter to power the dedicated UPS load. Advanced units operate the UPS load directly on the AC grid. The S1 switch is closed in this mode of operation. A bi-directional inverter is used to charge the battery in grid-connected mode. During power outages, S1 opens very quickly and the UPS load is supplied by the bi-directional inverter now working with reverse energy flow (DC to AC). S1 is an electronic switch and must open fast enough

not to cause problems with computer power supplies. When S2 is closed, PV electricity is used to charge the battery. In standard operation mode the PV system operates as any other grid-connected PV unit.

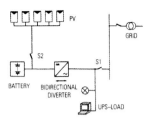

Fig 21 Advanced UPS system supported by photovoltaic energy
(Project UBS-Suglio)
Source: Peter Toggweiler, Switzerland

A direct connection between a standard UPS system and a PV generator was used for the first time in the Swiss project UBS Suglio. The PV array is connected to the DC bus bar of the battery with a relay switch (S2). When the PV array is connected to the battery the solar inverter is switched off because the battery absorbs all the PV electricity available. Depending on the weather conditions and the UPS load, the time of autonomy can be extended for several hours by PV.

PV UPS supply for a computer network

A similar UPS system is operated at the Helsinki University of Technology to supply a computer network at the research institute. Figure 22 shows the design of the 800 W unit.

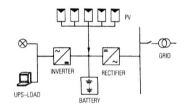

Fig 22 800 W UPS system supported by photovoltaic electricity
Source: HUT, Helsinki, Finland

Direct coupled PV systems / DC load

Any direct use application of DC electricity can be supported by photovoltaics. Beside the battery-based systems like UPS or telephone exchanges, there are many means of public transport that are operated with DC, including tram, trolley bus and subway. These DC networks can absorb a lot of PV energy. For example a 75 kW PV array is mounted atop the Neufeld park-and-ride facility in the City of Bern, Switzerland. The plant supplies the solar-generated DC current directly to the grid of the public transportation system. The annual energy production is 76,000 kWh. This is about 0.5 per cent of the total electricity demand of the tram and trolley bus network of SVB Bern.

System reliability

The reliability of a photovoltaic system may be defined as its ability to perform according to specifications over a given period of time. Limitations on performance due to design factors such as shading or perhaps unusually poor weather should be excluded from this definition. Although grid-connected photovoltaics is a relatively new technology, there is now a history of more than 30 years of off-grid photovoltaics use on which to base estimates of reliability. In fact, manufacturers of photovoltaic modules typically offer warranties of 20 or more years, based on this record.

In the last few years, the number of grid-connected photovoltaic installations, including building-integrated systems, has increased dramatically. Large well-monitored programmes, such as the German '1000 Roofs' programme, provide a large enough sample group to make results statistically meaningful. Large grid-connected projects are also underway internationally and IEA Task 7 has undertaken to survey some of these projects in member countries.

The following summarises the limited experience with utility interactive photovoltaic systems:

Photovoltaic array support structure

Support structures hold the photovoltaic modules in place; modules generally do not blow off the roof. The main concern for this component occurs in the original design. Photovoltaic arrays should be installed to ensure maximum solar exposure for the active photovoltaic area while minimising wind loadings on the array surface. At roof edges and eaves, wind loads may surge by a factor of three compared to more central sections. This requires a higher resistance against pulling forces. Also, for retrofit of photovoltaic systems on existing buildings, the addition of the support structure on the roof should not compromise the structural integrity or weather seal of the existing roof.

Photovoltaic modules

Photovoltaic modules produce electricity whenever the sun shines and if they perform well for the first year, they are likely to continue to perform for a very long time. While perhaps not a reliability issue, one reason for reported poor system performance has been the overrating of module power by the manufacturers. So far, no system has been in operation longer than the expected lifetime of the photovoltaic module. If there is a future problem, it is likely to occur in the module junction box. Because this box is very exposed to the elements and mounted on the back of a glass surface, it experiences higher than ambient temperatures. Evidence of corrosion may show after 10 to 15 years of operation. Corrosion may be delayed if cables are always introduced from the bottom and drainage openings for water condensation are provided.

Reliability issues for photovoltaic modules relate almost entirely to vandalism and theft. Photovoltaic modules must, by their inherent function, be exposed. Also, in the case of non-building installations such as street lighting, they are often located in isolated areas. They are therefore vulnerable to damage by vandals. As the general public becomes aware of the uses of photovoltaic systems, the modules also become targets for theft. Vandalism and theft issues can only be minimised during the system design. Generally, the photovoltaic modules should be kept out of reach. They can also be installed with theft-proof fasteners.

Reliabililty of electrical components

Fuses, breakers and switches normally function as required and are likely to function according to specifications for the life of the photovoltaic systems. Their reliability may reflect their passive role as well as the maturity of the electrical industry. Array string blocking diodes have failed in some systems due to lack of heat dissipation. IEA study IEAPVPS T5-03:1999 suggests that, with proper wiring, both the string diodes and the string fuses can be eliminated, resulting in a simpler and more reliable system.

Reliability of inverter (power conditioner)

For the small inverters used in BiPV systems, often used as string inverters in large arrays, the system designer should consider ease of unit removal, shipping to supplier and replacement until such time that the industry is able to supply a very reliable product. AC modules, again a relatively new technology, place the inverter in an extremely hostile environment. Several European installations have experienced problems during the first weeks of operation. Aside from high failure rates, accessibility and replacement of defective units after array installation could be an issue for this technology. The quality of this unit is the key factor in determining overall system performance. It accounted for two-thirds of reported troubles according to IEA study IEAPVPS T5-03:1999.

In today's market, the prevailing size for an inverter is in the 1–5 kilowatt range. About 30,000 units per year are now being produced. They should therefore have been thoroughly tested in the factory. With simple, straightforward installation, no start-up problems should be expected from these units. Nevertheless, the photovoltaic industry is relatively new and component designs are largely unproven. Failures have occurred with the smaller inverters, usually within several weeks of actual field operation. The problems can usually be attributed to control circuitry that has been recently developed but lacks actual field experience. Failure to protect itself from utility irregularities or user abuse usually leads to destruction or at best, failure of the control to bring the unit into operating mode.

Standards

As part of an electrical system, a building-integrated photovoltaic power supply must meet established rules and regulations. The components of the system must meet applicable product standards to ensure safe and reliable operation. The installation of the system must meet applicable electrical codes, again to ensure safe operation. In addition to these requirements, a utility-connected photovoltaic system may be required to meet additional criteria established by the local electrical power authority to ensure that its system and public safety are not compromised by the connection of this power supply.

Product standards

As discussed earlier, photovoltaics is a relatively new technology. Some components of a photovoltaic system have only recently been introduced to the market, while some designs are still evolving. The development of the technology has been followed by the development of standards. However the development, approval, and publishing of standards, especially international standards, is a slow process and with a rapidly advancing technology the standards inevitably lag behind the technology.

PV standards have been developed by a number of countries and standards organisations. This has resulted in a large number of standards, far too many to include in this chapter. Many of the early standards and qualification procedures were developed as part of the procurement process for national and international PV demonstration programs. For example in the U.S., the Jet Propulsion Laboratories in California developed standards in the 1970s to evaluate modules for safety and structural integrity as part of a block purchase by the U.S. Department of Energy. In Europe, the Joint Research Center (JRC) of the European Commission developed similar standards for the European Commission demonstration programs. However, in this age of globalisation, there is a concerted effort to harmonise these standards. This is especially true for product standards where the products have to be sold worldwide. Fortunately many of the standards are fairly compatible as they have been generated from similar sources and documents. The International Electrochemical Commission (IEC) develops international standards for renewable energy technologies. IEC Technical Committee 82 (TC 82) is responsible for solar photovoltaic energy systems. TC 82 has now published a series of initial basic standards and is in the process of developing further standards. Details of the standards and how to obtain them are available at www.iec.ch.

The most important and unique components in a building-integrated photovoltaic system are the photovoltaic modules. There is a well-established standard for PV modules that is normally referenced when purchasing modules for PV systems. This is IEC 61215 (1993–04), *Crystalline silicon terrestrial photovoltaic (PV) modules – Design qualification and type approval*, which lays down requirements for the design qualification and type approval of terrestrial photovoltaic modules suitable for long-term operation in general open-air climates. IEC 61646 (1996–11) does the same for thin-film photovoltaic modules.

The other main unique component of a building-integrated PV system is the inverter. Again individual countries have developed their own standards for this component. In Canada, the Canadian Standards Association has simply added sections pertaining to unique photovoltaic and utility requirements to an existing *Standard 107.1 for General Use Power Supplies*. In the U.S., Underwriters Laboratories Inc. has produced a completely new independent document, UL1741 *Standard for Static Inverters and Charge Controllers for Use in Photovoltaic Power Systems*. IEC TC 82 has developed two standards relevant to photovoltaic inverters: IEC 62109, *Electrical safety of static inverters and charge controllers for use in photovoltaic (PV) power systems* and IEC 62116, *Testing procedure – Islanding prevention measures for inverters used in grid connected photovoltaic (PV) power generation systems*. Other components used in a building-integrated photovoltaic system may be considered as variations of conventional electrical products to which established standards apply. However, where conventional components are used in the system, their application (such as bi-directional power flow in breakers), may mean that existing standards need revision.

In Australia, grid connection of energy systems via inverters is required to be supplied and installed to meet AS 4777. PV modules require a minimum 10-year warranty, and other components and designs need to meet relevant Australian Standards. Refer to Australian Standard AS 4777 at www.standards.com.au.

Installation codes

In addition to components of a system meeting safety and performance standards, the manner in which these components are installed must also conform to prescribed installation practices. Each country has evolved its own set of standards based largely on local characteristics and requirements. Working with a recently developed technology, installers have had to conform to myriad different and sometimes conflicting electrical standards. For example, the European community favours ungrounded circuits in the photovoltaic arrays, whereas U.S. regulations require these source circuits to be grounded.

There is a move towards international harmonisation with the drafting of IEC 62234, *Safety Guidelines for Grid Connected Photovoltaic Systems Mounted on Buildings*, which is being produced by IEC TC 82. Another IEC Technical Committee, TC 64, is responsible for all electrical components within buildings and has produced a draft addition to its existing standard, *Electrical Installations: Requirements for special installations or locations – PV power supply systems* (IEC 60364-7-712). There is also an existing standard that gives guidance on overvoltage protection for both stand-alone and grid-connected photovoltaic power generating systems – IEC 61173 (1992-09), *Overvoltage protection for photovoltaic (PV) power generating systems – Guide*.

Utility connection

Almost all building-integrated photovoltaic systems will be connected to the local utility distribution system. They will be classified as distributed resources or embedded generators, depending on the terminology used in a rapidly changing electricity business. Throughout the world, the structure of electric utilities is changing from those with large central generators and radial distribution systems to networks of loads and generators spread throughout their service territories. Utility-connected photovoltaic systems will be joining other small independent generators such as small co-generation plants, wind turbines and fuel cells.

Many utilities are reluctant to adopt this new mode of operation. They feel that a network incorporating many small independent generators will result in a loss of control of their own distribution systems and compromise their ability to guarantee the safety and power quality of the distribution system. The main concern is 'islanding', a condition in which a generator feeds power back to part of a utility distribution system that is experiencing a power outage. Part of the distribution line would then be live and could present a hazard to utility service personnel, interfere with normal power restoration and feed electricity not meeting utility power quality requirements to surrounding loads. To avoid this problem, standards generally require that inverters have means for detecting utility outages and isolating themselves from the distribution system. The international standard IEC 61727 (1995–06), *Photovoltaic (PV) systems – Characteristics of the utility interface*, addresses the interface requirements between the PV system and the utility, and provides technical recommendations. In the U.S. the IEEE has recently completed IEEE P1547, *Standard for Distributed Resources Interconnected with Electric Power Systems*.

Monitoring system performance

Procedures for monitoring the energy performance of PV systems are well established. In Europe, the Joint Research Center (JRC) of the European Commission developed guidelines for use in EC-funded projects. These were widely used internationally and then used as the basis for developing IEC 61724 (1998–11), *Photovoltaic system performance monitoring – Guidelines for measurement, data exchange and analysis*. This recommends procedures for the monitoring of energy-related photovoltaic (PV) system characteristics, and for the exchange and analysis of monitored data. In the U.S., IEEE P1547.3 *Guide for Monitoring, Information Exchange, and Control of Distributed Resources Interconnected with Electric Power Systems* is referenced.

Conclusion

Building-integrated photovoltaics, by their very nature, are electrical components that bring together DC and AC power within buildings. This creates inherent challenges in terms of electrical performance design and configuration, installation procedures, and the safe interconnection with utility grid systems. This chapter provides an overview of core electrical concepts, reliability issues and best practice standards and compliance requirements, many of which continue to evolve as the industry matures and the number of BiPV installations increases. There is no doubt that electrical elements of BiPV applications require careful preparation, ongoing monitoring and expertise from accredited practitioners.

CHAPTER SEVEN

NON-TECHNICAL ISSUES AND MARKET DEPLOYMENT STRATEGIES

Introduction

Building-integrated photovoltaics (BiPV) represents a significant technical innovation in the electricity and construction industries. In order to realise the vision of the widespread commercialisation of BiPV, collaboration is necessary among key stakeholders. Qualified design professionals, engineers, glaziers, roofers, general contractors, utility companies, bankers and financiers, realtors, policy makers and other government officials, building inspectors, legislative bodies, and knowledgeable building owners make up the team that will bring this technology to commercialisation in the marketplace. Each one of these stakeholders represents one or more of the institutions that hold the keys to unlock the gateway of this sustainable solution for architecture. Just as any building needs a frame or skeleton to provide the essential structural support, the technical innovation of BiPV requires the necessary institutional support to fulfil its promise of being a viable technology and a sustainable solution. The solar industry has demonstrated that the technology works, with hundreds of thousands of installed systems around the world. The architectural world has created award-winning, elegant solar buildings. Utility companies and municipalities have adopted this technology to augment their infrastructure and electricity services network. It is widely recognised that the potential for BiPV is significant; however, institutional barriers can slow its deployment.

This chapter provides insights into the economic and other non-technical aspects of BiPV. It presents an overview of the current institutional issues related to the introduction and commercialisation of photovoltaic (PV) power systems into the built environment. This overview includes an economic analysis, discussion of values, an assessment of the existing barriers in the marketplace and an overview of how to remove these barriers by adopting proper deployment strategies.

The economic performance of BiPV

BiPV systems generally have low operating expenses because of avoided fuel costs; however, the initial system purchase and installation costs make them capital-intensive and economically prohibitive when using first-cost analysis. Hence, economic incentives (interest rate buy-downs, utility rebates, metering programmes, tax advantages, depreciation allowances and programmes for financing new BiPV construction and renovations) may be needed to encourage their use. Conventional energy systems may initially be less expensive, but may have higher long-term costs, because of higher recurring fuel costs. When PV technology is adapted and used as a building component, as exemplified in BiPV, its economic costs and benefits may be shared between the occupant and the utility company. For a building owner, the added costs of installing and operating a system to generate electricity may be offset by the avoided costs of purchasing electricity, or by selling surplus electricity to the utility company.

Costs associated with the integration, design and installation should also be evaluated in comparison with the traditional construction products and systems, in order to determine the marginal cost of the BiPV system. Figure 1 presents indicative costs versus other cladding material.

Building element	Cost US$/m²
Concrete or terracotta roof tile/battens	15–25
Colourbond decking	30
Faced brick cavity wall	100
Pre-cast concrete panels	150
Window walling	400+
Curtain walling	500–800
Polished stone (marble, granite)	2,400–2,800
PV (mono- or poly-crystalline silicon)	650–1450
PV (thin film)	400–450

Fig 1 PV versus conventional cladding material
Source: BP Solar

The cost of a BiPV system can also be compared to using a standoff PV system added to an existing building. When evaluating the expense of BiPV technology, the following must be considered in addition to the cost of the materials: marginal added cost, labour costs, maintenance costs, utility interconnection costs, and costs associated with building permits. Until BiPV becomes a mainstream technology, there are added labour costs associated with specialised architectural design, engineering design, and installation. However, with technical supervision, traditional building tradesmen such as glaziers, roofers, sheet metal workers, and electricians can install BiPV systems.

Regular maintenance costs are typically low. Manufacturers recommend periodic system checks and cleaning as part of a preventive maintenance routine. This includes regularly clearing away any debris and cleaning the PV surfaces exposed to the environment. Rain or water from a simple garden hose is often sufficient to keep the system clean. As a rule of thumb, visual inspection of essential components, based on an inspection checklist provided by the manufacturer and/or the installer, should be made every six months. The utility meter and bill can be reviewed monthly to determine whether output from the system is dropping (adjusting for seasonal or other mitigating factors). Further investigation is warranted if this simple screening indicates poor system performance.

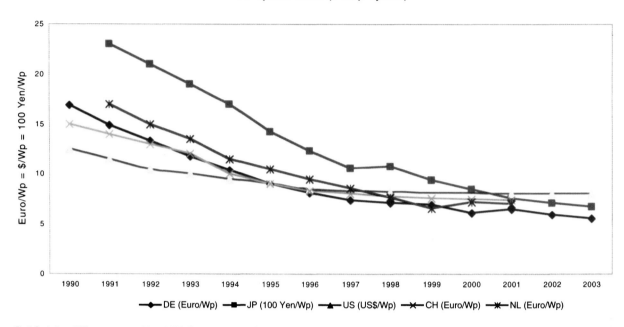

PV system costs (3 kWp system)

Fig 2 Evolution of PV system costs 1990–2003 in Germany, Japan, USA, Switzerland and the Netherlands; excluding value added tax (VAT) or goods and services tax (GST)
Source: Reinhard Haas

Utility interconnection costs are associated with the specific requirements determined by each local authority. Public utility companies have widely varying attitudes towards additional requirements. Costs can include interconnection fees, net metering tariffs, metering calibration charges, engineering study fees, and standby charges. Additional requirements for liability insurance, property easement, legal indemnity, record-keeping of all operation and maintenance (O & M) costs, and additional protection equipment will contribute to even greater utility interconnection costs. The relative cost of meeting these requirements can be much greater for small systems than it is for larger systems. Furthermore, building permits may be required before any construction, addition, moving, or altering of a building takes place. Electrical permits are required for new, remodelled, or upgraded structures. Some of the costs include fees for land disturbance, residential or commercial building permit fees, and re-inspection fees. Building permit fees vary, and are often based on the estimated cost of construction or building floor area. Therefore, permit fees may be increased by the addition of a BiPV system. Installers must contact local land-use and building design officials to identify specific permit requirements.

Price trends

The cost trends for small grid-connected PV systems in the built environment, in countries where comprehensive promotion activities have taken place, are depicted in figure 2. It is of interest that system prices dropped substantially between 1990 and 1996. Yet since 1996, with the exception of Japan, no remarkable price reductions have been achieved. Germany has recognised this and through government programmes and market uptake, system costs are beginning to fall.

In recent years, non-module costs such as inverters, design and planning, assembling, construction work and installation, have been reduced to a larger extent than module costs, as illustrated in figure 3. This shows the importance of pursuing further reductions both in module prices and in design and installation costs.

EUROPE

1990 total costs: 15 Euro/Wp

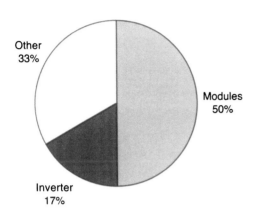

Other
33%

Modules
50%

Inverter
17%

2003 total costs: 5.7 Euro/Wp

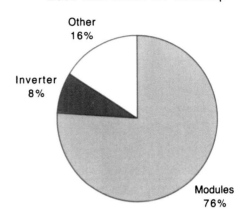

Other
16%

Inverter
8%

Modules
76%

JAPAN

1990 total costs: 2500 Yen/Wp

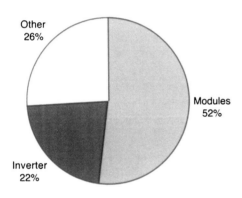

Other
26%

Modules
52%

Inverter
22%

2003 total costs: 679 Yen/Wp

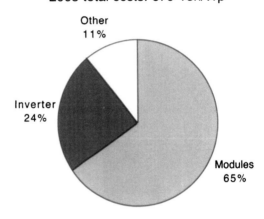

Other
11%

Inverter
24%

Modules
65%

USA

1990 total costs: US$12/Wp

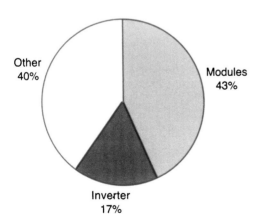

Other
40%

Modules
43%

Inverter
17%

2003 total costs: US$8/Wp

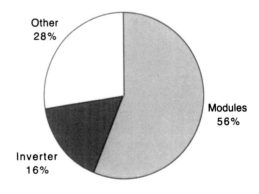

Other
28%

Modules
56%

Inverter
16%

Fig 3 Share of modules, inverters and other costs on total PV system
prices 1990–2003 in Europe, USA, and Japan; excluding VAT or GST
Source: Reinhard Haas

Methods of economic evaluation

Despite conventional wisdom, economics is not the only assessment criterion that needs to be taken into account, although the high (monetary) investment costs of PV systems are the major impediment to wider market penetration.

The (conceivable) environmental benefits of residential PV electricity should, in principle, be compared with the production costs of PV electricity. Using a traditional first costs economics approach, the cost of a kWh of PV electricity (p_{PV}) can be calculated by considering the investment costs (C_{PV}), the capital recovery factor (**CRF**), the yearly solar insolation (Q_{sol}), the ratio of total radiation on the optimal tilted plane to that on the horizontal surface **R** the performance factor φ^2, and the efficiency of the PV system η_{PV} as described in the following equation:

$$p_{PV} = \frac{C_{PV} CRF(r, LT)}{Q_{sol} \, R \, \eta_{PV} \, \varphi}$$

where r is the interest rate and LT the lifetime of the system. Currently, the PV module efficiency is by-and-large the same all over the world, whereas the investment costs and the solar insolation may vary tremendously between different countries and locations making them important variables.

Figure 4 elucidates this point, with costs varying a great deal from country to country. While price per kilowatt hour is a key determinant in evaluating the value of PV system generation, the true value will be affected by relative retail electricity costs. Figure 5 represents the convergence of conventional electricity tariffs and PV-derived electricity for different IEA countries. Japanese residential electricity tariffs are amongst the most expensive and hence, the competitive price gap that PV electricity has to bridge

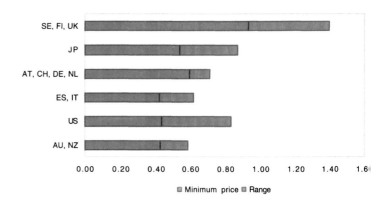

Cost of PV electricity 2002/2003
(USD/kWh = EUR/kWh = 100 JPY/kWh)

Fig 4 Range of costs per kWh of PV electricity in different OECD countries in 2002/2003 (related to small systems of 3 kWp)
Source: Reinhard Haas

is smaller. If a building uses BiPV-generated electricity directly, its value is equivalent to the conventional retail price. In other circumstances, where the PV is exported to the grid, purchase price may be determined by the wholesale electricity market. Some countries operate green energy markets, with an inflated value for renewable energy sourced electricity, or higher utility buy-back rates. Given the likelihood of conventional electricity prices increasing over the longer term, the value of PV generated power will also appreciate.

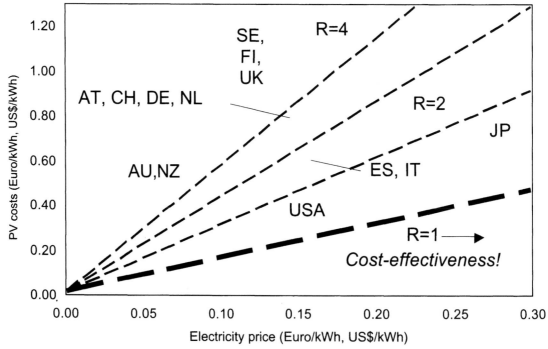

Fig 5 PV versus conventional electricity convergence costs for various IEA countries (R = ratio of PV generation costs/retail electricity price)
Source: Reinhard Haas

Evaluating the added value of BiPV

The benefits of BiPV systems can directly affect the decision-making process. These benefits can be identified and evaluated based on direct economic impact, indirect economic impact, and qualitative value; they include energy, environmental, architectural and socio-economic aspects. Beneficiaries include utilities, building owners and occupants, the design, engineering and construction industries and society as a whole (Watt, 2001). Moreover, various other added values have been identified from different points of view, as discussed by Haas (1995) and Watt (2001) and summarised in table 1. For further in-depth reading, the report by Watt (2001) is recommended.

From a PV customer's point of view, the most important values are that PV systems are available in every size, due to their inherent modularity, and that they are fast and relatively easy to install. Further, PV systems with storage increase consumer independence from central supply by utilities.

From society's point of view, environmental benignity – no emissions, no pollution in operation – is the most important added value, along with its decentralised applicability and the virtually complete absence of acceptance problems to date. Further socio-economic values are that decentralised PV systems foster employment in general as well as gradually enhancing local employment in both urban and rural areas. Decentralised PV applications could impact on both supply-side and demand-side issues. Consumers are more likely to be aware of their electricity meter and their consumption habits, leading to energy conservation and the purchase of energy-efficient appliances and other renewable energy-using equipment. Finally, increased security of supply due to decentralised generation may, in some countries, also be of value.

Benefit/Value	Remarks	Weight
PV CUSTOMERS:		
Modularity	PV systems can be constructed in any size and can be expanded over time	+ + +
Fast and easy to install	This value is subject to further improvements!	+ + (+)
Increases independence (with storage)	Storage adds system complexity and cost	+ + +
SOCIETY:		
Environmentally benign	No emissions or pollution in operation	+ + +
No acceptance problems	This may occur with large-scale deployment and/or unsympathetic urban design	+ + +
Decentralised generation facility	Adds to regional energy self-sufficiency and strengthens local grids	+ + +
Enhances general and local employment	A range of skills required	+ +
Indirect effect: triggers energy conservation	Enhances awareness of energy issues	+
Increase supply security	In some countries with a low developed grid or high summer peak demands, PV may contribute to increased supply security	+
Local and community choice and control		+
Educational device	PV is probably the best example for teaching energy supply and electricity generation	+
UTILITIES:		
Peak shaving	Summer peak loads are an increasing issue in many countries	+
Low transaction costs and short lead-time compared to building a power station	Especially important in a liberalised electricity market!	+
BUILDING INDUSTRY:		
For architects: Innovative multi-functional building construction element	Needs education of architects to ensure proper application	+ +
For developers: An attractive new marketing concept with a green image	Need support materials to be able to show benefits to customers	+

Table 1 Survey on values added by PV systems
Source: Watt (2001), Haas (2002)

The values of PV for utilities are that transaction costs and lead times for building a PV power station are low (this may become important especially in liberalised electricity markets) and that in some countries with daytime peak electricity demands in summer, PV may contribute substantially to 'peak-shaving'. The effective load-carrying capacity (ELCC) of PV can be especially high for commercial customers, with typically good matching between peak PV output and daytime air-conditioning load (Watt, 2001).

Moreover, PV may also provide values for other sectors in the market. Architects may benefit from PV as a new and innovative multi-functional building construction element. It also allows the designer to create environmentally benign and energy efficient buildings, without sacrificing comfort, aesthetics or economy, and offers a new and versatile building material (Watt, 2001). Builders and developers may use the positive 'green' image of PV for marketing purposes.

Figure 6 illustrates the process of developing cost-effective PV systems, from the current stage where incentives are used to buy down the effective cost, through increasing acceptance and quantification of the added values PV can offer. These added values can serve to reduce the net cost of PV-generated electricity and eventually lead to situations where PV is more attractive than conventional energy sources.

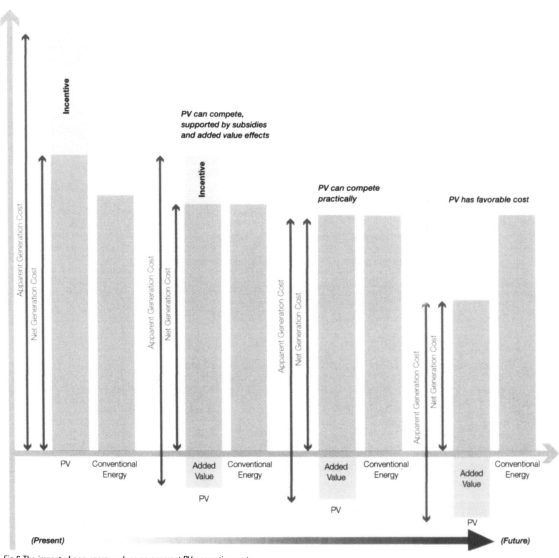

Fig 6 The impact of non-energy values on apparent PV generation costs
Source: Konno, (1999)

Barriers associated with the broader dissemination of BiPV

Photovoltaic power systems face a number of barriers to their entry into the mainstream energy and building markets. These include high capital costs and associated financing, plus administrative, architectural, communication, marketing and environmental issues. As discussed previously, some of these barriers can be overcome by assessing both the energy and non-energy benefits that can be provided by a photovoltaic power system, thus making it a cost-effective option even with current costs and energy prices.

Photovoltaic suppliers have indicated that they need a bigger market to acquire the necessary price reductions, but conversely, price reductions are needed to accelerate market growth. Government incentive programmes, such as buy-downs and tax incentives, help offset the high initial costs and encourage customers to make the investment in PV. However, some policy and incentives designed to stimulate the PV industry may actually hinder PV investments. The short-term duration of many regulations and government funding programmes creates obstacles for investors and developers, in that the rules and procedures that apply to today's project may not be in place for future projects. In the financial sector, a lack of confidence or awareness on the part of investors makes it difficult for individuals to secure financing for renewable energy projects. Investors lack information about technology, costs, performance and planning. These factors create barriers based on false premises, inaccurate suppositions and a lack of specialised knowledge.

Architects, project developers, and building contractors encounter special challenges when designing and building PV structures. These problems are primarily a result of a lack of standardised PV manufacturing as well as building design issues, and the associated paradigm changes in thinking needed. One of the main architectural issues is that PV products do not correspond with building dimensions. This mismatch results in the necessity for non-PV building materials to be used as filler materials around PV modules, creating a 'filled-in' appearance. General public reaction to the aesthetics of BiPV applications is not always positive. Whilst PV manufacturers have made an effort to produce BiPV modules that resemble building materials (such as solar shingles), architects have for some time been selecting building materials that mimic PV modules, especially glass.

Architects perceive three problems with the supply of BiPV product information:

- it is prepared for engineers and PV specialists and is therefore inappropriate for architects;
- it is not available through information avenues normally used by architects; and
- the routes for obtaining the information are not known; the information is spread out across a wide range of sources.

Communication barriers to the growth of BiPV in the built environment are based on a lack of knowledge, experience, and information among the parties involved. While pockets of intense activity exist at some municipal or national government levels, they are inconsistent from region to region. Vast areas with few or no government PV programmes exist in regions throughout the world, with accompanying vast knowledge voids. One suggested problem in communications is that the target groups do not know enough about PV and are prejudiced against it by their lack of knowledge. The building sector, for instance, does not readily accept new innovations in some cases, lacks awareness of the potential and possibilities of PV, and therefore may be reluctant to implement PV.

Within the energy sector there appears to be a focus on supply-side issues, a lack of experience and insight into the customer's needs, and a strong resistance to PV among the utilities. Utilities have insufficient understanding of the economics of PV, and are sceptical about PV, believing that it requires a high level of insolation or that it will interfere with their networks. Among the communications barriers with the public, a lack of trust towards developers and planners stands out. Many appear to lack awareness and knowledge of PV and have a critical attitude towards it. There is a lack of independent reliable information sources and educational programmes, due to the broad and interdisciplinary nature of renewable energy. The marketing barriers relate to problems within the PV industry and the energy sector. Lack of standardisation within the PV industry makes it difficult to develop comparative analysis of PV products, designs, and applications. This makes it difficult to develop new PV markets or penetrate existing markets. Manufacturers' product literature is often lacking in information on PV components' weight, attachment options, visual details, durability, certification and warranties, maintenance, repair and cleaning procedures, installation practices, costs, and embodied energy data.

In general, the PV industry may not fully understand who its customers are, the size of its market, or what kind of products are needed. It may also lack adequate understanding of the utility market and other possible niche markets. Additionally, the PV industry operates in an international market and may not be able to focus on local needs. The energy sector shares similar challenges. It needs a better understanding of customers and PV application potential. The energy sector also displays little knowledge of niche markets, or of which markets are most amenable to early PV exploitation. The gap between the utility marketplace and PV technology is caused by a lack of utility-related distribution channels or third parties that can deliver services such as maintenance, installation and products. Due to a lack of sales and service infrastructure, buyers must turn to the utility when they experience problems or have questions, which places an extra burden on the utilities.

International market deployment strategies

Market strategy considerations

There are five important steps to be identified before launching dissemination programmes (figure 8):

- Investigate the benefits of a technology, for example, environmental benignity, load-shaving, modularity.
- Estimate potential: is it possible that the technology contributes seriously to solving a problem, such as meeting energy demand?
- Identify barriers: what are the major impediments for a broader market penetration (such as lack of technical reliability, high investment costs, no social acceptance).
- Define target areas (the market, the technology) and target groups (private individuals, PV industry, architects, governments) for actions to be undertaken.
- Derive and assess possible strategies to overcome barriers such as financial incentives, information and education campaigns.

Fig 8 Issues associated with BiPV marketing
Source: Reinhard Haas

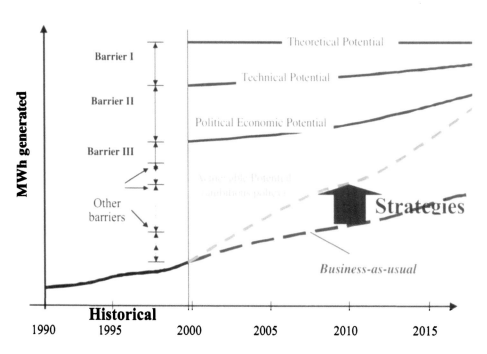

Fig 9 Interaction between potentials, barriers and strategies
Source: Reinhard Haas

Figure 9 depicts how potentials, barriers and strategies are linked in principle. The electricity generated is shown depending on the time. We start with the historical development of PV in a certain country and identify different potentials. Various barriers exist, which impede the practical achievement of the potentials. If no policy strategies are implemented, the lower broken line will be achieved, the so-called business-as-usual scenario. If an ambitious policy launches the proper strategies, the upper broken line will be achieved. The first important target area is the technology itself. Technical issues such as standardisation and reliability are of high relevance in a successful dissemination strategy. Next are market aspects such as competition, transaction costs and market transparency. Moreover, the individual preferences of customers and their willingness to pay (WTP) represents a very important role. Finally, the status and the acceptance of the technology in society has to be considered.

o increase the market penetration of PV and to enjoy the corresponding environmental benefits, very different strategies have been implemented in various European countries in recent years. In this section a survey of different types of promotion strategies, and how they work in principle is provided. These strategies are illustrated by reference to specific programmes. In addition, an overview of the flow of money as well as legal aspects is given. As already mentioned in the introduction, only dissemination strategies that focus on a non-specified target group are dealt with.

With respect to a classification of strategies, the following questions are of interest:

- Who launches a strategy? Governments? NGOs? Commercial companies such as 'Green' utilities?
- Who is addressed? PV generators? PV electricity users? Or a specific target group such as architects, teachers, electricians?
- Is it a voluntary strategy or based on regulation?
- Are financial incentives provided or not?
- Is the investment influenced, for example, by subsidies for the kWh generated?

Table 2 gives an overview of the most important strategies with their core features, categorised by the criteria described above.

In the history of PV implementation, various types of dissemination strategies have been launched, as shown in Table 3. As early as 1982 in Massachusetts, USA the first net metering programme for PV (and other renewables) was introduced. While this programme addressed only very few customers, in 1991 in Germany the first comprehensive dissemination strategy was launched – the German '1000 roofs' programme. In the early 1990s the municipal utility of Sacramento in California – SMUD – introduced various new strategies. Most important were green pricing models, which later also attracted attention in other countries, and the PV pioneer programme, which was a type of contracting process introduced in 1993. At about the same time, the first rate-based incentives programmes were launched in Switzerland and Germany (Burgdorf and Aachen model).

Solar stock-exchange models for PV electricity became popular in the city of Zurich in 1996. By the end of 2000, more than half of the Swiss households had access to 'Solarstrom'. Since about 1998, programmes for the promotion of renewables have focused on liberalised electricity markets. Examples are public purpose programmes as introduced in California, followed by other states in the USA, and tradable certificates. Other developments include soft loans introduced in the German '100,000 roofs' programme, green pricing programmes with labels and NGO initiatives. As can be seen clearly from Table 3, the widest variety of different types of strategies undoubtedly exists in Germany.

	FEATURE					
STRATEGY	Financial incentive	Marketing and targeted information	Improved technical performance	Regulatory conditions	Social event/public awareness	Creating market transparency
I. GOVERNMENTAL TARGET PROGRAMMES:						
1. Voluntary national targets	N	Y/N	Y/N	N	Y/N	Y/N
2. Mandatory national targets	Y/N	Y/N	Y/N	Y	Y/N	Y/N
II. REGULATORY FINANCIAL INCENTIVE INVESTMENT FOCUSED						
3. Governmental rebates	Y	Y/N	N	Y	?	Y/N
4. Financing, soft loans	Y	N	N	Y	?	Y/N
5. Tax incentives	Y	N	N	Y	Y	Y/N
III. REGULATORY FINANCIAL INCENTIVE, GENERATION-BASED						
6. Net metering	Y	N	Y/N	Y/N	Y	N
7. Enhanced feed-in tariffs	Y	N	Y/N	Y	Y	N
8. Rate-based incentives	Y	N	Y	Y	Y	N
9. Environmental pricing (eg CO_2-taxes)	Y	N	N	Y	N	N
IV. VOLUNTARY FINANCIAL INCENTIVE INVESTMENT FOCUSED						
10. Contracting	Y	Y	N	N	Y	N
11. Green shareholder	Y	Y	N	N	Maybe	N
12. Contribution	Y	Y	N	N	Maybe	N
13. Bidding	Y	N	N	N	Maybe	Y
V. VOLUNTARY FINANCIAL INCENTIVE, GENERATION-BASED						
14. Green tariffs	Y	N	N	N	Y	Y
15. Green-power marketing	Y	N	N	N	Y	Y
16. Solar stock exchange	Y	N		Y/N	Maybe	Y
VI. OTHERS						
17. NGO marketing	N	Y	N	Y/N	Y	Maybe
18. Selling green buildings	Y	Y	N	Y/N	Y	Maybe
19. Retailer alliances	N	Y	N	Y	Y	Maybe
20. Commercial financing programmes	N	N	N	Y	Y	N
21. Public building programmes such as schools, leisure centres, sports facilities	Y	Y	Y/N	Y	Y	Maybe

Table 2 Major types of strategies for dissemination of small PV systems and their core features

Year	Country	Type of strategy	Programme name	Remarks
1982	US	Net metering		
1987 – present	DE	Rebate	REN	
1991 – 1995	DE	Rebate	1000-Dächer-Programm	
1991 – 2000	CH	Voluntary target programme	ENERGIE 2000	
1992 – 1994	AT	Rebate	200 kW PV-Programm	
1992 – 1999	DE, CH, AT	Regulated rates	Kostendeckende Vergütung	
1993 – 1997	US	Contracting	SMUD PV pioneer I	
1994 – present	JP	Rebate	Residential PV promotion programme	revised 1998
1994 – present	ES	Regulated rates	P.A.E.E.	revised 1998
1994 – present	DE	Rebate, contribution	Sonne in der Schule, SONNEonline	launched by various utilities and government institutions
1996 – present	DE	Green pricing	RWE Umwelttarif	
1996 – present	CH	Bidding/Green pricing	Solarstrombörse	
1997 – present	NL	Voluntary target programme	Heading into the Solar age together	
1997 – present	CH	Green pricing	Solarstrom vom E – Werk	
1997 – 2001	DE	Bidding	Solarbörse Berlin	
1998 – present	DE	Labelling	Golden and Silver label (EUROSOLAR)	
1999 – present	AT	Shareholder	Sonnenschein	
1999 – present	DE	Soft loans	100,000 Dächer – Programm	
1999 – present	NL	NGO initiative	SOLARIS	
1999 – present	US (CA)	Rebates	California's emerging renewables buydown programme	
1999 – present	AUS	Rebates	PV Rooftop Programme	For grid and off-grid buildings, revised 2000
2000 – present	DE	Enhanced feed-in tariff	Neues Einspeisegesetz (EEG)	
2000 – present	DE	Rebate, contribution	Kirchengemeinden für die Sonnenenergie	

Table 3 A history of the most important promotion strategies for grid-connected PV electricity

AT = Austria; AUS = Australia; CH = Switzerland; DE = Germany; NL = the Netherlands;
UK = United Kingdom; US = United States (CA = California); JP = Japan; ES = Spain

National target programmes

In some countries, programmes have been launched which focus on a certain target market share or on installed capacity of PV or renewables in general. In principle, both regulatory and voluntary strategies are possible. Such programmes have been introduced and are still under way in Switzerland (ENERGIE 2000), the Netherlands (NOVEM programme), and the USA (President's 'Million Solar Roofs' programme).

The Swiss 'ENERGIE 2000' programme

In Switzerland in the early 1990s the 'ENERGIE 2000' programme was launched by government institutions. Within this programme, various energy-conserving and solar energy converting technologies were promoted and specific goals for market penetration by the year 2000 were set. The ENERGIE 2000 programme is now over, but a follow-up programme 'EnergieSchweiz' began in 2000. For PV, a capacity of 50 MWp was planned. At the end of 1999, about 14 MWp were installed. Until 1999, the PV systems were subsidised to 30 per cent.

The Dutch 'NOVEM' programme

In 1994 in the Netherlands, various organisations under the leadership of the Ministry for Energy and Environment (represented by NOVEM) launched a cooperative programme for a broader market dissemination of decentralised PV systems, the 'NOZ-pv programme'. Under the Dutch 'PV introduction plan' (NOVEM 1997) it was planned to install 7.7 MWp of PV capacity by the year 2000 and 500 MWp by 2010 (see Schoen 2000). The first target of 7.7 MWp was surpassed, with 9.2 MWp installed by the end of 1999 and 12.5 MWp installed by the end of 2000. The new PV covenant is anticipated to aim at a target of 300 MWp by 2010 and 1,400 MWp by 2020. In addition, it is intended to reduce the investment costs to 2.75 NLG/kW by 2010.

US 'Million Solar Roofs' initiative

In 1997, President Clinton announced the 'Million Solar Roofs' initiative. Within this programme, one million roofs in the USA were to be equipped with a PV system and/or a solar thermal system for water heating, pool heating, or space heating by 2010. With respect to PV, the Team-Up initiative with its 'friendly PV programmes' is of special relevance. Tax credits up to US$2,000 were made possible for individual systems.

Rebate programmes and soft loans

By 2002, three large rebate programmes had been introduced worldwide:

The Japanese residential PV promotion programme

The largest worldwide dissemination programme so far was launched in Japan in 1994. In the following years the number of small grid-connected systems skyrocketed. This programme was combined with low-interest consumer loans and comprehensive education and awareness activities for PV. The programme makes blocks of funds available to PV system retailers in a competitive bidding process. In 1997, the 'New Energy Promotion Law' was introduced, with subsidies for PV and a target of 400 MWp by 2000 and 4,600 MWp by 2010. While in 1994 'only' about 540 systems had been installed, by the end of the year 2003 about 52,000 small grid-connected systems with an average capacity of about 3.6 kWp had been installed. This led to 200 MW installed in FY2003 and to a cumulated installed capacity of 600 MW in Japan by the end of FY2003 (Ikki, 2004).

A major question is whether the Japanese programme has brought down the PV price substantially. Currently, it appears that it has. In the Japanese programme, rebates were decreased

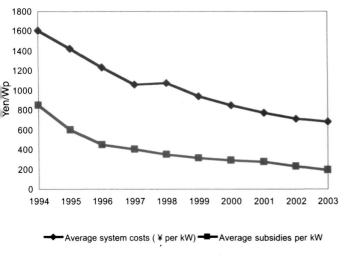

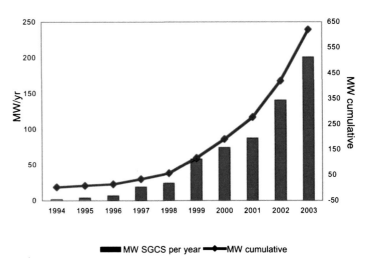

Fig 10 Japanese residential PV promotion programme: average system costs, average subsidies, and development of installed capacity, 1994–2003
Source: Reinhard Haas

continuously over time, from 50 per cent of the total investment costs in 1994 to less than 30 per cent in 2003. The upper limit for rebates has been reduced from ¥900,000 in FY1994 to ¥500,000 in FY1996 to ¥190,000 in FY2003 (see figure 10).

Figure 10 depicts the development of investment costs and subsidies in Japanese Yen over the time period of the promotion programme (Ikki, O, 1998, 2000, 2003). A result of this effort is that Japan is now the world leader in the development of grid-connected systems. This success is the direct result of a conscious policy to promote PV technology, both for reasons of national energy security (Japan imports most of its fuels) and for reasons of economic development (Japan aims to dominate PV manufacturing to the same extent as it dominates the production of electronic equipment).

The German '1000 Roofs' and '100,000 Roofs' programmes

The first comprehensive international dissemination programme was the '1000 Roofs' programme launched in Germany in 1989. This programme was completed in 1994. Some 2,250 German roofs were equipped with PV systems of an average size of 2.6 kWp and a total capacity of about 6 MWp. Average system costs were US$15,000/kWp and average subsidies were 70 per cent of the investment costs. During this dissemination programme, and also in its aftermath, comprehensive investigations on technical and sociological aspects took place. The major results of this programme were that PV systems reached a certain standard of technical reliability, that PV system costs dropped, and that the acceptance of the technology increased considerably. Moreover, experiences gained in this programme were also used for similar activities in Austria and Japan.

From 1999 to 2003, a new financial approach has been pursued in Germany with the '100,000 Roofs' programme. Within this programme, very attractive credits (soft loans) were provided to the public. Initially the interest rate was 0 per cent (in 1999) over a ten-year repayment period. The loan has been repaid in eight instalments from years three to ten, and the last instalment in year ten is cancelled if the system is still operating. The response to this programme in the first year (1999) was disappointing. Only about 3,000 new projects (about 9 MWp) were approved. This was only half of the planned capacity of 18 MWp. In March 2000 the German Law for the Priority of Renewable Energy came into force with the accompanying introduction of a substantial feed-in tariff of 99 Pfennig/kWh (0.5 Euro/kWh) for PV in March 2000. This was a major reason for the boost in approvals in April 2000 (figure 11). The programme was then stopped for almost six months and the initial plan was revised. The interest rate was raised from 0 per cent in 1999 to 1.8 per cent in June 2000. The target for 2000 was raised to 50 MWp/year, increasing to 95 MWp/year in the year 2003. Unfortunately, the second target in the year 2000 was also missed (figure 11). In 2000, 41.7 MWp were installed instead of the target of 50 MWp. For 2001, 65 MWp were planned, in addition to the 9 MWp missed from the first two years' targets. The targets for 2001 and 2002 were missed by only a small amount. Finally, in 2003, the growth was beyond the planned target, with 350 MW being installed (figure 11).

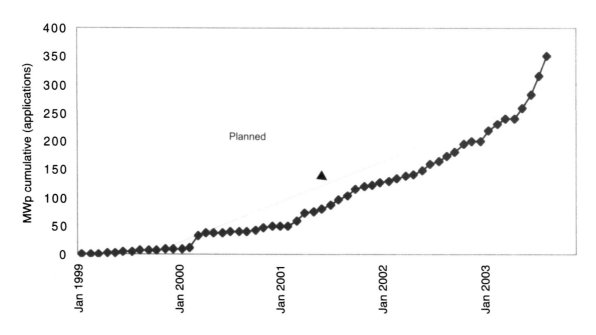

MWp cumulative

Fig 11 Soft Loans: German '100,000 Roofs' programme: cumulative applications 1999–2003
Source: Reinhard Haas

Regulated rates

In recent years other types of financing models have been introduced in various European countries that are based on regulated rates. The most important models are:

Net metering

Since the early 1980s, especially in the US, the 'net metering' financing model has gained attention for PV (and other renewable and CHP electricity). Within this approach the Net Excess Generated Electricity (NEG) is refunded by the utility at about the same price as the retail price of electricity.

Rate-based incentives

In the early 1990s, almost at the same time in Switzerland and in Germany, the idea of full production cost rates was launched. This meant that the public utility has to buy back PV electricity at (almost) the full production cost. This idea gained attention mainly in cities where municipal utilities are responsible for power supply and where local politicians have the power to put these full cost rates into practice. They have gained special attention in Germany, Switzerland, and Austria. They vary currently between about US$0.7 and US$1.0. Figure 12 shows the installed PV capacity per capita in some cities with rate-based incentives.

Utility programmes

Utility programmes are in general voluntary and usually provide financial incentives for a PV generator. These financial incentives may be collected from the utility's electricity consumers. Of course, it is also possible that the utility itself is the generator. Utility programmes also include possible 'Green brokers' and 'Green utilities'.

Contracting programmes

The first, and so far most popular, contracting programme worldwide for PV was launched in 1993 by SMUD (Sacramento Municipal Utility District) in California – the 'PV Pioneer I programme'. Under this programme the system was purchased, installed, owned, and operated by SMUD (Osborn, 2000). It feeds its power directly into the SMUD electric grid. SMUD residential customers volunteer to share in this effort through a form of 'green pricing' and by providing the roof area to place the SMUD PV power plants, each about 4 kWp. The PV Pioneer I customer pays a $4 per month 'green' premium in addition to their normal utility bill in order to participate.

The programme was aimed at developing the experience needed to successfully integrate PV as distributed generation into the utility system and developing long-term market and business strategies. In addition, its role was to stimulate the collaborative processes needed to accelerate the cost reductions necessary for PV to be cost-competitive in these applications by about the year 2003. This effort resulted in about 8 MWp of PV systems installed in Sacramento by the end of 2000, distributed over some 700 installations.

SMUD gained experience in the installation, operation, maintenance, pricing strategies and other aspects of residential PV systems, and obtained low-cost 'power plant sites'. With little marketing undertaken, SMUD has been adding about 100 PV Pioneer I systems each year. Finding customers willing to pay has not been difficult. PV Pioneer marketing normally consists of just one or two leafleting promotions a year, door-to-visits in neighbourhoods with a predominance of 'good roofs', and as a result, free media coverage.

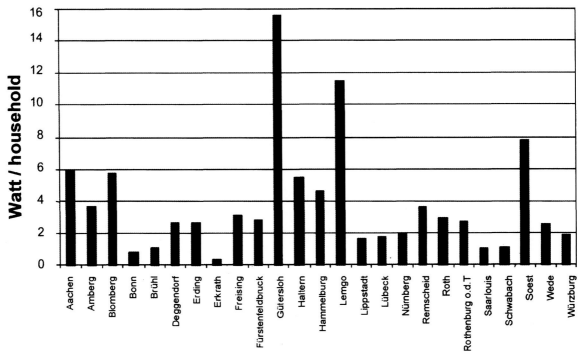

Fig 12 Installed PV-capacity in watt per capita in German cities, with rate-based incentives.
(Note that in some cities the limit of the price cap increases has already been reached)
Source: Reinhard Haas

Green tariffs

The major feature of this type of financing programme is that participants pay a special price premium per kWh over regular rates. This type of financing programme has gained attention up to now in Switzerland, Germany, Australia, the USA, Austria, and the Netherlands. Within this programme type, utilities offer 'green' electricity – that is, electricity generated by wind turbines, biomass, small-scale hydro, and PV – at a price that by-and-large meets the generation costs.

In recent years, labels have become more and more important to prove the content of the product. Green tariffs, with respect to PV, are most important in Germany, Switzerland and the US. In Australia a wide variety of Green tariffs exist and most utilities have installed some PV, although it is typically a very small portion of total green power requirements (Weller, 2000). With respect to the promotion of BiPV systems, it has to be stated that mostly larger plants are built from the revenues of the 'Green tariff' programmes.

Table 4 describes the major features of the most popular Green tariffs in different countries. Only those programmes that contain a substantial amount of PV are listed.

Private shareholders/participation shares

Another concept that has attracted attention, mainly in Germany, is the sale of shares in a PV plant to private customers, for example in blocks of 100 W. The customer thus becomes a shareholder in a renewable power station. An example of this programme type is the 'Bürger für Solarstrom' model of the 'Bayernwerke'.

Contribution programmes

Within contribution or donation programmes, customers can contribute to a fund for renewable energy projects. Usually these funds are managed by electric utilities. It is an approach that focuses mainly on promotion of PV systems to the public, for example, schools. The projects developed are unrelated to the customers' electricity usage.

Utility (country)	Start year	Product/ Label (PV share)	Premium/ price	Number of participants (year)	Participation rate (%)	PV capacity installed (kW)	PV electricity generated (MWh/yr)
RWE (DE)	6/1996	Umwelttarif (26% PV)	USD 0.102/kWh (Mix)	15,800 (1998) 12,500 (2000)	0.5	1,050	800
ENBW (DE)	2/1997	Umwelttarif grün (1% PV)	USD 0.41/kWh (Mix)	2,070 (2000)	0.12		
	2/1997	Umwelttarif solar (100% PV)	USD 0.818/kWh	230 (2000)	0.013	62	59
Elektra Basel Liestal (CH)	1992	Solar-strom für alle	CHF 1.40 (1998) 1.30 (2000)/kWh	333 (2000)	n.a.	104	84
Elektra Birseck München-stein (CH)	1994	EBM-Solar	CHF 1.40/kWh	440 (2000)	n.a.	71	85
NUON (NL)	1996	Natuurstroom (0.5 % PV)	NLG 0.09/kWh	52,000 (2000)	n.a.	1,000	702
Energy Australia	1997	Pure Energy/ GreenPower	AUD 0.137/kWh (40% premium for 100% greenpower)	15,500 (2000)	n.a.	600	877
Arizona Public Service (USA)	1997	Solar Partner	USD 0.176/kWh	1,600 (2000)	n.a.	500	

Table 4 Green tariff schemes with PV
Source: Authors' investigations

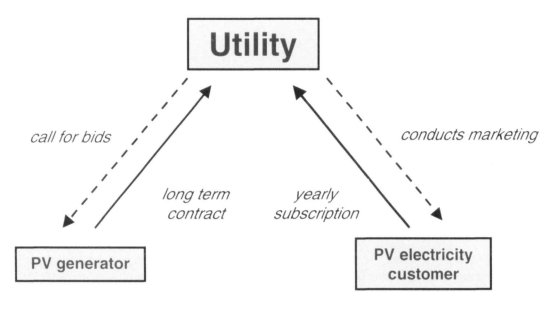

Fig 13 The principle of 'Solarstrombörse' applied by various utilities in Switzerland

Solar stock exchange

Another idea of providing financial incentives for the construction of PV systems is the 'Solar stock exchange' where electricity is generated by privately-owned PV systems and fed into the public grid. Other customers may buy this electricity and pay rates corresponding to the PV production costs. On the supply-side only, the most cost-effective projects are selected by a bidding process. The utility acts as a 'power exchange'. That is, it organises the balance between supply and demand, launches calls for tenders for new PV capacities and signs long-term contracts with the generator. On the demand-side, marketing activities are conducted and the customer may subscribe on a yearly basis or longer (figure 13). In most cases, the utility bears the administration costs but has no other expenses. The customers choose how much solar electricity they want to buy. The minimum order is usually 100 kWh/year. The price was around USD 0.63–0.75/kWh (CHF 1.00–1.20/kWh) in 2002. It has to be recognised that the system's costs and the price for customers has decreased continuously since 1996 (from about CHF1.40 in 1996 to CHF 0.85/kWh in 2002).

The advantages of this strategy are:

- customers' willingness to pay is fully exhausted;
- efficient operation is encouraged;
- private 'green' PV owners ensure that only the best examples for PV will be constructed;
- a 'green label' with high credibility ('pure solar electricity') may be associated with this type of strategy.

This idea was first developed for the city of Zurich in Switzerland and has since attracted attention in other cities. At the end of 2001 about 2 MWp had been installed in Switzerland. With respect to the system size, it has to be stated that in most instances larger plants are built from the revenues raised by the 'Solarstrombörse'.

Non-government organisation (NGO) initiatives

Aside from green tariffs, private shareholder and donation projects have been launched by different types of organisations, specifically NGOs. The most important programmes are summarised below.

The Sonnenschein programme

An example of a very successful shareholder programme is the Sonnenschein campaign in the Austrian province of Vorarlberg, where about 150 kWp of decentralised systems were installed between 1998 and 2001. This programme is still running.

The Solaris programme

Another successful example is the Solaris programme, launched by Greenpeace in 1997 in the Netherlands. About 15,000 applicants were registered and 3,000 systems were installed up to the end of 1999 (Schoen, 2000). Within this programme, no financial incentives are provided for residential customers.

Core issues

This review of marketing strategies for PV shows that there is a wide range of possibilities for increasing their dissemination with real success stories. Yet, there are considerable differences in these strategies with respect to technical and economic efficiency as well as their success in triggering a substantial number of new installations. The most important conclusions of this analysis are reported below, split into three major categories:

1 What are the core issues for successful dissemination strategies, regardless of which strategy is chosen?

2 Under which conditions are different types of strategies successful?

3 Which activities are required now with respect to different target groups?

Regardless of which strategy is chosen, the following basic requirements apply for success:

- Predictability and continuity over time: avoid 'stop and go' strategies!
- Comprehensive associated information and education activities are important.
- Pure cost-effectiveness is not crucial, with affordability being the key. However, over the next five years the costs have to come down substantially, to a level close to residential retail electricity prices.
- High environmental credibility of the institution/ company that launches the strategy.
- It has to be ensured that after the programme is terminated, a sustainable development of the PV industry is likely and that the market does not collapse.
- With respect to financial incentives, it is important that they exhibit a decreasing trend over time, are designed dynamically, and have justifiable benefits to society.
- It is of high relevance that efficient promotion programmes depend on consumers' 'Willingness To Pay' (WTP). It is very important that financial incentives provide at the most only the difference between the system costs and the WTP for PV. The incentives in most programmes up to now have not been optimally designed. Consumers' WTP for PV is higher than expected by programme designers. With the same amount of total subsidies it would have been possible to promote more PV systems.
- Minimise administrative and transaction costs.

To be successful, it is necessary to design strategies in a way that ensures the cooperation of governments, utilities, customers and potential investors.

Under which conditions are different types of strategies successful?

The conditions required for the success of different types of strategies are:

- National targets: national target programmes work if they are pursued seriously, accompanied by information and education activities, and if clearly defined and achievable targets per year exist. Moreover, it is of high relevance that a carefully conducted progress report is provided.
- Rebates: in principle, rebates work as a dissemination strategy; they are more effective if they are accompanied by comprehensive information and education activities. Rebates have to decline continuously over time, as the suppliers would not reduce the system costs to the same extent, but rather earn an extra profit. If rebates on investments are provided, it is of high relevance that they are linked to performance standards and that they are accompanied

by comprehensive information, education and technical monitoring activities. Otherwise, the typical 'fly-by-night' effect of suppliers may take place, as happened in the early 1980s in the USA with respect to solar thermal collectors. In addition, the purchase of a PV system leads to changes in consumer behaviour. To accelerate the cost reduction of PV systems, it is important that rebates are a fixed amount per kWp capacity, or even better, of kWh generated, and not a percentage of the investment costs.

- Financing/soft loans: this type of instrument works in a similar way to rebates, although up to now the experience available has been rather limited. This being the case, the lessons learnt so far from the German '100,000 roofs' programme are nevertheless very encouraging.
- Regulated rates: regulated rates are preferable to rebates, since they are based on PV system performance and have lower transaction costs and bureaucracy; virtually all programmes where the regulated rates were close to the production costs and guaranteed over a period of about 15 years were successful.
- Green pricing/solar stock exchanges/green shareholder: green pricing programmes work only if they are launched by a utility with high credibility, are marketed, and have an attractive label which, of course, has to include PV.

Which activities are required now with respect to different target groups?

The actions required now with respect to different target groups in the four relevant areas of activity are:

Technical issues

Technical barriers still exist and many technical issues are not yet solved satisfactorily. The most pressing problems concern:

- system optimisation;
- a need for an increase in standardisation/simplicity/compactness (i.e 'sun-in-the-box, 'plug-and-power');
- safety;
- utility interface.

It is a high priority to solve these problems before any widespread dissemination strategy is triggered.

Markets

In principle, we have to differentiate between countries and regions where there is already a mature market and where there is not.

To make any programme work, and to reduce the subsidies or the premium paid, it is important to have a good infrastructure and a competitive market. Yet, currently in most countries – except maybe in Germany and Japan – no competitive and transparent market exists.

Society

National and local governments must be convinced that the technology brings about societal benefit by means of:

- mitigating the environmental burden;
- increasing local employment;
- enhancing supply security;
- introducing environmental pricing, such as taxes or tradable Green certificates linked to a quota for PV.

Customers

With respect to private households, the most important aim is to increase their Willingness To Pay. This can be achieved by:

- proof of environmental benignity;
- personal identification;
- labelling of green power;
- simple purchase conditions;
- simple technical installation;
- affordable systems at reasonable prices;
- introduction of education programmes for architects and housing companies;
- provision of financing programmes for commercial companies.

Conclusion

This chapter has identified eight key factors for successful dissemination strategies of small grid-connected PV systems. These are illustrated in figure 14 and can be enlarged upon as follows:

- Provide a minimum financial incentive that allows exhaustion of customers' WTP.
- Improve the market: ensure that the competitiveness and the transparency of the PV system market, as well as of the market for electricity (eg by means of a power content label) is enhanced. Moreover, ensure continuity of the strategy over time, and sustainable growth of the industry.
- Strive for a guaranteed technical performance, and increases in standardisation and efficiency.
- Try to make the programme a social event and address the public as well as the mass media.
- Strive for setting the correct regulatory conditions from society's point of view. Remove barriers for access to the grid and introduce environmental pricing.
- Minimise the costs for the public. Strive for low administration and transaction costs and minimise monetary financial support to reach a certain amount of PV capacity.
- Provide comprehensive, detailed and targeted information for the potential programme participants.
- Conduct marketing. Who are the potential customers and what are their needs?

It must be recognised that there is a tremendous variety in strategies, programmes and dissemination ideas. If the most valuable lessons learnt from these widespread activities are summarised and extracted, the groundwork will be set for continuously increasing the dissemination of BiPV systems. It is also important, however, that the market continues to transform so as to reach new customers with well-performing, affordable systems.

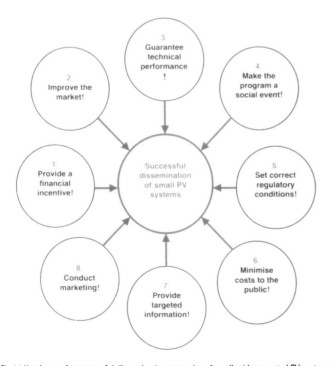

Fig 14 Key factors for successful dissemination strategies of small grid-connected PV systems

APPENDICES

GLOSSARY

Alternating current (AC): Electric current in which the direction of the flow is reversed at frequent intervals. In Europe and Australia, this occurs 100 times per second (50 cycles per second, i.e. 50 Hertz [Hz]) and 120 times per second in the USA. This is the opposite of direct current (DC).

Ancillary services: Resources used to maintain power supply quality such as reliability, voltage and frequency stability and waveform purity.

Anti-reflection coating: A thin coating of a material that reduces the light reflection and increases light transmission, applied to a photovoltaic cell surface.

Azimuth: Angle between the north direction and the projection of the surface normal into the horizontal plane, measured clockwise from north. As applied to the PV array, 180-degree azimuth means the array faces due south.

Balance of system (BOS): The parts of the photovoltaic system other than the PV array including switches, controls, meters, power-conditioning equipment, the supporting structure for the array and storage components, if any.

Building-integrated photovoltaics (BiPV): The harnessing of solar power technologies as a part of, or attached to, the external building skin.

Cogeneration: The simultaneous production of electricity and heat, usually for commercial or industrial use.

Direct current (DC): Electric current in which electrons are flowing in one direction only. This is the opposite of alternating current (AC).

Distributed resources: Small-scale generating, storage or demand management plant, sometimes referred to as micropower and typically connected into the electricity distribution, rather than transmission, network. These can include photovoltaic power systems, wind generators, batteries or other storage devices and appliances, such as solar water heaters, which reduce electrical load on the distribution network.

Embodied energy: All of the energy invested in bringing a material to its final product. This includes the transportation, maintenance, repair, restoration, refurbishment or replacement of materials, components or systems during the lifetime of, for example, a building.

Emissions trading: A mechanism to control the increase in greenhouse gas emissions by setting emission limits, allocating permits and allowing emitters to trade permits amongst themselves as a means of achieving the lowest cost emission reductions overall.

Energy payback time: The time required for any energy producing system or device to produce as much useful energy as was consumed in its manufacture and construction. For PV the energy payback time is approximately two to four years.

EVA: Abbreviation for Ethylene Vinyl Acetate, used as an encapsulant between the glass cover and the solar cells in PV modules. EVA is durable, transparent, is resistant to corrosion and is a flame retardant.

Final annual yield: Total photovoltaic energy delivered to the load during one year per kilowatt of power installed. Unit: kWh per annum per kW installed.

Fossil fuels: Energy sources derived from ancient plant and animal matter trapped on the earth's surface over geological time. These include coal, oil and natural gas, all of which are non-renewable over any human timeframe.

Greenhouse gas emissions: Emissions of gases that collect in the atmosphere and contribute to the Earth's 'greenhouse' effect. Increasing concentrations of gases, such as carbon dioxide, methane and nitrous oxide are currently producing an enhanced greenhouse effect because they are accumulating at a rate faster than they can be dispersed. The combustion of fossil fuels is considered to be a major cause of this enhanced effect, which in turn is expected to contribute to higher average global temperatures over the next century.

Grid-connected distributed photovoltaic power system: System installed on consumers' premises, usually on the demand side of the electricity meter. This includes grid-connected domestic photovoltaic power systems and other grid-connected PV power systems on commercial buildings, motorway sound barriers, etc. These may be used for support of a utility's distribution grid.

Insolation: A term that relates to direct and diffuse sunlight from incident solar radiation.

Installed power: Power delivered by a photovoltaic module or a photovoltaic array, under standard test conditions (STC). See below for an explanation of STC. Also called STC output power. Unit: Watt peak (Wp).

Insulation: Not to be confused with insolation. Insulation is a material characteristic that reduces the transfer of heat from, for example, the interior of a home, from the outside environment or vice versa.

Inverter: Device that converts direct current (DC) into alternating current (AC).

I-V curve: A graphical presentation of the current (I) versus the voltage (V) from a photovoltaic cell as the load is increased from the short circuit (no load) condition to the open circuit (maximum voltage) condition. The shape of the curve characterises cell performance.

Junction box: A PV generator junction box is an enclosure on the module where PV strings are electrically connected and where protection devices can be located, if necessary.

kWh: Symbol of kilowatt-hour, unit of energy (power expressed in kW multiplied by time expressed in hours).

Line-commutated inverter: An inverter that is tied to grid power. The commutation of power (conversion from DC to AC) is controlled by the power line, so that, if there is a failure in the power grid, the PV system cannot feed power into the line.

Load: The amount of electric power that is being consumed at any given moment. Also, in an electrical circuit, any device or appliance that is using power. The load for a utility company varies greatly with time of day and to some extent with season.

Maximum power point (MPP): The point on the current–voltage (I–V) curve of a module under illumination, where the product of current and voltage is maximum. For a typical silicon cell panel, this is about 17 volts for a 36 cell configuration.

NOCT: Abbreviation for Nominal Operating Cell Temperature. The solar cell temperature at a reference environment defined as 800 W/m^2 irradiance, 20 °C ambient air temperature, and 1 m/s wind speed with the cell or module in an electrically open circuit state.

Off-grid domestic photovoltaic power system: Systems installed in households and villages that are not connected to a utility's distribution network (grid). Usually a means to store generated electricity is used, which is most commonly lead-acid batteries. Also referred to as stand-alone photovoltaic power systems or remote area power supplies (RAPS).

Passive solar: Technology that utilises a structure to capture and store the sun's heat. In a passive system, there are usually no moving parts (fans, pumps, etc.). The primary use for passive thermal technologies is in homes and buildings for space heating.

Peak power: PV modules are rated by their peak power output. The peak power (or nominal power) is the amount of power output a PV module produces at standard test conditions (STC). Unit: Watt peak (Wp).

Performance ratio: Ratio of the final yield to the reference yield calculated on the annual or monthly or daily performance. The reference yield is the theoretically available energy on an annual, monthly or daily basis per kilowatt of installed power.

Photovoltaic power system: A system including photovoltaic modules, inverters, batteries (if applicable), and all associated installation and control components, for the purpose of producing solar photovoltaic electricity. Also commonly referred to as PV or photovoltaics.

PV: Abbreviation of photovoltaics and depending on the context, can refer to cells, modules or systems.

PV-T: Abbreviation for PV Thermal which refers to a photovoltaic system that, in addition to converting sunlight into electricity, collects the residual heat energy and delivers both heat and electricity in usable form.

Pyronometer: An instrument for measuring total hemispherical solar irradiance on a flat surface, or 'global' irradiance; thermopile sensors have been generally identified as pyranometers, however, silicon sensors are also referred to as pyranometers.

Renewable energy: Energy sources recently derived directly or indirectly from the energy of the sun, the earth's core, or from lunar and solar gravitational forces that are renewable over short timeframes. These include solar, wind, biomass, tidal, wave, hydro and geothermal energy.

Solar thermal: Term used to describe the generation of heat (rather than electricity) from the sun. Examples are solar swimming pool heaters and household domestic water heaters.

Standard test conditions (STC): The testing conditions used to measure photovoltaic cells' or modules' nominal output power where the irradiance level is 1000 W/m², with the reference air mass 1.5 times the solar spectral irradiance distribution and a cell or module junction temperature of 25 °C.

Vertically integrated utilities: Where generation, transmission, distribution and retailing of electricity are combined in a single organisation. A vertically integrated utility could also consist of generation, electricity transport (transmission and distribution) with no retail functions, or of only electricity transport.

Watt (W): SI unit of power. Symbol is W. One Watt is equal to one Joule per second, with 1 kWh equaling 3.6 MJ. Multiples like kW (1000 W) or MW (1000 kW) are also used. In this publication, it is understood to be power output under standard test conditions (STC). Also written Wp (watt peak) by PV professionals to mean peak power at STC.

REFERENCES

CHAPTER 1: BUILDING DESIGN AND ENVIRONMENTAL CONCEPTS

Alsema, EA 1998, 'Energy Requirements and CO2 Mitigation Potential of PV Systems', presented at the *BNL/NREL Workshop on PV and the Environment 1998,* 23–24 July, Colorado, USA.

Duffie, JA & Beckman, WA 1991, *Solar Engineering of Thermal Processes,* John Wiley & Sons, Inc., New York.

Green, MA 1992, *Solar Cells – operating principles, technology and system applications,* Prentice-Hall, New York.

Honsberg, CB & Bowden, S 1999, *Photovoltaics – devices, systems and applications,* CD-ROM, UNSW Photovoltaics Centre, Vol. 1, Version 1.

IEA–PVPS, 2000/4, *Task 1 Workshop on the Added Value of PV Systems,* Glasgow, 5 May 2000.

Lloyd-Jones, D 2000, 'Effective use of building integrated photovoltaic waste heat: Three projects', *Proceedings of the 2nd World Solar Electric Buildings Conference,* 8–10 March, Sydney, Australia, pp. 29–34.

Lovins, A & Lehmann, A 2000, *Small is Profitable: The Hidden Benefits of Making Electrical Resources the Right Size,* (Rocky Mountain Institute, Boulder, Colorado, USA, forthcoming), cited in Dunn, 2000.

Munro, D, Ruyssevelt, P, Knight, J 2000, 'Photovoltaic Building Integration Concepts and Examples', *Proceedings of the 2nd World Solar Electric Buildings Conference,* 8–10 March, Sydney, Australia, pp. 115–120.

Outhred, H & Watt, M 1999, 'Prospects for Renewable Energy in the Restructured Australian Electricity Industry', presented at *World Renewable Energy Congress,* 10–13 February, Perth, Western Australia, Australia.

Perez, R et al 1997, 'PV as a Long-Term Solution to Power Outages, Case Study: The Great 1996 WSCC Power Outage', *Proceedings of the ASES Annual Conference,* Washington, DC, USA.

Perez, R 1998, 'Photovoltaic Availability in the Wake of the January 1998 Ice Storm', *Proceedings of Renew 98,* NESEA, Greenfield, Massachusetts, USA.

Prasad, DK, Schoen, TNJ, Hagemann, I, Thomas, PC 1997, 'PV in the Built Environment – An International Review', *Proceedings of the ISES 1997 Solar World Congress,* 24–30 August, Expo Science Park, Taejon, Korea.

Prasad, DK and Snow, M, eds 2000, 'IEA PVPS Task VII Photovoltaics in the Built Environment – part of the Renewable Energy for the New Millennium', *Proceedings of the 2nd World Solar Electric Buildings Conference,* 8–10 March, Sydney, Australia.

Sick, F & Erge, T, eds 1996, *Photovoltaics in Buildings, design handbook for architects and engineers,* IEA Task 16 Solar Heating and Cooling Energy Systems, James and James, Glasgow, UK.

Snow, M & Prasad, DK 2002, 'Architectural and Aesthetic experiences for Photovoltaics (PV) in the Built Environment', *Proceedings of Passive & Low Energy Architecture (PLEA) Conference,* Toulouse, France.

Sørensen, H & Munro, D 2000, 'Hybrid PV/Thermal collectors', *Proceedings of the 2nd World Solar Electric Buildings Conference,* 8–10 March, Sydney, Australia.

Watt, ME, Kaye, RJ, Travers, DL, MacGill, I 1997, 'Assessing the potential of PV in Buildings', *14th European Photovoltaic Solar Energy Conference,* 30 June–4 July, Barcelona, Spain.

Watt, ME, et al 1999, *Opportunities for the Use of Building Integrated Photovoltaics in NSW,* Report to the NSW Department of Energy, SERDF, PV Special Research Centre, Sydney, Australia (see PV Centre website for reports).

Watt, ME 2000, *Added Value of PV Systems,* Report IEA–PVPS, 09:2001, International Energy Agency, University of NSW, Australia.

Wenham, SR, Green, MA & Watt, ME 1994, *Applied Photovoltaics,* Prentice-Hall, New York.

CHAPTER 2: TECHNOLOGIES AND INTEGRATION CONCEPTS

Bazilian, MD, Leenders, F, Van Der Ree, BGC & Prasad, DK 2001, 'Photovoltaic Cogeneration in the Built Environment', *Solar Energy,* Vol. 71, No. 1, pp. 57–69.

IEA-PVPS T7-03 2000 'Photovoltaic Building integrated concepts', *Proceedings of the IEA PVPS Task 7 Workshop,* 11–12 February 1999, EPFL, Lausanne, Switzerland; Paris, France.

Meike, W 1998, 'Hot Climate Performance between Poly-Crystalline and Amorphous Silicon Cells connected to a Utility Mini-Grid', *Proceedings of 36th ANZSES Solar Energy Conference,* 25–27 November, Christchurch, New Zealand, pp. 464–470.

Reijenga, T 2000, 'Architectural quality of building integration of solar energy – case studies in the Netherlands', *Proceedings the 2nd World Solar Electric Buildings Conference,* 8–10 March, Sydney, Australia, pp. 42–47.

Wren, C & Barram, F 2000, 'Solar Integration on Commercial Buildings', *Proceedings of the 2nd World Solar Electric Buildings Conference,* 8–10 March, Sydney, Australia, pp. 68–74.

CHAPTER 3: CASE STUDIES

AUSTRALIA: SYDNEY OLYMPIC VILLAGE

Deo Prasad, Mark Snow (www.fbe.unsw.edu.au)

SOLARCH Group, University of NSW, Sydney NSW 2052, Australia

Tel: +61 2 9385 4868 Fax: +61 2 9385 6735

Additional references:

Collins, R, Davenport, T, & Schach, M 2000, 'Recent experiences in building integrated photovoltaics', *Proceedings of the 2nd World Solar Electric Buildings Conference,* 8–10 March, Sydney, Australia, pp. 95–98.

Prasad, DK & Snow, M 2000, 'The shiny side of gold – Solar Power in Sydney 2000', *Renewable Energy World,* July, pp. 77–87.

Prasad, DK & Snow M 2000, 'Sydney Olympics 2000: A solar power showcase', *REFOCUS,* ISES International Solar Energy Society Magazine, Sept/Oct, pp. 22–24.

AUSTRIA: ENERGIEPARK WEST

Karin Stieldorf (www.tuwien.ac.at)

Department for construction and building design 270/3, Faculty of Architecture, Vienna University of Technology

Karlsplatz 13, A-1040 Vienna, Austria

Tel: +43 1 588 01-0 Fax: +43 1 58801-40199

CANADA: TORONTO HIGHRISE ROOF

Per Drewes

Sol Source Engineering

Newmarket Ontario L3Y 1R7, Canada

Tel: +1 905 898 0098 Fax: +1 905 898 1668

CANADA: WILLIAM FARRELL BUILDING

Ljubisav Stamenic (www.bcit.ca)

BCIT Technology Centre

3700 Willingdon Avenue, Burnaby BC V5G 3H2, Canada

Tel: + 1 604 451 6934 Fax: + 1 604 436 0286

DENMARK: BRUNDTLAND CENTRE

Henrik Sørensen (www.esbensen.dk)

Esbensen Consulting Engineers

Vesterbrogade 124 B, DK-1620 Copenhagen V, Denmark

Tel: + 45 33 26 73 00 Fax: + 45 33 26 73 01

GERMANY: FRAUNHOFER ISE

Hermann Laukamp (www.ise.fhg.de)

Fraunhofer-Institut fuer Solare Energiesysteme

Heidenhofstrasse 2, D-79110 Freiburg, Germany

Tel: + 49 761 4588 5275 Fax: + 49 761 4588 5217

GERMANY: MONT-CENIS ACADEMY

Ingo Hagemann (www.rwth-aachen.de)

Architekturbüro Hagemann

Annuntiatenbach 43, D-52062 Aachen, Germany

Tel: + 49 (0) 241 34530 Fax: + 49 (0) 241 30547

Additional references:

Hagemann, IB 2001, 'Gebäudeintegrierte Photovoltaik. Architektonisch sinnvolle Integration der Photovoltaik in die Gebäudehülle', Verlagsgesellschaft Rudolf Müller. Köln, Germany.

Susa-Verlag, Ed 1999, 'Akademie Mont-Cenis Herne', *Baudokumentation 84*, Hannover, Germany.

Pilkington Solar International GmbH, Ed, 1998, *Megawattstark*, (company brochure), Köln, Germany.

Entwicklungsgesellschaft Mont-Cenis GmbH, Ed 1998, 'Mont-Cenis. Fortbildungsakademie Herne, Stadteilzentrum Herne-Sodingen', Herne, Germany.

Dassler, Friedrich et al 1999, 'Solarzeitalter. Fortbildungsakademie Mont-Cenis in Herne', *Intelligente Architektur* No 19, Eine Spezialausgabe der AIT, pp. 29–43.

EMC – Entwicklungsgesellschaft Mont-Cenis, Ed 2000, '...auf Mont-Cenis', Pressemappe, Herne, Germany.

www.akademie-mont-cenis-herne.nrw.de

www.sma.de/de/photovoltaik/veröffentlichung/elekronik/191999/inhalt.html.

'Solarkraftwerk mit modularem Aufbau', *Elektronik*, Sonderdruck aus Heft 19/1999, Franzis Verlag, Germany.

ITALY: **THE CHILDREN'S MUSEUM OF ROME**

Cinzia Abbate (www.aevarchitetti.it)

Abbate &Vigevano Architetti

Piazza S. Anastasia 3, 00186 Rome, Italy

Tel: +39 06 6796498 Fax: +39 06 69783038

Additional references:

Wetzel, T, Baake, E & Muelbauer, A, *La Stampa,* 14 February 2001.

JAPAN: **NTT DoCoMo BUILDING**

Tadashi Ito

Kajima Technical Research Institute, Kajima Corporation

19-1 Tobitakyu 2-Chome Chofu-shi, Tokyo 182-0036, Japan

Tel: +81 424 89 7090 Fax: +81 424 89 7128

JAPAN: **J-HOUSE**

Jiro Ohno

Senior Project Manager, Architectural Design Division

Nihon Sekkei, Inc.

29th floor, Shinjuku I-LAND Tower, 6-5-1, Nishi-shinjuju, Shinkuju-ku, Tokyo 163-1329, Japan

Tel: +81 3 3344 2311 Fax: +81 3 5325 8821

JAPAN: **SBIC EAST BUILDING**

Jiro Ohno

Senior Project Manager, Architectural Design Division

Nihon Sekkei, Inc.

29th floor, Shinjuku I-LAND Tower, 6-5-1, Nishi-shinjuju, Shinkuju-ku, Tokyo 163-1329, Japan

Tel: +81 3 3344 2311 Fax: +81 3 5325 8821

THE NETHERLANDS: **ECN BUILDINGS 31 AND 42**

Tjerk Reijenga (www.bear.nl)

Co-authors: Astrid Schneider, Henk Kaan (ECN)

BEAR Architecten bv

Postbus 349, NL-2800 AH Gouda, The Netherlands

Tel: +31 182 529 899 Fax: +31 182 582 599

THE NETHERLANDS: **LE DONJON**

Tjerk Reijenga (www.bear.nl)

Co-author: Astrid Schneider

BEAR Architecten bv

Postbus 349, NL-2800 AH Gouda, The Netherlands

Tel: +31 182 529 899 Fax: +31 182 582 599

THE NETHERLANDS: **NIEUWLAND**

Tony JN Schoen (www.ecofys.nl)

Ecofys

PO Box 8408, 3503 RK Utrecht, The Netherlands

Tel: +31 30 2808 300 Fax: +31 30 2808 301

Additional information supplied by: Adriaan Kil, Edith Molenbroek (Ecofys);
Ingmar Gros, Frans Vlek (REMU); Cinzia Abbate.

SOUTH KOREA: **KIER SUPER LOW ENERGY BUILDING**

Jongho Yoon (www.hanbat.ac.kr)

Department of Architectural Engineering, Hanbat National University

Taejon, South Korea, 305-719

Tel: +82 42 821 1126 Fax: +82 42 821 1115

EJ Lee, MW Jung (www.kier.re.kr)

Korea Institute of Energy Research

Taejon, South Korea, 305-343

Tel: +82 42 860 3514 Fax: +82 42 860 3132

SPAIN: **UNIVER**

Jorge Aguilera, Gabino Almonacid (www.ujaen.es)

Grupo Jaén de Técnica Aplicada, Universidad de Jaén

23071 Jaén, Spain

Tel: +34 953 002434 Fax: +34 953 002400

SWITZERLAND: **ABZ APARTMENT BUILDINGS**

Daniel Ruoss (www.enecolo.ch)

Enecolo AG

Lindhofstrasse 52, CH 8617 Mönchaltorf, Switzerland

Tel: +41 1 994 9001 Fax: +41 1 9940 9005

SWITZERLAND: **STUDENT HOUSING**

Christian Roecker, François Schaller (www.epfl.ch)

EPFL (Swiss Federal Institute of Technology), LESO – ITB

CH 1015 Lausanne, Switzerland

Tel: +41 21 963 45 45 Fax: +41 21 963 27 22

UK: **JUBILEE CAMPUS NOTTINGHAM UNIVERSITY**

Prof. John Berry (www.arup.com)

Ove Arup & Partners

13 Fitzroy Street, London W1T 4BQ, UK

Tel: +44 (0)20 7 636 1531 Fax: +44 (0)20 7 465 3673

UK: **SOLAR OFFICE, DOXFORD INTERNATIONAL BUSINESS PARK**

David Lloyd Jones (www.studioe.co.uk)

Studio E Architects Ltd

Palace Wharf, Rainville Road, London W6 9HN, UK

Tel: +44 (0)20 7385 7126 Fax: +44 (0)20 7381 4995

USA: **4 TIMES SQUARE**

Kiss + Cathcart Architects (www.kisscathcart.com)

44 Court Street, Tower C, 12th Floor, Brooklyn NY 11201, USA

Tel: +1 718 237 2786 Fax: +1 718 237 2025

Additional references:

World Architecture 83, '4 Times Square: Breaking New Ground,' February 2000, pp. 56–63.

Eiffert, P & Kiss, G 2000, *Building-Integrated Photovoltaic Designs for Commercial and Institutional Structures: A Sourcebook for Architects,* U.S. Department of Energy, Oak Ridge, Tennessee.

CHAPTER 4: **NON-BUILDING PV STRUCTURES**

Clavadetscher, L & Nordmann, Th 1999, '100 kWp Grid-Connected PV-Installation along the A13 Motorway in Switzerland Plant Monitoring and Evaluation – Operation and Maintenance', *Annual Reports for the Swiss Federal Office of Energy,* 1990–1999, Project No. 32046.

Nordmann, Th, Frölich, A, Reiche, K, Kleiss, G & Götzberger, A 1998, 'Integrated PV Noise Barriers: Six Innovative 10 kWp Testing Facilities – A German Swiss Technological and Economical Success Story!', *2nd World Congress and Exhibition on Photovoltaic Solar Energy Conversion,* Vienna, Austria.

Nordmann, Th & Götzberger, A 1994, 'Motorway Sound Barriers: Recent Results and New Concepts for Advancement of Technology', *IEEE First World Conference on Photovoltaic Energy Conversion,* December, Hawaii, USA, pp. 766–773.

Nordmann, Th & Götzberger, A 1995, 'Motorway Sound Barriers: The Bifacial North/South concept and the Potential in Germany', *13th European Photovoltaic and Solar Energy Conference and Exhibition,* 23–27 October, Nice, France, pp. 707–709.

Snow, M & Prasad, DK 2000 'Power without the noise – opportunities for integrating photovoltaic noise barriers (PVNB) under Australian conditions', *ANZSES Solar Energy Conference,* Griffith University, 29 Nov–1 Dec, Brisbane, Australia.

TNC Energy Consulting, GmbH Freiburg DL, ENEA IT, ISE Freiburg DL, Utrecht University NL, NPAC UK, PHEBUS France, TNC Consulting AG Switzerland, 1999, *EU PVNB POT – Evaluation of the Potential of PV Noise Barrier Technology for Electricity Production and Market Share,* EU Thermie B Project, 1990–1999, Project Number: 32046.

CHAPTER 5: BIPV POTENTIAL AND DESIGN TOOLS

ETSU 1997, *Photovoltaics in Buildings – a survey of design tools*, ETSU Report No. S/P2/00289/REP.

Lee, T, Oppenheim, D & Williamson, T 1995, *Australian Solar Radiation Data Handbook*, ERDC.

Little, AD, Inc. 1995, *Building-integrated photovoltaics (BiPV) – Analysis and US Market Potential*, report to US Department of Energy, Office of Building Technologies.

Nowak, S, Gutschner, M, Toggweiler, P, Ruoss, D 2000, *Potential for Building Integrated Photovoltaics, Solar-Yield-Differentiated Area Potential for Some Selected IEA Countries*, NET Nowak Energy & Technology, Ltd., Ursen, Switzerland; Enecolo Ltd., Mönchaltorf, Switzerland.

Snow, M, Jones, P, Lannon, S, & Prasad, DK 2000 'Application of a GIS-building integrated photovoltaic (PV) modelling approach for cities. A case example Newcastle, Australia', *Architecture, City, Environment – Proceedings of PLEA 2000*, Cambridge, UK, pp. 684–689.

Snow, M, Travers, DL, Jones, P, Prasad, DK 1998, 'Building integrated photovoltaics (BIPV) – Developing a modelling tool for cities', *ANZSES Solar Energy Conference*, 22–25 November, Christchurch, New Zealand.

CHAPTER 6: ELECTRICAL CONCEPTS, RELIABILITY AND STANDARDS

Ambo, T 1997, 'Islanding Prevention by Slip Mode Frequency Shift', *Proceedings of the IEA PVPS Workshop on Grid-interconnection of Photovoltaic Systems*, 15–16 September, Zurich, Switzerland.

Begovic, M, Ropp, M, Rohatgi, A, Pregelj, A 1998, 'Determining the Sufficiency of Standard Protective Relaying for Islanding Prevention in Grid-Connected PV Systems', *Proceedings of the 2nd World Conference and Exhibition on Photovoltaic Solar Energy Conversion*, 6-10 July, Vienna, Austria.

von Bergen, Ch 1999, *Convert 2000 with LonWorks-Interface*, Sputnik Engineering AG, 12/1999.

Bonn, R, Ginn, J, Gonzalez, S 1999, *Standardized Anti-Islanding Test Plan*, Sandia National Laboratories.

Bower, W, et al 2000, 'Testing To Improve PV Components and Systems', *Proceedings of the 16th European PV Solar Energy Conference and Exhibition*, 1-5 May, Glasgow, UK.

Burges, K 2000, 'Prototyping of AC Batteries for Stand Alone PV Systems', *EU PV Conference*, Glasgow, UK.

Eclipse PV/E, Energy management tool for PV systems, Ecofys, Utrecht, Netherlands.

Gonzalez, S 2000, 'Removing Barriers to Utility-Interconnected Photovoltaic Inverters', *Proceedings of the 28th IEEE PV Specialists Conference*, Anchorage, Alaska, USA.

He, W, Markvart, T, Arnold, R 1998, 'Islanding of Grid-connected PV Generators: Experimental Results', *Proceedings of the 2nd World PV Solar Energy Conference*, Vienna, Austria.

Häberlin, H, et al 1992, 'PV Inverters for Grid Connection: Results of Extended Tests', *Proceedings of the 11th PV Solar Energy Conference*, Montreux, Switzerland, p. 1585 ff.

Häberlin, H & Graf, JD 1998, 'Gradual Reduction of PV Generator yield due to pollution', *Proceedings of the 2nd World Conference and Exhibition on Photovoltaic Solar Energy Conversion*, Vienna, Austria.

Häberlin, H & Graf, JD 1998, 'Islanding of Grid-connected PV Inverters: Test Circuits and Some Test Results', *Proceedings of the 2nd World Conference and Exhibition on Photovoltaic Solar Energy Conversion*, Vienna, Austria, pp. 2020–2023.

IEA–PVPS Task V 1997, *Proceedings of the IEA PVPS Task V Workshop on Utility Interconnection of PV Systems*, Zurich, Switzerland.

IEA–PVPS V-1-03 1998, *Grid-Connected Photovoltaic Power Systems: Status Of Existing Guidelines And Regulations In Selected IEA Member Countries* (Revised Version), Task V Internal Report.

IEA–PVPS T5-01: 1998, *Utility Aspects of Grid Interconnected PV Systems*, IEA–PVPS Report.

International Electrochemical Commission 2001, *Safety guidelines for grid connected photovoltaic systems mounted on buildings*, IEC 62234, from www.iec.ch.

Kälin, T, *Solar PV Power Plant P + R Neufeld*, Bern, Electrowatt Engineering, Zürich, Switzerland.

Kern, G, Bonn, R, Ginn, J, Gonzalez, S 1998, 'Results of Sandia National Laboratories Grid-Tied Inverter Testing', *Proceedings of the 2nd World Conference and Exhibition on Photovoltaic Solar Energy Conversion*, 6-10 July, Vienna, Austria.

King, DL, Hansen, B & Boyson, WE 2000, 'Improved Accuracy for Low-Cost Irradiance Sensors', *Proceedings of the 28th IEEE PV Specialists Conference*, Anchorage, Alaska, USA.

King, DL, Kratochvil, JA & Boyson, WE 1997, 'Temperature Coefficients for PV Modules and Arrays: Measurement Methods, Difficulties, and Results', *Proceedings of the 26th IEEE PV Specialists Conference*, California, USA, pp. 1183–1186.

King, D, Kratochvil, J, Boyson, W & Bower, W 1998, 'Field Experience with a New Performance Characterization Procedure for Photovoltaic Arrays', *Proceedings of the 2nd World Conference and Exhibition on Photovoltaic Solar Energy Conversion*, 6–10 July, Vienna, Austria.

Kobayashi, H, Takigawa, K 1998, 'Islanding Prevention Method for Grid-interconnection of Multiple PV Systems', *Proceedings of the 2nd World Conference and Exhibition on Photovoltaic Solar Energy Conversion*, 6–10 July, Vienna, Austria.

Kroposki, B et al 1996, 'Photovoltaic Module Energy Ratings Methodology Development', *Proceedings of the 25th IEEE PV Specialists Conference*, Washington, DC, USA.

Laukamp, H et al 2000, 'Reliability Issues in PV Systems – Experience and Improvements', *Proceedings of the 2nd World Solar Electric Buildings Conference*, Sydney, Australia, pp. 88–94.

Laukhamp, H 2001, *Reliability of Photovoltaic Systems*, Final Report Draft rev. 30.08.01, Task 7 Report IEA–PVPS.

Lund, P, contributions to the report 'New Electrical Concepts', Helsinki University of Technology.

Pacific Gas and Electric Co., 1993, *Independent Power Producers' Interconnection Handbook*.

Panhuber, C, *Economical and Reliability Optimisation of a 2 kW Grid-Coupled Inverter*, Fronius KG, Wels, Austria.

Pellis, J 1997, *The DC Low-Voltage House*, ECN Report ECN-C-97-58.

Request for Proposal, Utility Scale Photovoltaic Power Systems (US-2), Kerman, 'Photovoltaics for Utility Scale Applications', PVUSA Project, Jan 1992.

Ropp, M, Begovic, M, Rohatgi, A 1999, 'Prevention of Islanding in Grid-connected Photovoltaic Systems', *Progress in Photovoltaics Research and Applications*, John Wiley & Sons, Vol. 7, No. 1.

Spooner, E 2000, 'Standards Development for BIPV', *Proceedings of the 2nd World Solar Electric Buildings Conference*, 8–10 March, Sydney, Australia, pp. 82–87.

Stanton, A, et al 2000, *Primer of Applied Regression and Analysis of Variance*, McGraw Hill Professional.

Stevens, J, Bonn, R, Ginn, J, Gonzalez, S, Kern, G 2000, *Development and Testing of an Approach to Anti-islanding in Utility-Interconnected Photovoltaic Systems*, SAND2000-1939, Sandia National Laboratories, Albuquerque, New Mexico, USA.

Toggweiler, P 2000, Contributions to the report, *New Electrical Concepts*, Enecolo AG.

Toggweiler, P, Meyer T, 'AC Modules', joint development with FHG ISE and ETH, Zurich.

Watt, ME 1993, *Environmental and Health considerations in the production of cells and modules*, Centre for Photovoltaic Devices and Systems, UNSW, Sydney, Australia, Centre Report No. 1993/02.

Wilk, H 1998 'Grid Connected PV Systems in Austria', *Plenary Presentation, 2nd Conference on PV Solar Energy Conversion*, Vienna, Austria.

Wilk, H, Schauer G/Verbund, 1997, 'Operational Behaviour of Small Power Conditioners for Grid-Interactive PV-System', *14th EU PV Conference*, Barcelona, Spain.

Wilk, Schauer/Verbund, Harich, Enders (Arsenal Research) 1998, 'Testing Inverters for Utility Interactive Operation', *2nd Conference on PV Solar Energy Conversion*, Vienna, Austria.

Wilk, H 2000, *Grid Connected PV-Systems in Austria, Lessons Learned*, internal report, presented at Task VII expert meeting, Sydney, Australia.

CHAPTER 7: NON-TECHNICAL ISSUES AND MARKET DEPLOYMENT STRATEGIES

olinger, M, Wiser, R 2002, *Customer-Sited PV: A survey of Clean Energy Fund Support*, LBNL Report 49668.

rasil, T 2002, *Emerging Renewables Buydown Program Past and Future*, PV Alliance, San Ramon, CEC.

ouncil of the European Union, 2001, *On the promotion of electricity from renewable energy sources in the internal lectricity market*, proposal for a Directive of the European Parliament and the council, Brussels, Belgium.

COFYS, 2000, *Financing PV in the Netherlands*, mimeo.

iffert, P, Leonard, G and Thompson, A 2001, *Guidelines for the Economic Analysis of Building Integrated Photovoltaic ystems*, National Renewable Energy Laboratories, Golden, Colorado, USA.

iffert, P & Kiss, GJ 1999, *Building integrated photovoltaics for commercial and institutional structures – A sourcebook for Architects*.

iffert, P, Leonard, G, Thompson, A 2001, *Guidelines for the Economic Analysis of Building Integrated Photovoltaic Systems*, NREL.

TSU, 1998, *The Value of electricity generation from Photovoltaic Power Systems in Buildings*, S/P2/00279/REP.

European Commission, *Energy for the future: renewable sources of energy*, White paper for a community strategy and action plan, COM (97) 599, Brussels 1997.

Farhar, BC & Houston, AH 1996, 'Willingness to Pay for Electricity from Renewable energy', *Proceedings of 1996 ACEEE Summer Study on Energy Efficiency in Buildings*, ACEEE.

Farhar, BC, Roper, M 1998, 'Understanding Residential Grid-tied PV Customers and Their Willingness to Pay Today's Costs: A Qualitative Assessment', *Proceedings of 1998 ACEEE Summer Study on Energy Efficiency in Buildings*, ACEEE.

Frauenfelder, S, *Erfolgsrezepte für das Solarstrom-Marketing*, Bulletin SEV/VSE 10/2000.

Genennig, B & Hoffmann, VU 1996, *Sozialwissen-schaftliche Begleituntersuchung zum Bund-Länder-1000 Daecher Photovoltaik Programm*, Umweltinstitut, Leipzig.

Green, J, Plastow, J 2001, *Options of labelling Green Electricity in Europe* Report of the project ELGREEN co-financed under the 5th framework programme of the European Commission, IT Power, UK.

Gregory, A, Bahaj, AS & Stainton, RS 1994, 'Stimulating Market Success for Solar Energy: A Global Perspective', *The Yearbook of Renewable Energies*, pp. 222–227.

Groenendal, BJ, et al 2000, *Critical Success Factors for the Large-Scale Introduction of grid-connected PV Systems*, ECN-Report 00-086, Petten, Netherlands.

Haas, R 1995, 'The Value of Photovoltaic Electricity for Society', *Solar Energy*, 54 (1), pp. 25–31.

Haas, R 1997, 'Successful dissemination programs for residential PV applications – an international survey', *Proceedings of 26th IEEE Photovoltaic Specialists Conference*, 30 September–3 October, Anaheim, USA.

Haas, R 1998, 'Residential Photovoltaics Applications: The Relevance of Non-Technical Issues', *The International Journal of Solar Energy*, 20 (1), pp. 37–55.

Haas, R 2000, *Marketing Strategies for PV Systems, an International Survey*, Institute of Energy Economics, Vienna University of Technology, Vienna, Austria.

Haas, R, Ornetzeder, M, Hametner, K, Wroblewski, A, Hübner, M 1999 'Socio-Economic Aspects of the Austrian 200 KWp-Photovoltaic rooftop programme', *Solar Energy*, 66 (3), pp. 183–191.

Haas, R 2002, 'Market deployment strategies for PV systems in the built environment: An evaluation of incentives, support programmes and marketing activities', *IEA Photovoltaic Power Systems Programme Report* IEA–PVPS T7-06:2002.

Haberland, U, Stuhlweissenburg, P 2000 'Das 200 Dächer-programm der HEW', *Proceedings of 16th Symposium Photovoltaische Sonnenenergie*, Staffelstein, Germany.

Hirshman, WP 2001, 'Australia's PV rebate program proves too successful', *PHOTON International*, 2/2001.

Hoffmann, VU 1995, 'Sozialwissenschaftliche Begleitforschung zum 1000-Daecher-PV-Programm', *Elektrizitätswirtschaft*.

Huber, C (ed) 2001, *Action plan for a green European electricity market*, Report of the project ELGREEN, co-financed under the 5th framework programme of the European Commission, Energy Economics Group, Vienna University of Technology, Vienna, Austria.

IEA-PVPS, *Trends in Photovoltaic Applications in selected IEA countries between 1992 and 2001*, IEA Report, PVPS T1–11:2002.

ISE, '1000-Daecher-Meß- und Auswerteprogramm', *Fraunhofergesellschaft*, ISE Freiburg/Leipzig 1994–2000.

Kiefer, K, Hoffmann, VU, Erge, T 2000, 'Gesicherte Erträge von netzgekoppelten Photovoltaik-Anlagen', *Proceedings of 16th Symposium Photovoltaische Sonnenenergie*, Staffelstein, Germany.

Maycock, PD 2000, 'The world PV market 2000: shifting from subsidy to "fully economic"?', *Renewable Energy World*, 3 (4), July–August, pp. 59–74.

Maycock, PD 2001, 'The PV boom: Where Germany and Japan lead, will California follow?' *Renewable Energy World*, 4 (4) July–August, pp. 145–163.

National Renewable Energy Laboratory (NREL) 1998, US Department of Energy, Office of Energy Efficiency and Renewable Energy, www.eren.doe.gov/greenpower/netmetering.html.

National Renewable Energy Laboratory (NREL) 1998, US Department of Energy, Office of Energy Efficiency and Renewable Energy, www.eren.doe.gov/greenpower/market_brief_2 /policies.html.

Nordmann, T 1996, 'Successful Solar Marketing and Financing in Europe: How to get from here to there in Photovoltaics?', *Proceedings of EUROSUN '96*, Freiburg, Germany.

NOVEM, 1997, *Heading into the Solar Age Together*, Amsterdam, Netherlands.

Osamu, I, *PV Activities in Japan*, various issues, 1998–2002.

Osamu, I & Ohigashi, T 2001, 'Progress and Future Outlook of PV market in Japan', *Proceedings of the 17th European PV Conference*, Munich, Germany, (forthcoming).

Osborn, DE 1997, 'Commercialization of utility PV distributed power systems', *Proceedings of Solar 97*, American Solar Energy Society (ASES), Washington, USA.

Osborn, DE 2000, 'SMUD PV programmes', *Renewable Energy World*, 3 (5), September–October.

Painuly, JP 2001, 'Barriers to renewable energy penetration: a framework for analysis', *Renewable Energy*, no. 24, pp. 73–89.

Petrovic, L 2000, *The greener option*, PEI.

Rezzonico, S, Nowak, S 1997, *Buy-back rates for grid-connected Photovoltaic power systems*, Report IEA PVPS TASK 2, St. Ursen.

Rocky Mountain Institute, 1991 'Photovoltaics – Clean Energy now and for the future', *RMI Newsletter*, VII (1), pp. 5–7.

Rogers, EM 1993, *Diffusion of Innovations*, Free Press, New York/London.

Ruoss, D 2000, 'A Green Pricing model in Switzerland – The "Solarstrom stock exchange" from the Electric Utility of the City of Zurich', *Proceedings of the 2nd World Solar Electric Conference*, 8–10 March, Sydney, Australia.

Scheer, U 2000 'Photovoltaikförderprogramm "Münchner Solarstrom" der Stadtwerke München', *Proceedings of 16th Symposium Photovoltaische Sonnenenergie*, 14–16 March, Staffelstein, Germany.

Schoen, T 2000, 'BIPV Overview & Getting PV into the Marketplace in the Netherlands', *Proceedings of the 2nd World Solar Electric Buildings Conference*, 8–10 March, Sydney, Australia.

Skorka, I, Kotzerke, C, Bohlen, M, Hoffmann, VU, Kiefer, K 1998, 'SONNEonline – The Photovoltaic School Program', *Proceedings of the 2nd World Conference Photovoltaic Solar Energy Conversion*, 6–10 July, Vienna, Austria.

Swezey, B & Bird, L 2001, *Green Power Marketing in the U.S.: A status report*, National Renewable Energy Laboratory, Golden, Colorado, USA.

US Department of Energy, 1998, *The borrower's guide to Financing Solar Energy Systems*, Washington, DC, USA.

Watt, ME, Ellis, M, O'Regan, S, Gow, S, Fisher, B and Fowkes, R 1996, *Study of local government regulations impacting on the use of remote area power supply systems and other renewable energy technology*, Department of Primary Industries and Energy, Canberra, Australia.

Wattanapong, R, Herbert, W, O'Donoghue, J 1998, 'Photovoltaic application in the 8th National, Social and Economic Development plan of Thailand (1997–2001)', *Proceedings of the 2nd World Photovoltaic Solar Energy Conference*, Vienna, Austria.

FURTHER RESOURCES: WEBSITES

About solar power:	www.greenhouse.gov.au/yourhome/consumer/cg7b.htm www.howstuffworks.com/solar-cell.htm www.eere.energy.gov/solar/
Australian Business Council for Sustainable Energy:	www.bcse.org.au
BiPV simulation tools:	PVSyst: www.pvsyst.com PV design Pro-G: www.mauisolarsoftware.com PV*Sol: www.valentin.de Solar Pro: www.lapsys.co.jp/pro/pro_english.html
European Photovoltaic Industry Association (EPIA):	www.epia.org
IEA Task 7 PV in the Built Environment:	www.task7.org
International Energy Agency (IEA) PVPS programme and Task reports:	www.iea.org www.iea-pvps.org
International Solar Energy Society (ISES):	www.ises.org
Japan Photovoltaic Energy Association (JPEA):	www.ipea.gr.jp
NCPV virtual library:	www.nrel.gov/ncpv/libbody.html
PV databases:	www.pvdatabase.com www.demosite.ch
PV information weblinks:	www.pvresources.com/en/literature.php
PV manufacturers, index:	www.solarbuzz.com www.pvportal.com www.pvresources.com
PV software:	www.pvresources.com/en/software.php
PV standards:	International Electrotechnical Commission (IEC): www.iec.ch European Commission (EC) Joint Research Centre (JRC): www.jrc.it Institute of Electrical and Electronic Engineers (IEEE): www.ieee.org
Solar Energy Industries Association (SEIA):	www.seia.org
Solar power in the built environment:	www.fbe.unsw.edu.au
University of NSW PV Centre weblink page:	www.pv.unsw.edu.au/solpages.html
US National Renewable Energy Laboratory (NREL):	www.nrel.gov/
US NREL - National Center for Photovoltaics (NCPV):	www.nrel.gov/ncpv/

INDEX

ACKNOWLEDGMENTS

ASSOCIATE PROFESSOR DR DEO PRASAD

Professor Deo Prasad is the Director of the SOLARCH Group: University of NSW Centre for a Sustainable Built Environment, Sydney, Australia. He has qualifications and experience in architecture and engineering and has an international reputation for his work in the field of sustainable buildings. Deo was a sub-task leader of Task 7 in the IEA PVPS programme 'PV in the Built Environment'. He is also director of the International Solar Energy Society (ISES).

DR MARK SNOW

Dr Mark Snow is a senior researcher at the SOLARCH Group, University of NSW and previously worked with a leading energy environmental consultancy, Mandix, in Cardiff. His doctorate, completed at the Welsh School of Architecture, Cardiff University was in the field of modelling BiPV and solar access in urban environments. Mark consults in this area and has particular interest in urban sustainability research and education.

The principal editors acknowledge and greatly appreciate the assistance of Professor Ray Cole, School of Architecture, University of British Columbia, Canada and Dr Muriel Watt, Centre for Photovoltaics Engineering, UNSW in reviewing manuscripts.